Noble Gas Geochemistry

Noble Gas Geochemistry discusses the fundamental concepts of using noble gases to solve problems in the earth and planetary sciences.

The discipline of noble gas geochemistry has become a major branch in the earth and planetary sciences. It offers a powerful and unique tool in resolving problems such as the origin of the solar system, evolution of the planets, Earth formation, mantle evolution and dynamics, atmospheric degassing and evolution, ocean circulation, dynamics of aquifer systems, and numerous applications to other geological problems. This book gives a comprehensive description of the physical chemistry and cosmochemistry of noble gases before leading on to applications for problem solving in the earth and planetary sciences. Abundant tables and figures for noble gas elemental and isotopic data in the solar system (terrestrial, meteoritic, solar, etc.) are included.

There have been many developments in the use of noble gases since publication of the first edition of this book in 1983. This second edition has been fully revised and updated. The book will be invaluable to graduate students and researchers in the earth and planetary sciences who use noble gas geochemistry techniques.

Minoru Ozima is Emeritus Professor in the Department of Earth and Planetary Sciences, University of Tokyo. As well as the first edition of this book, Professor Ozima has published *The Earth: Its Birth and Growth* (1981, Cambridge University Press) and *Geohistory* (1987, Springer Verlag, Heidelberg).

Frank A. Podosek is a professor in the Department of Earth and Planetary Science at Washington University in St. Louis. He is currently chief editor of *Geochimica et Cosmochimica Acta*. Professor Podosek is a Fellow of the Meteoritical Society.

Noble Gas Geochemistry

Second Edition

Minoru Ozima
University of Tokyo

Frank A. Podosek
Washington University

PUBLISHED BY THE PRESS SYNDICATE OF THE UNIVERSITY OF CAMBRIDGE
The Pitt Building, Trumpington Street, Cambridge, United Kingdom

CAMBRIDGE UNIVERSITY PRESS
The Edinburgh Building, Cambridge CB2 2RU, UK
40 West 20th Street, New York, NY 10011-4211, USA
10 Stamford Road, Oakleigh, VIC 3166, Australia
Ruiz de Alarcón 13, 28014 Madrid, Spain
Dock House, The Waterfront, Cape Town 8001, South Africa

http://www.cambridge.org

© Cambridge University Press 2002

This book is in copyright. Subject to statutory exception
and to the provisions of relevant collective licensing agreements,
no reproduction of any part may take place without
the written permission of Cambridge University Press.

First published 2002

Printed in the United States of America

Typefaces Times New Roman 10.25/13.5 pt. and Joanna   *System* QuarkXPress [BTS]

A catalog record for this book is available from the British Library.

Library of Congress Cataloging in Publication Data
Ozima, Minoru.
    Noble gas geochemistry / Minoru Ozima, Frank A. Podosek. – 2 ed.
        p. cm.
    Includes bibliographical references and index.
    ISBN 0-521-80366-7
    1. Gases, Rare. 2. Geochemistry. I. Podosek, Frank A. II. Title.
QE516.R23 O98 2001
551.9–dc21
                                                                00-054674

ISBN   0 521 80366 7   hardback

Contents

Preface to the Second Edition   ix

Preface to the First Edition   xiii

Chapter 1 **Introduction**   1

1.1   Retrospect and Prospect   1
1.2   Geochemical Characteristics of the Noble Gases   4
1.3   Constants and Conventions   5
1.4   Nomenclature   7
1.5   Air   10
1.6   Atmosphere   12
1.7   Nuclear Chemistry   15
1.8   Three-Isotope Diagrams   25

Chapter 2 **Physical Chemistry**   30

2.1   Introduction   30
2.2   Adsorption   33
2.3   Solution   42
2.4   Crystal-Melt Partitioning   52
2.5   Trapping and Implanting   55
2.6   Clathrates and Fullerene C_{60} 60
2.7   Diffusion   63
2.8   Stepwise Degassing   69
2.9   Isotopic Fractionation   74

Chapter 3 **Cosmochemistry** 81

3.1 Cosmic Abundances 81
3.2 Solar Noble Gases 83
3.3 Planetary Noble Gases 87
3.4 Extrasolar Noble Gases 93

Chapter 4 **Water** 98

4.1 Introduction 98
4.2 Solubility Data 103
4.3 Seawater 106
4.4 Meteoric Water 110
4.5 Juvenile ^{4}He 112
4.6 Juvenile ^{3}He 113
4.7 Paleotemperature: Noble Gas Record 119

Chapter 5 **Crust** 123

5.1 Introduction 123
5.2 Sediments 123
5.3 Cosmic Dusts 129
5.4 Noble Gases in Aquifer System 138
5.5 Cosmogenic Noble Gases 140
5.6 Crustal He and Ne – Nucleogenic Components 147
5.7 The ^{4}He/^{40}Ar Ratio 155
5.8 Cyclo-Silicate Minerals 157

Chapter 6 **Mantle** 160

6.1 Introduction 160
6.2 Mantle-Derived Volcanic Rocks 161
6.3 Diamonds 165
6.4 Volcanic Gases, Fumaroles, and Hydrothermal Waters 167
6.5 Mantle Xenoliths 168
6.6 Mantle Noble Gas (Abundance) 169
6.7 Mantle Noble Gas (Isotopic Ratios) 178
6.8 Degassed, Less-Degassed, and Regassed Mantle 187
6.9 Mantle Degassing Mechanism 190
6.10 Mantle Degassing Chronology 197
6.11 Excess $^{131-136}$Xe: ^{244}Pu or ^{238}U Derived? 203
6.12 The Mantle He Flux 205
6.13 ^{4}He/Heat and Mantle Dichotomy 212

Chapter 7	**Noble Gases in the Earth** 217
7.1	Overview 217
7.2	Primordial Noble Gas in the Earth 218
7.3	SW, Q, and Terrestrial Noble Gases 227
7.4	Ne Fractionation 230
7.5	Missing Xe 233
7.6	Radiogenic ^{129}Xe* and Fission Xe 235
7.7	^{244}Pu/^{238}U-^{129}I/^{127}I Systematics 238
7.8	Origin of Terrestrial Noble Gas 241
7.9	Inventory 249

References   253

Index   281

Preface to the Second Edition

When the first edition of *Noble Gas Geochemistry* was published in 1983, this discipline was still comparatively underdeveloped, and few people seemed to expect that this apparently arcane subject would become one of the major tools of geochemistry. But noble gases have become mainstream, spoken of in the same breath as Pb, Sr, and Nd. Due to unique properties such as extreme scarcity in nature and (almost) perfect lack of chemical interaction, noble gases are now being used as a geochemical tracer to address a variety of problems in the earth and planetary sciences in ways that other tracers cannot. In this light, we thought that the time was ripe to revise the first edition in accordance with recent developments.

Current noble gas geochemistry deals with very broad subjects ranging from the origin and evolution of the earth and the solar system to local geological problems. A single-volume monograph cannot deal comprehensively with all these issues. In this revised edition, we, therefore, decided to concentrate on the more fundamental aspects of noble gas geochemistry, necessarily forcing us to give short shrift to many specific geological applications. Considerable space is devoted to a general discussion of the physics and chemistry of noble gases. In the last decade, much laboratory work has led to progress in understanding adsorption, absorption, and diffusion of noble gases in melts and in solids. These subjects are directly relevant to earth science; hence, in Chapter 2 we present extensive discussions with as much recent data as possible. Chapter 6 is essentially new and constitutes the longest chapter, in accordance with rapid development in this field. Mantle evolution is one of the most successful examples of noble gas geochemistry in recent application, and topics such as mantle dichotomy, in noble gases as in other systems, and the evolution of solid earth-atmosphere system are central issues attracting heated debate.

Since the discovery of nontrivial amounts of extraterrestrial noble gases in meteorites by Gerling and his colleagues in the early 1950s, noble gas measurements have been integral to meteorite studies and have contributed greatly to our understanding not only of the origin of meteorites but also of the planetary system as a whole. Because the scope of this book has to be limited, however, we have had to be content with a relatively concise review of this subject, concentrating on aspects with special relevance to noble gas *geo*chemistry (Chapter 3). Recent developments in meteorite studies clearly suggest that a deeper understanding of terrestrial noble gases can only be gained in the framework of noble gas *cosmo*chemistry. Hence, in Chapter 7 we discuss terrestrial noble gases in the general context of cosmochemistry. Interaction of cosmic rays with terrestrial materials is another specific example that shows the importance of nonterrestrial noble gases in noble gas geochemistry. This new field is discussed in Chapter 5.

The second edition is essentially a new book except for Chapter 4 (Water), which is somewhat modified, and Chapter 2, which maintains the original flavor but with some significant new material.

Thanks to the advent of commercial high-precision mass spectrometers for the analysis of extremely small quantities of noble gas in the mid-1980s, a large amount of high-quality noble gas data are now available in numerous publications. This accessibility has made a direct comparison of experimental data from different laboratories much easier and has also greatly aided meaningful interpretation of the data. In this spirit, we have tried to accommodate as much basic data for terrestrial noble gases as possible in table and figure formats, mostly taken from recent publications. In dealing with these data, however, we have focused on our own views while trying to attend to and illuminate diverse interpretations. Readers seeking more extended advocacy of alternative views will readily find them in the liberal literature citations herein.

Finally, we owe much to too many colleagues to do justice to them in a reasonable space. But we acknowledge the special help of Bob Walker, Kazuo Saito, Tom Bernatowicz, Takuya Matsumoto, Rainer Wieler, and Jun-Ichi Matsuda. We also give special thanks to the McDonnell Center for the Space Sciences at Washington University for its assistance in helping the authors get together.

Minoru Ozima and Frank A. Podosek

During the copyediting stage of this book we learned of the death of John H. Reynolds, and it seems fitting to add some remarks here. Although everybody has forebears, many view John Reynolds as the "father" of the study of the natural occurrence of noble gases; for decades he was the scientific leader of this discipline. John built the machine – the one that graces the cover of this book – that was truly useful for analyzing the tiny amounts of noble gases that commonly characterize natural materials, using principles that even today lead people to speak of Reynolds-type mass spectrometers. It was not enough to build and further develop the instrument, however. For many years John applied this capability to scientific investigations, asking many questions, answering some, but most importantly opening doors, first in cosmochemistry and then in terrestrial geochemistry, and in large measure establishing the agenda of this discipline. He was rigorous, he was intuitive, he was careful, he was bold, and, amazingly often on important issues, he was right. Perhaps even more important, many of the people now called noble gas geochemists or cosmochemists got their skills and scientific passion in Berkeley, as John's students, postdocs, or visitors, or from those who were there. We learned much from him, we owe him much, and of John Reynolds it can truly be said that noble gas geochemistry just wouldn't be the same without him.

Preface to the First Edition

In addition to being the collective name for the elements in the rightmost column of the periodic table of the elements, the term "noble gases," also "rare gases" or occasionally "inert gases," is a convenient label for the branch of the scientific enterprise concerned with studying the occurrence and distribution of these elements in nature, particularly in the earth and the terrestrial planets. This research area is usually considered part of geochemistry, although other labels would serve as well.

The most familiar and most widely practiced area of geological research involving noble gases is, of course, geochronology, especially K–Ar dating. There are many good books about geochronology and it is not our intention to try to add another. Noble gas geochemistry, the subject of this book, will mean here what it usually suggests in the geological and planetary science community: the study of the natural occurrence of noble gases and what may be learned thereby, other than determination of the ages of rocks.

In the last two decades the discipline of noble gas geochemistry has prospered, at least by the measure of getting its own sessions at scientific conferences and attracting practitioners in sufficient numbers that, regrettably, they no longer all know each other or are even familiar with each other's work. In spite of its fruits, however, noble gas geochemistry often seems to non-practitioners to have much the air of the secret society and its dark art. Among the deplorable consequences of this circumstance are too many cases where good science and important scientific results obtained from noble gases are widely ignored and, on the other hand, too many cases of uncritical acceptance of noble gas results and interpretations which are ambiguous, misleading, out of context, fatuous, or outright nonsense.

Among the reasons which could be advanced to explain why noble gas geochemistry has not been very well integrated into the larger scientific community of which it is part is, basically, unfamiliarity. Many of the traditional academic disciplines are involved in the study, and there are no courses in noble gas geochemistry in graduate school. There is also no book which covers the field, and we hope that this one will fill that need. The general style is that of the review paper: The intent is for professional level and currency and, within the limitations of scope, comprehensiveness, with sufficient background and introduction to encourage access by the student of the geosciences. A major aim of this book is that it not only be useful to our colleagues in the field but that it help those in other specializations form a critical appreciation of "noble gases."

Our intent is to take the "geo-" in geochemistry literally, i.e. our focus is on terrestrial noble gases. Still, even more than in most specializations, it is impossible to develop an appreciation of terrestrial noble gases without a corresponding appreciation of the features which have emerged from the study of noble gases in extraterrestrial materials, notably meteorites. Accordingly we have included a chapter on noble gas "cosmochemistry." This is background material, however, and not an attempt to treat this subject as thoroughly as we hope to have treated the terrestrial subjects; such an attempt would require a considerable widening of the scope of this book.

The writing of this book was planned and initiated while one author, on sabbatical leave, was a guest in the other's home institution. This was made possible by support from the Japan Society for the Promotion of Sciences, and we are very glad to express our appreciation to the JSPS. We would also like to thank the McDonnell Center for the Space Sciences at Washington University for a variety of kinds of support which greatly facilitated the preparation of this book.

M. Ozima
F. A. Podosek

Chapter 1

Introduction

1.1 Retrospect and Prospect

In view of their scarcity and failure to form chemical compounds, it is not surprising that noble gases remained unknown until relatively late in the history of chemistry. The first known experimental indication of their existence was a persistent gaseous residue after chemical removal of nitrogen and oxygen from air, as noted by Cavendish in 1784; the residue was small, however, "not more than 1/120th part of the whole," presumably attributed to experimental error, and in any case subsequently ignored. The first definitive identification came when several observers found a previously unknown line in the spectrum of the solar chromosphere during the 1868 eclipse; this was quickly recognized to belong to a new element, not yet known on earth, which was named helium ($\eta\lambda\iota o\nu$: sun); of course, no chemical characterization was possible.

The actual "discovery" of the noble gases came principally from the work of Rayleigh and Ramsay in the late nineteenth century. In 1892 Rayleigh reported that nitrogen prepared from ammonia was consistently less dense (by 0.5%) than "nitrogen" prepared from air (by removal of oxygen, carbon dioxide, and water). Both Rayleigh and Ramsay, working in collaboration, followed up this experimental clue; pursuing the possibility that the density difference reflected admixture of a heavier gas in air, they, like Cavendish, found a residue when chemically reactive species were removed from air. This residue was clearly denser than air and spectroscopic examination (by Crookes) gave a spectrum unlike that of any known element. Rayleigh and Ramsay announced the discovery of a new element in a joint paper in 1895. The discovery attracted interest beyond that due a new element because they also found it to be chemically unreactive, an unprecedented property that led to the name argon ($\alpha\rho\gamma o\nu$, negative of $\varepsilon\rho\gamma o\nu$: not working, idle, inert, lazy). They further found argon to be a monatomic gas of atomic weight 40. This result was puzzling since the concept of the periodic table was then well established, and it was surprising to find an element whose weight was that of calcium, with no apparent place for it in the sequence potassium, calcium, scandium.

Further discoveries proceeded rapidly. In seeking possible compounds of argon, Ramsay suspected that "nitrogen" previously reported to be liberated in acid treatment of a uranium mineral was actually argon rather than nitrogen; he prepared gas in this way (in 1895) and indeed identified argon in its spectrum. He also noted another line in this spectrum; in examining this gas Lockyer and Crookes quickly identified it as the helium line previously found in the solar spectrum. Helium was found to be monatomic, of atomic weight 4, and like argon, chemically inert. By 1897 Ramsay was able to conclude that helium and argon represented a new column in the periodic table and predicted that at least one additional element, of intermediate atomic weight, remained to be found. In 1898 Ramsay and coworkers conducted fractional distillation experiments on liquid air, which led to the identification of three new elements: neon ($\nu\varepsilon o\nu$: new), krypton ($\kappa\rho\upsilon\pi\tau o\nu$: hidden), and xenon ($\zeta\varepsilon\nu o\nu$: alien). The identification included spectroscopic characterization, measurement of atomic weight, and confirmation of chemical inactivity. The noble gas column of the periodic table was completed in 1900–1903 by identification of gaseous emanation of thorium and radium as a noble gas, radon ($\rho\alpha\delta\iota\upsilon\sigma$: ray).

Since their discovery more than a century ago, the unique properties of the noble gases have been the subject of much research in theoretical chemistry and physics. These gases have also found many applications as tools for scientific research and many commercial, industrial, and even medical applications as well. These uses are well known, and our further discussion will focus on the role of noble gases in geochemistry.

The first such application was in geochronology. Even while they were studying the phenomenon of radioactivity of uranium and thorium, Rutherford and Soddy noted its potential value as a geochronometer. The first radiometric age measurement, by the uranium-helium method, was reported by Rutherford at the International Congress of Arts and Science in St. Louis in 1904. Uranium-helium dating was subsequently pursued as a major methodology in geochronology and was later supplemented by (and largely supplanted by) the potassium-argon method (Wasserburg & Hayden, 1955), particularly the ^{40}Ar-^{39}Ar variant (Merrihue & Turner, 1966). Geochronological study of terrestrial, meteoritic, and lunar samples continues to be pursued very actively, and to account for much of the research in noble gas geochemistry.

Appreciation of the relevance of noble gases to broader aspects of geochemistry was somewhat later in developing. A noteworthy step was von Weizäcker's (1937) recognition that atmospheric argon was radiogenic; the corollary inference of gas transport from the solid earth to the atmosphere suggests the possibility of the quantitative evaluation of the degassing history of the earth, a subject that has received much attention and in which radiogenic noble gases play the principal role. Subsequently, Brown (1952) used noble gas abundances to argue that the Earth's atmosphere as a whole was secondary rather than primary, that is, the present atmosphere was formed by outgassing the interior (cf. Chapter 6). The discovery of (juvenile) He in sea water (Clarke, Beg & Craig, 1969) showed that this degassing process is incomplete and continues today. Considerable recent work has been done on developing the use of noble gases as geochemical tracers and indices of tectonic activity. More generally, characterization of both primordial and radiogenic noble gas abundances in the mantle is a major present concern in noble gas geochemistry because of the profound implications for both the formation of the Earth and its thermal, tectonic, and chemical evolution. Such developments and prospects are the principal subject of this book.

As with other scientific disciplines, advances in noble gas geochemistry can be seen to follow not only from prior progress in research but also from exploitation of technological and methodological developments. Early work measured gases by volume, using the handling techniques developed by Ramsay and Rayleigh and progressively refined to the point that uranium-helium ages could be determined for ordinary surface rocks rather than only for unusual and scarce uranium and thorium ores (cf. Dubey & Holmes, 1929). The present character of the discipline, however, is set by the research tool which is its sine qua non, the mass spectrometer. The first demonstration of multiple isotopic forms of a stable element (neon!) was made in the "parabolic" mass analyzer by Thomson (1912), and the first focusing mass "spectrographs" suitable for isotopic abundance measurements were constructed by Dempster (1918) and Aston (1919). During the 1930s and 1940s, the basic instrumental configuration for isotope ratio mass spectrometry evolved to the form still most commonly used today: a monoenergetic ion source (with ions of gaseous samples produced by electron bombardment), mass separation by single-focusing magnetic sector analysis, and electrical (rather than photographic) ion detection, usually with an electron multiplier for small samples. A particularly significant development for noble gases was Reynolds' (1956) introduction of techniques and methodologies for static rather than dynamic analysis. This finding permitted the accurate isotopic analysis of very small quantities of gas, which characterizes noble gas geochemistry today and justly may be said to have begun its modern era.

In the early 1980s, an ultra-high-vacuum (pressure below 10^{-10} torr can be maintained in a static condition for hours) mass spectrometer was made commercially available; the machine is capable of analyzing isotopic ratios of an extremely small amount of noble gases. With the introduction of these commercial machines, high-quality noble gas isotopic data on a variety of samples, especially those of mantle-derived materials, are now emerging, widening the scope of noble gas geochemistry and offering a powerful tool to earth science.

1.2 Geochemical Characteristics of the Noble Gases

Although this whole book is about noble gas geochemistry, two important general characteristics should be noted here. These features are commonly known by workers in the field but deserve explicit attention from the nonpractioner.

First, the noble gases are, obviously, noble (i.e., chemically inert). This does not mean that they completely fail to interact with anything (cf. Chapter 2), but that the interactions are of the van der Waals type, much weaker than normal chemical reactions. In practical terms, this means that the interactions are less complicated, and we may be more optimistic that they can be understood in terms of fairly simple ideas. (That the relevant chemistry is still not very well understood reflects more the lack of empirical data than intrinsic complexity.) It also means that the noble gases constitute a reasonably coherent group so that geochemical parameters can often be seen to vary more or less smoothly from light to heavy gases. This coherence is the reason why it makes sense to consider their geochemistry as a group, as well as why, experimentally, they are naturally studied as a group.

Second, the noble gases are scarce, whence their common alias, the rare gases. This feature is a consequence of their inertness. The noble gases are not actually rare in the cosmic sense (Chapter 3) nor in the solar system as a whole (i.e., the sun and presumably the major planets). At some point in its evolution, however, the solar system, or at least that part most familiar to us, passed through a stage in which some of its constituent elements were mostly in solid phases, whereas others, including the most abundant, were in a gas phase. The outstanding chemical characteristic of the terrestrial planets (including meteorites) is that they are made from the solids, to the virtually complete exclusion of the gases. Because of their chemical inertness, the noble gases were overwhelmingly partitioned into the gas phase and so are depleted in the earth and other accessible samples, often to a very striking degree, and in general more than any other elements.

Thus, the noble gases are trace elements par excellence. As an example, a not unreasonable value of Xe concentration in a rock is some 10^{-11} cm^3 STP/g (about 3×10^8 atoms/g), or 0.00006 ppb. It is nevertheless quite feasible to perform an adequate analysis on a 1-g sample of such a rock, in the sense of a sample to blank ratio in excess of 10^2, 5–10% uncertainty in absolute abundance, and 1% or less uncertainty in relative abundances of the major isotopes. Detection limits are much lower than this, and for the scarcer isotopes the blank and thus the quantity necessary for analysis are two to three orders of magnitude lower. It is worth noting that the reason why such an experiment is possible is the same reason why noble gases are so scarce in the first place: their preference for a gas phase and the ease with which they can be separated from more reactive species.

An important corollary of this scarceness is that many effects, which are not intrinsically characteristic of the noble gases, are best studied through them. These are effects in which small amounts of material are added to some reservoir. In many cases, such additions are undetectable against the background of material already there, but when the background is low, as it usually is for noble gases, the effects become observable and thereby interesting. An important subclass of such effects is that of nuclear transmutations, many of which are totally unobservable if the daughter nuclide is a "normal" element but which make prominent changes if the daughter is a scarce noble gas. Nuclear chemistry (Section 1.7) is thus a considerable part of noble gas geochemistry.

Many examples of the utility of scarcity will be found in this book, so one will suffice here. The nuclide ^{40}K is naturally radioactive and thus useful for geochronology. Some 90% of the decays produce ^{40}Ca but constitute an imperceptible perturbation on normally present Ca in all but a few highly unusual samples, and K-Ca dating is only marginally useful in geochemistry. The 10% of the decays that produce ^{40}Ar generally overwhelm any Ar initially present and are readily observable experimentally, and K-Ar dating is probably the most extensively practiced form of geochronology. By noble gas standards, the Earth is awash in radiogenic ^{40}Ar. Indeed, on the basis of atmospheric ^{40}Ar overabundance relative to other Ar isotopes, von Weizäcker (1937) predicted the decay of ^{40}K to ^{40}Ar even before K was known to be radioactive.

Finally, we should note that even though Rn is chemically a noble gas, geochemically it might as well not be. Rn has no stable isotopes and is found in nature only because it is part of U and Th decay schemes (Section 1.7). Because of its extremely low abundance even by noble gas standards, it is observed by different techniques (based on its radioactive decay) than the other noble gases. Most important for our purposes, the geochemical factors that determine the distribution of Rn seldom have any close relation to the factors that control the other noble gases. Accordingly, we make no attempt to present a comprehensive description of the geochemistry of Rn.

1.3 Constants and Conventions

A number of physical constants and conversion factors frequently useful in noble gas geochemistry calculations are collected in Table 1.1. These values have been used for all the calculations in this book.

In subsequent data tabulations in this book, primary experimental data are stated with their attendant uncertainties when it is important to do so. Most numerical data are stated without uncertainties, however, because experimental uncertainties are commonly minor in comparison with those originating in the failure of natural systems to conform to simple models or in the fuzziness of quantitative interpretations. In general, numerical values are stated with more significant figures than their precision justifies. This reflects not ignorance but the utility of many such data as starting points for calculations in which the excess figures help alleviate two annoying problems: the accumulation of rounding errors in calculations based only on

Table 1.1. *Physical constants and conversion factors*

Avogadro's number	N_A	6.02217×10^{23} molecules/mole
Gas constant	R	8.31434 J/K/mole, 82.0578 cm^3-atm/K/mole
Boltzmann's constant	k	1.38062×10^{-16} erg/K
Ice point (0°C)		273.15 K
Electron charge	e	1.60219×10^{-19} C
Planck's constant	h	6.6256×10^{-34} J-s
1 atm (760 torr)		1.01325×10^6 dyns/cm$^2 \equiv 1.01325 \times 10^5$ Pa
1 cm^3 STP		4.465×10^{-5} mole = 2.688×10^{19} molecules
1 cal		4.184 J
1 eV/molecule		23.06 kcal/mole
1 amu (^{12}C $\equiv$ 12)		1.66053×10^{-24} g
RT at 0°C		2.24141×10^4 cm^3-atm/mole
RT at 25°C		2.44655×10^4 cm^3-atm/mole

significant figures and the vexation of obtaining different values of the same quantity when calculated by different paths.

Presumably in continuity with early investigations in which gases were actually measured by volume, quantities of noble gases in modern geochemical literature are generally reported in terms of cm^3 STP. This has the advantage that cm^3 STP are proportional to number of atoms, which is the way gases are measured in the mass spectrometers responsible for most of the modern data. Numbers of atoms, rather than, say, weight percent or partial pressures, are fortunately also the units generally most convenient for calculations and interpretations. Reporting of gas quantities in moles, or actual numbers of atoms, would have the same advantage but is not customary. Unfortunately, none of these units (nor others that might be chosen) has the desirable feature of being comparable to a sufficiently extensive set of observed data that its use would permit abundance data to be remembered as integers rather than powers of ten.

A widespread convention in geochemistry, particularly for isotopic data, is use of values in which numerical values are expressed as permil (‰) deviations from some standard:

$$\delta_x = 1000(X/X_0 - 1) \tag{1.1}$$

where X is the quantity of interest and X_0 is the standard value. Other normalizations such as 10^2 for percent (%) deviations or 10^4 for epsilon (ε) deviations are also used (unfortunately, noble gas data cannot yet justify using the epsilon notation), but they do not have as consistent notation as δ for ‰ deviations. This convention has the advantage that small deviations from the standard value are more readily apparent. It has the more important advantage that if the standardization is based on a physical standard measured under the same circumstances as the sample, δ_x can be determined from the ratio of instrumental responses without the uncertainties involved in correcting either X or X_0 for instrumental isotopic discrimination. In high-precision "stable isotope" (O, C, N, S, etc.) geochemistry, δ values are thus the norm as primary experimental data, so variations of isotopic ratios from the standards are known to better precision than are the absolute ratios.

In noble gas mass spectrometry, comparisons using δ values are frequent, but primary experimental isotopic data are generally absolute ratios rather than δ values. To a large extent, the standardization is accomplished by calibrating instrumental performance by analysis of a widely accessible standard (air), but for practical reasons the calibrations are generally less frequent and employ less of a match of sample and standard conditions than is the rule in stable isotope geochemistry. Fortunately, this entails little difficulty in practice, since isotopic effects are frequently so large that minor discrepancies are unimportant or, for all practical purposes, nonexistent. For a few cases of effects near the border of observation, however, a more thorough adoption of δ-value normalization could be of benefit.

1.4 Nomenclature

Most specialized disciplines have their own specialized jargons, each a lexicon and syntax developed partly to make communication more efficient and partly by historical accident. For convenience, this section lists a number of words and phrases used elsewhere in this book and in the pertinent literature that require special attention. Some such terms will be obscure outside the circle of communicants, some have special meanings or connotations beyond their normal dictionary definitions, and some need particular attention because they are not used consistently in the literature.

Component: In isotopic geochemistry, and especially in noble gas geochemistry, this term is widely used to designate any compositionally well-defined and uniform reservoir. It is particularly useful to describe component resolution (i.e., the attempt to interpret analytical data as a superposition of previously known components), the reservoirs of which might have been sampled by the specimen under consideration. A component may have a single and well-understood origin (e.g., the collection of Xe isotopes produced by spontaneous fission of ^{238}U), but it need not. Thus, Xe in air is an ill-understood mixture of Xe contributed by a number of sources; it is nevertheless useful to consider air Xe itself as a component because it is a uniform reservoir that can be and is sampled by materials of interest. The term *component* usually denotes a reservoir of isotopes of a single element. In some cases, it may include more than one element, but it is then less useful. Thus, fission Xe from ^{238}U will be produced in proportion to ^{4}He, and together the Xe and He will constitute a component, but the Xe and He can be very easily separated (much more so than, say, ^{136}Xe and ^{134}Xe), and the notion of U-derived gas is less useful in component resolution than is U-derived Xe.

Nuclear Component: A nuclear component is one generated by nuclear transmutation. For example, Xe produced by fission of ^{238}U is a nuclear component; air Xe, in contrast, would generally not be considered a nuclear component, even though it contains a contribution from ^{238}U fission. Ultimately, all the isotopes have been produced in processes controlled by their nuclear properties, and so everything is a nuclear component, but it is still useful to make a distinction. In particular, it is common to designate as a primordial component all the nuclides present in the solar system when it was isolated from the rest of the galaxy (cf. Chapter 3), in contrast to a variety of nuclear components generated within the solar system since that time. Subdivision of this category is common (e.g., a radiogenic component is one generated by decay of natural radionuclides).

Trapped Components and In Situ Components: Noble gases in solid samples are grouped into two classes designated trapped and in situ. The *in situ* label designates a nuclear component that is still in the same locations in which it was generated. Gases that are not in situ are trapped. The distinction is operational (although it is sometimes difficult to determine experimentally) and only partly genetic. An in situ component is necessarily a nuclear component, but the origin of a trapped component need not be specified. Usually, we think of a trapped component as originating outside the specimen in question so that the specimen sampled some external reservoir; the gas in the reservoir may itself be primordial, a nuclear component, or a mixture. The origin is not necessarily external, however. As an example, a rock may contain some ^{36}Ar of external origin (trapped) and ^{40}Ar generated by decay of ^{40}K (in situ). If the rock is melted in a closed system and then solidifies again, the subsequently produced radiogenic ^{40}Ar will be an in situ component. The ^{40}Ar produced before the melting will no longer be in the same microscopic locations in which it was produced, so will be a trapped component. As specified, the distinction is operational and rests on the possibility, at least in principle, of separating trapped and in situ components by mechanical or chemical disassembly of the rock or by different diffusional behavior manifested in stepwise heating. (Hence, the distinction between trapped and in situ is not made for liquid or gaseous samples.) Thus, in the example, the first-generation ^{40}Ar, although radiogenic, will have become homogenized with the ^{36}Ar and no longer separable from it by any means short of a Maxwell's demon (or mass-dependent isotopic fractionation). The second-generation ^{40}Ar will be separable from these but not, say, from ^{39}Ar produced by ^{39}K(n, p) in a neutron irradiation: together these will constitute a single in situ component because they are sited in the same place (i.e., wherever the K is). The total ^{40}Ar is thus split between two components, one trapped and one in situ, and these components can be separated. This situation is not hypothetical: "excess ^{40}Ar," not uncommon in rocks, is radiogenic but trapped, not in situ.

Solar and Planetary: In addition to their normal usage in designating the sun and a planet, respectively, the terms *solar* and *planetary* are used to designate specific patterns of noble gas elemental abundances. These are discussed in Chapter 3.

Atmosphere: Air is the gas phase at the surface of the earth. In this book, as is common in the literature, *atmosphere* will be used in a more general sense to designate all the volatiles in surface reservoirs of the Earth, including air but also water and sedimentary rocks.

Juvenile: It is generally believed that the Earth's atmosphere was formed by degassing of volatiles from the interior. Volatiles that have remained in the Earth's interior, have never been part of the atmosphere, and are observed to be entering the atmosphere for the first time are designated as juvenile. *Atmospheric* is an antonym to juvenile.

Fractionation and Discrimination: Both terms are used in consonance with everyday meaning but also with special connotations. *Fractionation* connotes the process of selecting one element in preference to another or, more commonly in this book, one isotope in preference to another of the same element. For the latter usage in particular, it is supposed that in natural isotopic fractionation processes the only selection criterion is isotopic mass. In nature, isotopic fractionation effects are usually small and measured in per mil (in contrast to elemental fractionation, often measured in powers of ten) and in elements of three or more isotopes can be recognized by their smooth functional (often nearly linear) dependence on differences in isotopic mass. In the technical literature, the term *discrimination* is usually reserved for the specific case of instrumental rather than natural isotopic fractionation (i.e., that which arises as part of the measurement process). In contrast to thermal ionization mass spectrometers, noble gas mass spectrometers usually have rather constant discrimination, and the normal procedure is to calibrate this discrimination by analyzing standards and applying the corresponding correction to sample data. When small fractionation effects are inferred, it is sometimes questionable whether they are actually natural fractionations or perhaps simply artifacts (improper correction for instrumental discrimination). To our knowledge, no geochemical noble gas investigations have used the double-spike technique, which could distinguish between natural fractionation and instrumental discrimination. Even for geochronological applications, the use of a double spike (cf. Macedo et al., 1977) is very rare. However, a single spiking technique is now widely used in paleotemperature application of noble gas geochemistry, in that a high-precision quantitative analysis of noble gas amount is required (e.g., Poole et al., 1997).

Partition (Distribution) Coefficients: In describing the partitioning of a trace element among coexisting phases, we frequently use a partition (distribution) coefficient for a given element, defined as a concentration ratio C_2/C_1. Here C is concentration, and the subscripts identify the phases; often the normalizing phase is some convenient reservoir, such as a silicate melt, with which several other phases may equilibrate. For noble gases, it is often most convenient to normalize to a gas phase. If the concentrations are expressed in the same units, the distribution coefficient is dimensionless. It is conventional to cite noble gas concentrations in condensed phases in cm^3 STP/g, however, and to describe the gas phase by partial pressures, so most of our numerical citations of distribution coefficients will be in cm³ STP/g/atm.

1.5 Air

Air is a major reservoir for terrestrial noble gases. It is possible that air is *the* major reservoir for the Earth (i.e., that it accounts for most of the terrestrial noble gas inventory, cf. Section 6.9); it is often so assumed, but this assumption should not be made lightly. In any case, air is certainly the most conspicuous and accessible reservoir of noble gases, and its characterization is of fundamental importance in noble gas geochemistry. The geochemical considerations are taken up elsewhere in this book; here we focus on the presentation of basic data. It is noteworthy that these data are important not only in the direct sense of providing information about the Earth and its atmosphere but also in the sense of serving as a standard for noble gas geochemistry. Most data are produced by mass spectrometers whose performance, both sensitivity and mass discrimination, is calibrated by analyzing samples of air or of secondary standards whose parameters are ultimately determined by comparison with air. At least as far as noble gases (other than Rn) are concerned, we generally assume that air is compositionally uniform (not counting high altitudes and nuclear power plants, which are unimportant for total inventory). Actually, studies directed to this point are scarce, but there seems no reason to challenge this assumption (cf. Mamyrin et al., 1970).

Generally accepted data for elemental abundances and isotopic compositions are given in Tables 1.2 and 1.3, respectively. The air "concentrations" in the rightmost column of Table 1.3 are often used as a normalization for observed concentrations in samples. If the atmosphere actually does account for nearly the total terrestrial inventory, then indeed these values are near the average concentration of noble gases in the materials that accreted to from the Earth. Use of these data for normalization does not constitute endorsement of this proposition, however, and whether or not they represent the terrestrial inventory, they are a convenient data set with elemental ratios of air and absolute abundances of the same order of magnitude as many samples.

The primary experimentally determined data in Tables 1.2 and 1.3 are the isotopic ratios and volume fractions in air and the total mass of air; these are tabulated with their reported uncertainties. Quantities derived from these are shown without error limits in Tables 1.2 and 1.3; for reasons cited earlier, they are stated to more significant figures than are justified by the precision of the primary data from which they are calculated.

It seems likely that application of presently available information and technology could result in some improvement in the elemental abundance data (Table 1.2). At least from the geochemical viewpoint, however, relatively little advantage would result. High-precision absolute abundances, as needed, for example, in K-Ar dating or studies of gases dissolved in water, are generally isotopic dilution data calibrated

Table 1.2. *Elemental composition of dry air*

Gas[a]	Molecular Weight ($^{12}C \equiv 12$)	Volume Fraction[a]	Total Inventory grams	Total Inventory cm³ STP
Dry air[c]	28.9644	1	5.119×10^{21}	3.961×10^{24}
N_2	28.0134	0.78084 ± 0.00004	3.866×10^{21}	3.093×10^{24}
O_2	31.9988	0.20948 ± 0.00002	1.185×10^{21}	8.298×10^{23}
CO_2	44.0099	$(3.10 \pm 0.10) \times 10^{-4}$	2.45×10^{18}	1.248×10^{21}
He	4.0026	$(5.24 \pm 0.05) \times 10^{-6}$	3.707×10^{15}	2.076×10^{19}
Ne	20.179	$(1.818 \pm 0.004) \times 10^{-5}$	6.484×10^{16}	7.202×10^{19}
Ar	39.948	$(9.34 \pm 0.01) \times 10^{-3}$	6.594×10^{19}	3.700×10^{22}
Kr	83.80	$(1.14 \pm 0.01) \times 10^{-6}$	1.688×10^{16}	4.516×10^{18}
Xe[b]	131.30	$(8.7 \pm 0.1) \times 10^{-8}$	2.019×10^{15}	3.446×10^{17}

[a] Data from *U.S. Standard Atmosphere* (1962) and Mirtov (1961) as tabulated by Verniani (1966).
[b] Other gases more abundant than Xe are CH_4, H_2, N_2O, and CO; see Walker (1977).
[c] Verniani (1966), based on 0.33% H_2O in total air mass $(5.136 \pm 0.007) \times 10^{21}$ g.

volumetrically rather than by comparison with air. In other applications, concentration differences between air and other samples are seldom significant below 10%, more commonly not below a factor of two.

The opposite is true, however, for isotopic data (Table 1.3). Isotopic comparisons are of major concern in noble gas geochemistry, both terrestrial and extraterrestrial, and experience has shown that as error limits have shrunk in response to technological improvement, the arguments have followed the error limits down to progressively finer levels. In many instances involving comparison of air and sample compositions, or even only the use of air in instrumental calibration, the uncertainties in air isotopic composition are not inconsequential.

For quite some time the "industry standard" for air isotopic composition was the pioneering work of Nier (1950a, 1950b) and Aldrich and Nier (1948). For Ar, Nier (1950a) calibrated absolute instrumental discrimination by analyzing a spike mixture of separated isotopes whose composition was determined volumetrically. Absolute discrimination calibration for Ne, Kr, and Xe (Nier, 1950b) also used this spike: analysis of Ar^{2+} for discrimination in the Ne mass range and analysis of Kr^{2+} and Xe^{3+} in the Ar mass range to determine one isotope ratio, which was then used to determine discrimination for Kr^+ and Xe^+. Nier (1950a) is still the accepted standard for Ar. The more recent redeterminations of He and Ne also utilize volumetrically controlled mixtures of separated isotopes. For Kr and Xe, more precise data show persistent deviations from Nier's (1950b) ratios so that better air compositions are available, but no improved discrimination calibration has been made. Various laboratories (e.g., Podosek et al., 1971; Basford et al., 1973; Bernatowicz et al., 1979) have thus adopted compositions based on detector signal ratios corrected for discrimination so as to agree with Nier's (1950b) data, either in a specific ratio or to minimize overall mass-dependent trends. Such data are thus uncertain by an overall

Table 1.3. *Isotopic composition and abundance data for air*

Isotope	Isotopic Ratios[a]	Isotopic Ratios (alternate)	Percent atomic Abundance	Total abundance (cm³ STP/g)[b]
³He	0.000001399 ± 13		0.00014	
⁴He	≡1		100	3.473×10^{-9}
²⁰Ne	≡1	9.80	90.5	1.091×10^{-8}
²¹Ne	0.00296 ± 2	0.029	0.268	
²²Ne	0.1020 ± 8	≡1	9.23	
³⁶Ar	0.003378 ± 6	≡1	0.3364	2.083×10^{-8}
³⁸Ar	0.000635 ± 1	0.188	0.0632	
⁴⁰Ar	≡1	295.5[c]	99.6	
⁷⁸Kr	0.006087 ± 20	0.0199	0.347	
⁸⁰Kr	0.03960 ± 2	0.1297	2.257	
⁸²Kr	0.20217 ± 4	0.6623	11.52	
⁸³Kr	0.20136 ± 21	0.6597	11.48	
⁸⁴Kr	≡1	3.276	57.00	4.307×10^{-10}
⁸⁶Kr	0.30524 ± 25	≡1	17.40	
¹²⁴Xe	0.003537 ± 11	0.02337	0.0951	
¹²⁶Xe	0.003300 ± 17	0.02180	0.0887	
¹²⁸Xe	0.07136 ± 9	0.4715	1.919	
¹²⁹Xe	0.9832 ± 12	6.496	26.44	
¹³⁰Xe	0.15136 ± 12	≡1	4.070	
¹³¹Xe	0.7890 ± 11	5.213	21.22	
¹³²Xe	≡1	6.607	26.89	2.437×10^{-12}
¹³⁴Xe	0.3879 ± 6	2.563	10.430	
¹³⁶Xe	0.3294 ± 4	2.176	8.857	

[a] Errors refer to last digits listed. Data from Mamyrin et al. (1970), Eberhardt, Eugster, & Marti (1965), Nier (1950a, 1950b), Basford et al. (1973), and Nief (1960).
[b] Atmospheric inventory (Table 1.2) divided by the mass of the earth (5.976×10^{27} g).
[c] Nier's (1950a) primary data were stated as ratios to ^{40}Ar; the equivalent ^{40}Ar/^{36}Ar ratio is 296.0 ± 0.5. The figure 295.5 came into widespread use on the basis of his rounded percentage abundances, and Steiger & Jäger (1977) have recommended its adoption as a convention for geochronology.

mass-dependent fractionation, although this should not be too severe, probably not more than about 1‰/amu. The He, Ne, and Ar data in Table 1.3 are generally used for comparisons and calibration; no widely adopted alternative to Nier's (1950b) Kr and Xe compositions has yet emerged.

1.6 Atmosphere

In everyday life and in many scientific contexts the terms *air* and *atmosphere* are essentially synonymous. But in some geochemical contexts there are significant distinctions. *Air* is the more restricted term, referring to the gas phase at the surface of

1.6 Atmosphere

the Earth, and *atmosphere* is the more general term, including air and also the hydrosphere and some parts of the biosphere and lithosphere. Formally, the atmosphere is usually taken to be the near-surface inventory of excess volatiles, elements or their compounds present in amounts larger than those attributable to weathering of primary igneous rocks (Rubey, 1951). These include H, C, N, S, Cl, and the noble gases; the reactive elements are all volatile or form plausible volatile compounds. Most of the mass of some of these elements now resides elsewhere than air (e.g., H in the world ocean, C in carbonate sediments, S in sulfide and sulfate sediments), but under conditions not drastically different from those now prevailing, they could indeed be present primarily as gases (in which case we would have very dense air composed primarily of H_2O and CO_2). A few other elements are in the excess volatiles category, including B, F, Br, I, Sb, and Se (Horn & Adams, 1966); however, they are not as volatile and their supply by igneous rock weathering may not be as small. O is an interesting special case, an abundant and reactive element that forms both volatile and refractory compounds; in the form of H_2O, CO_2, carbonates, and the like, it is considered part of the atmosphere, although it is unclear whether it is present in significant excess of the weathering supply. The O_2 in air is a geochemical anomaly; the oxygen fugacity in the solar system at large, and in the Earth as a whole, is far too low to permit any significant equilibrium abundance of free O_2. In the past, minor but significant atmospheric O_2 may have been produced by photodissociation of H_2O in the upper atmosphere (followed by escape of the H, so that it could not recombine with the O). At present, atmospheric O_2 is supported by a constant resupply by biological photosynthesis, without which O_2 would be quickly consumed by a variety of chemical reactions (e.g., oxidation of organic C or of Fe^{2+} to Fe^{3+}).

It is generally believed that the Earth's present atmosphere (and those of Mars and Venus) are secondary, which means that the constituents of the atmosphere were originally part of the solids that accreted to form our planet. The contrast is with a primary atmosphere, which would be composed of gases captured, as gases, from the solar nebula by the gravitational field of an accreting solid planet, or perhaps even by nebular gases' self-gravity; Jupiter, for example, has a primary atmosphere in this sense (indeed, it is mostly atmosphere). A secondary atmosphere would be generated as part of the solid planet's evolution, by preferential transport of volatiles to the surface (degassing), for example, in a fluid phase (cf. Section 6.8 and 6.9).

The basic argument that the Earth's present atmosphere is secondary is quite simple: Its composition is wrong (Brown, 1952). Gravitational capture does not depend on chemical properties, and a primary atmosphere should have solar (cosmic) composition (cf. Section 3.2), perhaps fractionated by mass but not by chemistry. If the Earth ever had a primary atmosphere, it is now gone (and must have been gone throughout most of the Earth's history). The limiting factor turns out to be the noble gas Ne, the element most depleted on the Earth relative to solar abundance. If we supposed that Ne actually represented primary nebular gas, the corresponding quantities of other atmospheric gases would be minor or trivial compared to their atmospheric inventories, and would thus be inferred to be secondary. (Actually, good arguments can be advanced that the Ne itself is also secondary.)

The chemically reactive atmospheric volatiles typically undergo significant exchange among the various surface reservoirs (air, the hydrosphere, the biosphere, sediments), and their global geochemistry is discussed in terms of "budgets," which include fluxes as well as inventories. In contrast, we often assume, usually implicitly, that noble gas atmospheric inventories are chiefly in air (with only modest amounts dissolved in sea water or trapped in sediments), and furthermore that the atmosphere is conservative for noble gases (i.e., that when a noble gas is degassed from the solid Earth into the atmosphere, it stays there forever). This is not a trivial issue because keen interest attaches to the processes and rates of solid earth degassing, and many models rely on the atmosphere to provide a boundary condition in the form of the integral of prior degassing over geologic history. In some respects, conservation of atmospheric noble gases and residence primarily in air is a reasonable assumption, but there is one well-known exception, and the possibility of other exceptions or qualifications deserves attention.

The well-known exception is He, which escapes from the exosphere, partly by straightforward Jeans escape and partly by nonthermal (ionization) processes (cf. Section 7.9). (Other species also escape, but under present circumstances the escape is only significant in depleting the atmospheric inventory for He.) Atmospheric He must thus also be discussed in terms of a budget. Exospheric losses are balanced by source terms including degassing of the solid earth, solar wind capture, and nuclear production. The separate budgets for ^{3}He and ^{4}He are not closely linked; their sources are geochemically distinct, and their primary loss mechanisms are also different. Their atmospheric lifetimes are short on a geological time scale, of order 1 Ma for ^{4}He and of order half that for ^{3}He. In geochemical context, there is thus nothing particularly fundamental about the atmospheric ^{3}He/^{4}He ratio, and, although we have no evidence on the point, we can expect that this ratio varies significantly over geological time. There is even a claim for an anthrolopologic change in ^{3}He/^{4}He (Sano et al., 1989).

Again from the geochemist's perspective, the fact that He escapes from the atmosphere is, on the whole, a good thing. Because of its short lifetime, the abundance of He in air and in reservoirs that contain gases obtained from air (notably sea water) is at least a few orders of magnitude lower than it would be if the atmosphere were conservative for He. This circumstance has facilitated the identification of primordial He in terrestrial samples and of juvenile He degassing from the solid earth (see Chapter 6) simply because of low atmospheric contamination. If He accumulated in air, these discoveries probably would not have been made, nor would the other discoveries resulting from the investigations initially attracted by He; there would be fewer noble gas geochemists who would have much less to say; and the discipline of geochemistry at large would be significantly the poorer.

There is also the possibility of significant loss or flux of noble gases from air in the other direction (i.e., recycling back into the mantle by tectonic subduction or perhaps even just "hiding" of gas in some ill-studied reservoir, in quantities large enough to be important to inventory considerations). The issues here are more controversial, however, and there is no well-founded consensus on whether this effect is important or can safely be ignored. If loss, recycling, or hiding are important at all, they are most likely to be so for Xe, the least noble of the noble gases and the most likely to partition significantly into some reservoir other than air. These issues arise in a variety of contexts discussed elsewhere in this book (Sections 3.3, 6.8, and 7.5).

Caution of a different kind is in order in considering how far back in geological history it is safe to assume that conditions were fundamentally similar to those now prevailing. It is frequently noted, for example, that the overall level of tectonic activity is likely to have been greater (and perhaps qualitatively different) in the past than it is now simply because of greater heat production by the major radionuclides. Far enough back in time, of course, the Earth must have been quite different, but it is not clear just how far back this must be. In part, because of some aspects of atmospheric gases, for example, it can be and has been suggested that the early Earth had a primary or perhaps a much more extensive secondary atmosphere, one that was not conservative for noble gases; such an atmosphere, if it existed, is largely but not necessarily completely gone, and its dregs may be significant in the present atmosphere (Sections 3.3 and 6.8).

1.7 Nuclear Chemistry

For convenience of reference and familiarization with the isotopes of the noble gases and their nuclear neighbors, Figures 1.1(a)–1.1(f) display their places in the family of nuclides in standard N-Z format. Note, however, that the noble gases are identified as a coherent group on the basis of their chemical properties, not their nuclear properties. There is not a close correspondence between the structure of a nucleus and the electronic structure of an atom. Accordingly, there is nothing special about the physics of noble gas atomic nuclei taken as a group (except for the coincidence that the α particle, which plays a unique role in nuclear physics, also happens to be the nucleus of ^{4}He).

The noble gas isotopic abundances indicated in Figure 1.1 are those of air; it is noteworthy that the natural range of isotopic abundance variations of the noble gases is greater than that of most other elements, and also that air is a unique reservoir with a unique history, so that air composition is not necessarily representative of noble gas isotopic compositions in any given sample, or in the solar system or the cosmos at large. Since the synthesis of isotopic species is governed by their nuclear properties, there is nothing special about noble gas abundances on a cosmic scale (Section 3.3). The abundance and isotopic composition of He reflect conditions in the Big Bang, and the other noble gas nuclei reflect conditions inside stars. Interesting as the noble gas nuclei are in terms of cosmology, nuclear physics and astrophysics they have nothing to do with the fact that when these nuclides acquire electrons, the electrons form a closed outermost shell and become noble gas atoms.

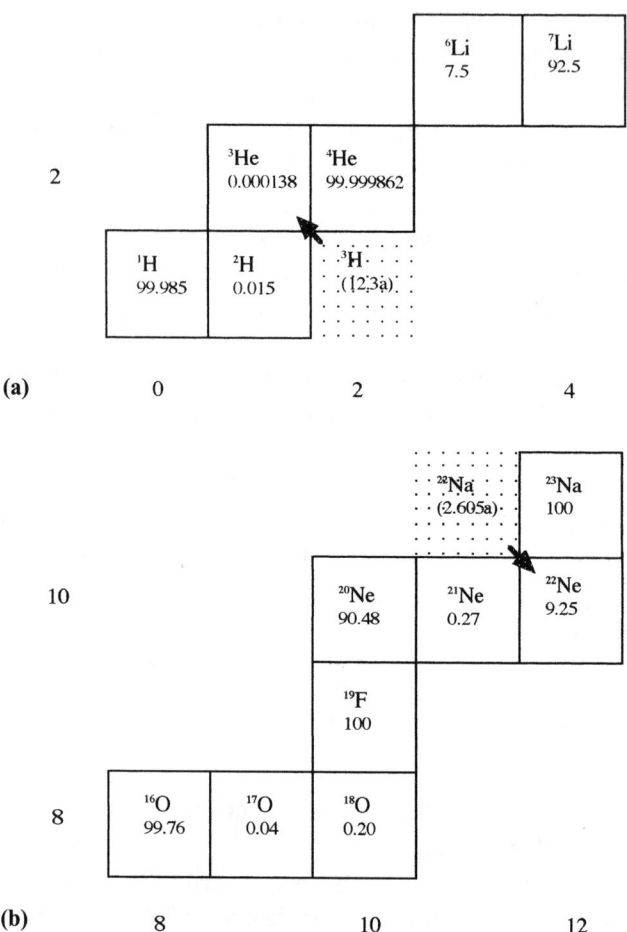

Figure 1.1 Display of the nuclei of noble gases and neighboring elements; neutron number is shown on abscissa, and proton number on ordinate. Stable nuclides and long-lived (extant) radionuclides are shown as solid boxes with percent normal abundances (from Walker, Parrington & Feiner, 1989); short-lived radionuclides are shown as shaded boxes, with half-lives in parentheses and the decay path indicated by thick arrows. Thin-line arrows indicate fission-fragment decay paths. Panels (a)–(f) illustrate the regions around He, Ne, Ar, Kr, Xe, and Rn, respectively.

If we understand *nuclear chemistry* to mean the study of the effects of nuclear transformations, especially of proton number (i.e., transformation of one element to another), then from the physicist's or chemist's viewpoint the nuclear chemistry of noble gases is neither more nor less interesting than any other group of elements, and there is no reason to distinguish noble gases from any other elements. From the geochemist's viewpoint, which defines the scope of this book, particular interest is attached to natural nuclear chemistry effects, which involve enough nuclear transformations to produce *observable* variations in elemental or, more commonly,

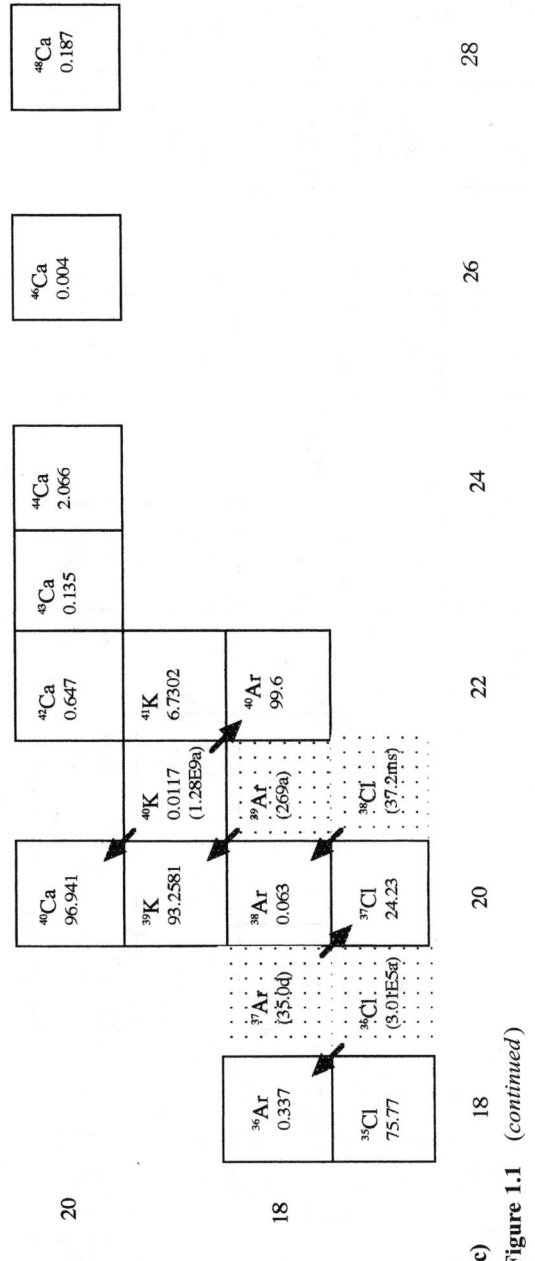

Figure 1.1 (continued)

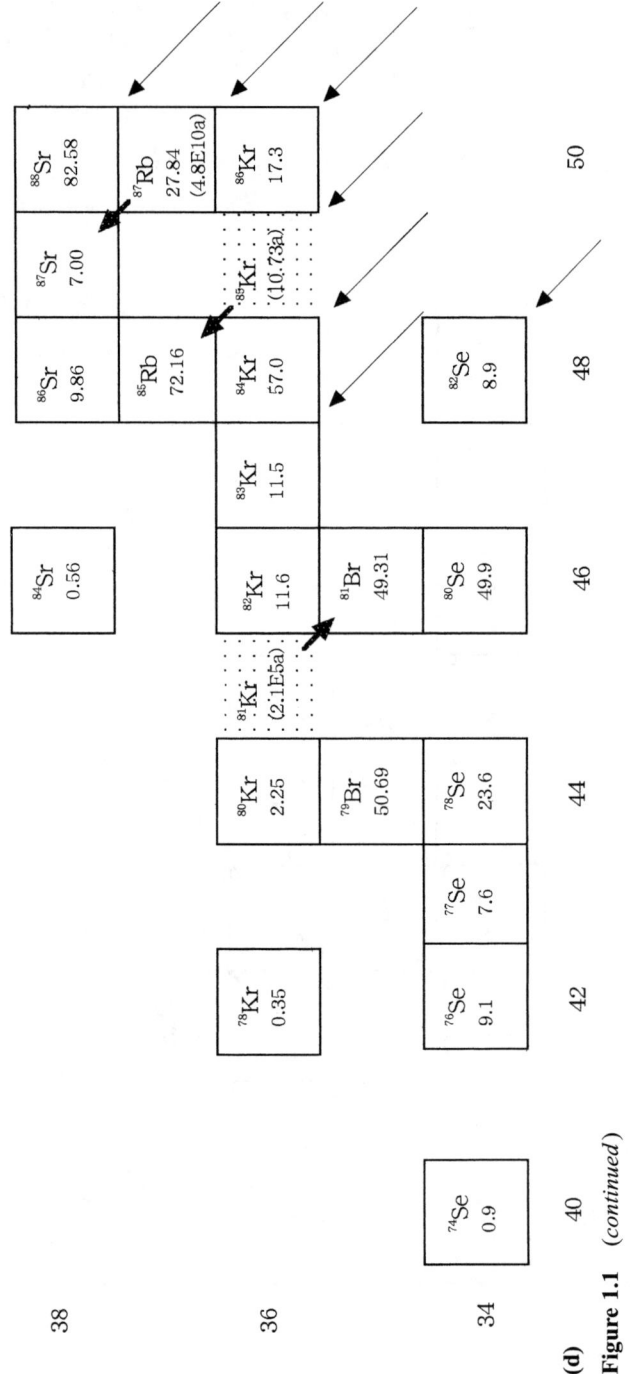

Figure 1.1 (continued)

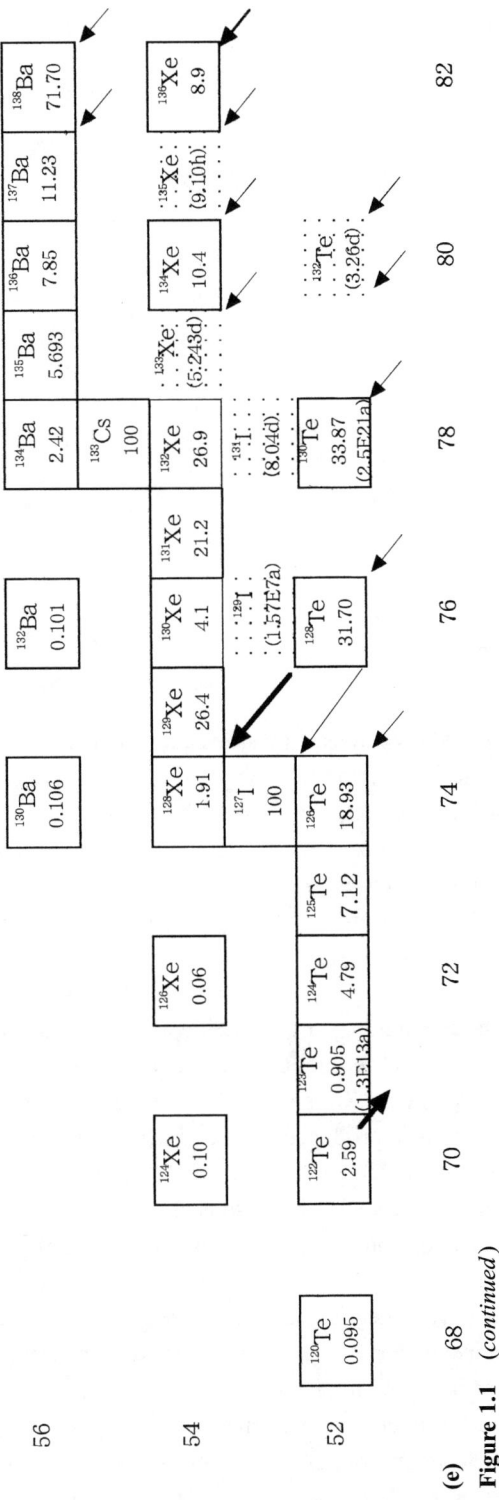

Figure 1.1 (continued)

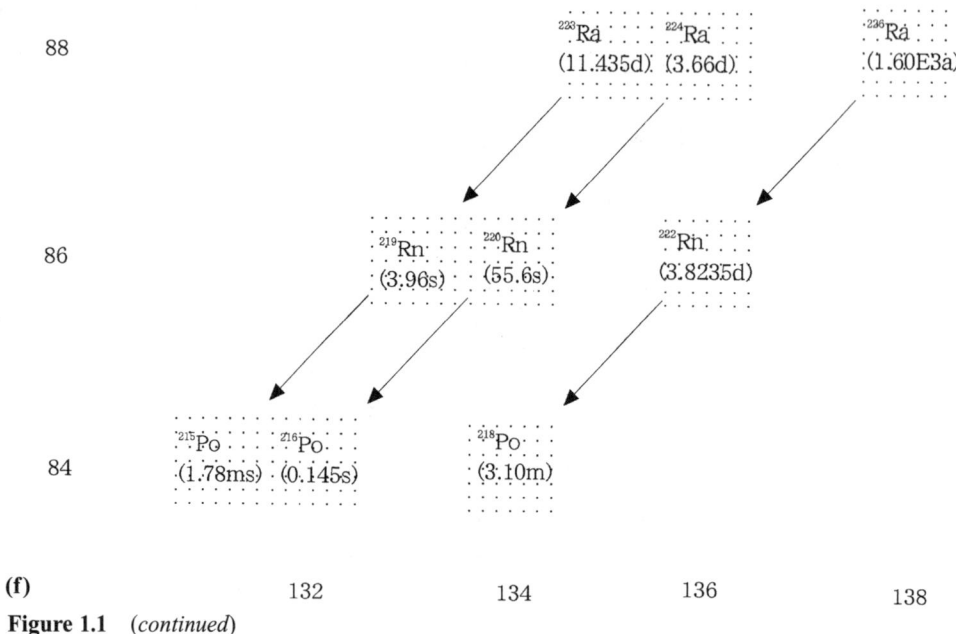

(f)
Figure 1.1 (*continued*)

isotopic abundances. Many nuclear transformations occur naturally, of which only a very small fraction lead to any effects large enough to be observable (but note that observability is a moving target, changing with the state of the analyst's art and also with advances in appreciation of which natural materials might display interesting effects). The relatively few cases in which effects *are* big enough to be observed are frequently exploited to great profit to geology (e.g., radiometric geochronology) and other sciences.

By the criterion of observability in geochemical context, there are very few cases of interest in which the parent or target nuclide is a noble gas nucleus; most of these concern decay schemes and lifetimes of radioactive species, and in the terrestrial environment only the special case Rn is important (Rn has no stable isotopes and occurs naturally only because it is produced in U and Th decay). In contrast, the range of interesting effects is very much wider when the daughter or product nuclide is a noble gas nucleus. The basic reason why this is so is chemistry, not physics. In most natural samples of the Earth or the other terrestrial planets, noble gas abundances are typically very low (Section 1.2), and small amounts of nuclear transformation products, which would have a negligible (i.e., unobservable) effect on the background abundance of a normal element, sometimes make a significant (i.e., observable) effect on the low background abundance of a noble gas.

One obvious class of nuclear transformations leading to noble gas daughters is the decay of naturally occurring radionuclides; prominent cases are collected in Tables 1.4 and 1.5. Generation of ^{4}He by decay of ^{238}U, ^{235}U, and ^{232}Th and of ^{40}Ar by decay of ^{40}K are well known. It is perhaps less well known, among terrestrially oriented

Table 1.4. *Natural radionuclides[a] decaying to noble gas daughters*

Parent	Half-life (Ga)	Decay	Daughter[b]	Yield
^{238}U	4.468	α	^{4}He	8
^{235}U	0.7038	α	^{4}He	7
^{232}Th	14.01	α	^{4}He	6
^{40}K	1.251	β	^{40}Ar	0.1048
^{82}Se	—[c]	$\beta\beta$	^{82}Kr	1
^{238}U	4.468	sf[e]	^{86}Kr	
^{244}Pu	0.082	sf[e]	^{86}Kr	
^{129}I	0.0157	β	^{129}Xe	1
^{128}Te	—[d]	$\beta\beta$	^{128}Xe	1
^{130}Te	—[d]	$\beta\beta$	^{130}Xe	1
^{238}U	4.468	sf[e]	^{136}Xe	
^{244}Pu	0.082	sf[e]	^{136}Xe	

[a] Parameters for U, Th, and K decay schemes as recommended by Steiger & Jäger (1977).
[b] Several other radionuclides decay by α emission but do not contribute significantly to global production rates.
[c] Half-life 1.0×10^{20} to 1.3×10^{20} yr (Kirsten et al., 1986; Lin et al., 1988).
[d] The half-life of ^{130}Te is controversial; arguments have been advanced for either end of the range 0.7×10^{21} to 2.9×10^{21} yr (cf. Manuel, 1991; Podosek et al., 1994; Takaoka et al., 1996); the discrepancy arises in geological interpretation rather than analytical uncertainties.
[e] Fission produces several isotopes of both Xe and Kr (those with no stable isobars of lower atomic number); see Table 1.5 for yields and compositions.

geochemists, that early in solar system history there were also present a handful of natural radionuclides with lifetimes so short that they are now extinct (e.g., Podosek & Nichols, 1997), among which ^{129}I and ^{244}Pu produce noble gas daughters. These two radionuclides are prominent in the study of meteorites, and their prior presence in the ancient Earth is also of considerable interest (Sections 6.11, 7.6, and 7.7). The equally obvious class of geochemical application of the decay of natural radionuclides to noble gases is the discipline of geochronology, including the workhorse methods of U-Th-He and K-Ar dating, but also less extensively practiced methods such as U-Xe (fission) dating and, in meteorites and even in some lunar samples, methods based on the short-lived radionuclides (e.g., Swindle & Podosek, 1988; Shukolyukov et al., 1994). The subject of geochronology as a scientific discipline, in the sense of obtaining ages of rocks, is outside the scope of this book and will not be an explicit subject for detailed exposition; this is because the types of information usually sought and obtained in geochronology are different from other aspects of noble gas studies and because the subject is both enormous and adequately treated in many other books.

Table 1.5(a). *Production of noble gases^a in fission*

Parent	Mode	Branching Ratio	Yield^{de} (%) ^{136}Xe	^{86}Kr	References
^{238}U	sf	5.45×10^{-7}	6.3 ± 0.4	0.95 ± 0.06	1, 2, 3, 7
^{244}Pu	sf	1.25×10^{-3}	5.6 ± 0.6	0.11 ± 0.03	4, 5, 6, 10
^{235}U	nf^c		6.47 ± 0.47	2.04 ± 0.02	7, 8, 9

^{a} Minor amounts of ^{4}He and ^{3}He are produced in ternary fission. Spontaneous fission of ^{232}Th has also been observed (Whetherill, 1953), as has induced fission of ^{239}Pu in the Oklo natural chain reactor (Drozd et al., 1974).
^{b} Except as noted, data from review/compilation by Hyde (1974).
^{c} Thermal neutron cross section 580 barns.
^{d} Isobaric yields; some precursors have significant half-lives.
^{e} The product of branching ratio and yield is more accurately determined than either factor separately; for ^{238}U Ragettli et al. (1994) give $\lambda_{sf}Y_{136} = (6.83 \pm 0.18) \times 10^{-18}$ yr^{-1}.

Table 1.5(b). *Xe and Kr compositions in fission*

Parent/Mode	Xe Composition (^{136}Xe $\equiv$ 1)				Kr Composition (^{86}Kr $\equiv$ 1)	
	^{129}Xe	^{131}Xe	^{132}Xe	^{134}Xe^a	^{83}Kr	^{84}Kr
^{238}U sf	<0.002	0.076 ± 0.003	0.595 ± 0.017	0.832 ± 0.012	0.03 ± 0.01	0.13 ± 0.02
^{244}Pu sf	0.048 ± 0.055	0.248 ± 0.015	0.893 ± 0.013	0.930 ± 0.005		
^{235}U nf	0.1	0.453 ± 0.013	0.677 ± 0.020	1.246 ± 0.036	0.27 ± 0.01	0.50 ± 0.02

^{a} In neutron irradiation, the yield of ^{136}Xe may be significantly enhanced due to neutron capture on ^{135}Xe (isobaric yield 6.5%, half-life 9 hr, thermal neutron cross section 3.6×10^6 barns).

Table 1.5 References: 1, Whetherill (1953); 2, Young & Thode (1960); 3, Segre (1952); 4, Fields et al. (1966); 5, Alexander et al. (1971); 6, Lewis (1975); 7, Walker et al. (1989); 8, Farrer & Tomlinson (1962); 9, Farrer, Fickel & Tomlinson (1962); 10, Hudson et al. (1989).

Tables 1.4 and 1.5 are not, and are not meant to be, exhaustive; they are instead conditioned by the criteria of observability and geochemical interest. There are, for example, several other (besides U and Th) long-lived radionuclides that decay by α emission, but their contributions of ^{4}He on a global scale, and in most rocks, are quite minor in comparison with U and Th. Long-lived ^{147}Sm (half-life 106 Ga) and short-lived ^{146}Sm (half-life 103 Ma), for example, both decay by α emission, and keen geochemical and cosmochemical interest attend each, but they are studied through their daughters ^{143}Nd and ^{142}Nd, respectively, rather than through ^{4}He. There are surely rare earth ores in which ^{4}He production from ^{147}Sm outweighs that from U and Th, but we are unaware of any relevant studies that exploit this. In contrast, ^{130}Te and ^{128}Te both have very long half-lives, and their double-beta decay production of ^{130}Xe and ^{128}Xe, respectively, are trivial on a global scale. Nevertheless, there are ores in which these daughters are prominent and are indeed the subject of interesting studies.

Another interesting class of nuclear transformations is that collectively termed spallation, consisting mostly of incidence of protons, neutrons, and alpha particles on some target nuclide at energies high enough to fragment (spall) the target without too much regard for the niceties of nuclear structure. The term *spallation* is usually reserved for cases in which the high-energy incident particles reflect, directly or indirectly, exposure of a sample to cosmic rays (usually galactic cosmic rays, but at relatively lower energies solar energetic particles can also be significant), in reflection of which the term *cosmogenic* is also frequently used to designate this class of reactions (cf. Section 5.5). With some liberty, more specific and lower energy reaction channels such as (p, n), (n, p), and (n, pn), which do not exactly constitute nuclear fragmentation, are lumped in with this category, as long as the incident particles ultimately arise from cosmic ray exposure.

In almost all relevant cases, the exposure to cosmic rays is not intense enough to change the composition of the target element directly; it is only the fragments that may become observable against a very low background abundance (i.e., those that happen to be noble gases, or not present at all save for recent production, such as the short-lived radionuclides ^{26}Al or ^{10}Be). The hallmark of spallation products is that the daughter fragments are distributed relatively evenly across significant mass ranges, without nearly as much structure as the compositions resulting from stellar nucleosynthesis. Spallation of Si targets, for example, will yield the three isotopes of Ne in roughly comparable abundances (also Mg, but that would not be observable); in contrast, the usual normal abundance of ^{21}Ne is much less than those of ^{20}Ne and ^{22}Ne, so that the manifestation of a spallation component added to Ne is elevation of the relative ^{21}Ne abundance. For the heavier elements (Kr and Xe), spallation products are typically enriched in the lighter isotopes (which also happen to be the scarcest isotopes of Kr and Xe) because of preferential removal of neutrons (relative to protons) from the target.

The characteristic range over which galactic cosmic ray production is attenuated is of order 1 m in most rocks (i.e., about $300 \, g/cm^2$) and, depending on trapped noble gas concentrations, becomes observable for exposure times typically of order 1 Ma. Isotopic effects due to cosmic ray exposure are endemic in meteorites. The Earth's atmosphere constitutes a column depth of about $1 \, kg/cm^2$ so that, even though cosmic rays are significantly attenuated at the surface of the Earth, the rate of generation of spallation products is not negligible, and noble gases can profitably be used to study surface exposures (Section 5.5), especially at high altitudes. In meteorites, the most sensitive noble gas indicator of cosmic ray exposure is usually ^{21}Ne, with ^{3}He a close second. In the terrestrial environment, the effects first become evident in ^{3}He, but ^{21}Ne may also be detectable.

Besides natural radionuclides and spallation, a handful of other interesting nuclear interactions lead to observable effects in the noble gases. Some of these are listed in Table 1.6. Most of these involve incidence of α particles or neutrons; in nature these are generated primarily by decay of U and Th, the α particles directly, and the neutrons as secondaries by a variety of (α, n) reactions. Again, the list in Table 1.6 is not intended to be exhaustive; there are many more reactions that have been found to be significant, in terrestrial samples as well as meteorites, in materials with unusual

Table 1.6. *Selected natural nuclear processes[a] generating noble gas daughters*

Product	Reaction	Reference
^{3}He	^{6}Li (n, α)	Morrison & Pine (1955)
^{21}Ne	^{18}O (α, n)	Wetherill (1954)
^{22}Ne	^{19}F (α, n)	Wetherill (1954)
^{38}Ar	^{35}Cl (α, p)	Wetherill (1954)
^{126}Xe	^{126}Te (μ, $\mu\pi\beta$)	Bernatowicz et al. (1993)
^{128}Xe	^{127}I (n, $\gamma\beta$)	Srinivasan, Alexander & Manuel (1971)
^{129}Xe	^{128}Te (n, $\gamma\beta$)	Hennecke, Manuel & Sabu (1975)
^{129}Xe	^{130}Te (μ, n$\beta\beta$)	Takagi, Hampel & Kirsten (1974)
^{131}Xe	^{130}Ba (n, $\gamma\beta$)	Srinivasan (1976)
^{131}Xe	^{130}Te (n, $\gamma\beta$)	Hennecke, Manuel & Sabu (1975)
^{132}Xe	^{131}Xe (n, γ)	Drozd et al. (1974)

[a] Other than radioactive decay (Table 1.4), fission (Table 1.5), or cosmic-ray-induced spallation (Tables 5.3 and 5.4).

histories or compositions. In this table only the most important (i.e., most readily observable) reaction of a given "kind," among a group of geochemically related processes, is listed; for example, ^{7}Li(α, ^{3}He)^{8}Be produces ^{3}He but at a lesser rate than ^{6}Li(n, α), and ^{17}O(α, n) produces ^{20}Ne but at a lesser rate than ^{18}O(α, n) produces ^{21}Ne. Among the reactions most broadly relevant, particular attention is called to those producing ^{3}He and the isotopes of Ne (the latter are sometimes called the Wetherill reactions, after Wetherill, 1954). α decay produces ^{4}He but not ^{3}He, so that, strictly speaking, "radiogenic" He designates monoisotopic ^{4}He. ^{3}He *is* produced indirectly, however, as a consequence of α decay, and, in broad terms, where there is ^{4}He, there is also likely some ^{3}He. It has thus become customary to lump the secondary ^{3}He in with the primary ^{4}He into a "radiogenic" He component. The ^{3}He/^{4}He ratio in radiogenic He is low but not vanishing (excepting short-lived radionuclides, it appears to be the most extreme naturally occurring isotopic ratio) and is useful in diagnosing the source of natural He (cf. Sections 5.6 and 6.7). Ne produced in nuclear reactions within the Earth, recognized by its diagnostic elevation of the relative abundance of ^{21}Ne and now commonly termed nucleogenic Ne, is sometimes quite prominent in crustal sources (Section 5.6) and has come to be seen as an important characteristic of "mantle" provenance (Section 6.7).

It is implicit in all the foregoing discussion that we are considering only natural nuclear reactions. The field of interest becomes much broader still if opened to consideration of anthropogenic effects, such as nuclear power plants or nuclear bombs. This topic is also mostly outside the scope of this book, but a few points are worth noting. First, the only such effect of global significance in noble gas geochemistry, as far as we know, is excess ^{3}He in meteoric water, originating from decay of bomb-produced ^{3}H (half-life 12 yr); this effect may be exploited in groundwater studies (e.g., Takaoka & Mizutani, 1987). The second is the so-called Oklo phenomenon, the only known instance of a *natural* fission chain reactor, and a unique exercise in nuclear chemistry of noble gases (e.g., Drozd, Hohenberg & Morgan, 1974; Shukolyukov et al., 1994) and other elements. Finally, many noble gas practitioners irradiate samples with neutrons (to quite high fluence, commonly of order 10^{19} neutrons/cm^2) in nuclear reactors, usually for the ^{40}Ar-^{39}Ar or ^{129}Xe-^{128}Xe (for meteorites) dating techniques. The desired reactions for these methodologies are ^{39}K(n, p)^{39}Ar and ^{127}I(n, $\gamma\beta$)^{128}Xe, respectively, but the neutron irradiation also induces a large number of other reactions that produce noble gases, some wanted and some not.

In concluding this section, note that nuclear chemistry has been an important and integral part of noble gas geo- and cosmochemistry ever since it grew into a recognizable discipline, contributing greatly to both its richness and its complexity. Nuclear chemistry is also responsible for frequent connections to other disciplines. As one example, noble gases have contributed significantly to fundamental physics in a variety of ways, providing, for example, a profitable avenue for study of double-beta decay (cf. Inghram & Reynolds, 1950; Bernatowicz et al., 1993; Takaoka, Motomura & Nagao, 1996). As one more, noble gases were responsible for observation of the first known short-lived radionuclide in the early solar system (^{129}I; Reynolds, 1960) and more recently for isolation of preserved presolar grains in meteorites (e.g., Anders & Zinner, 1993), both of which have had a major impact on the study of stellar nucleosynthesis and the formation of the solar system. The near ubiquity of nuclear effects in noble gases can be a curse or a blessing, sometimes both, depending on one's viewpoint and objectives. Spallation noble gases, for example, may complicate or preclude accurate characterization of trapped gases, but they are also the principal way to study cosmic ray exposure histories of meteorites and lunar samples. Often enough, several nuclear components are superimposed and require several layers of correction, correlation, and perhaps ancillary measurements before the wheat can be separated from the chaff. Often enough, unusual samples challenge the ingenuity of the practitioner to recognize effects never before encountered (e.g., Bernatowicz et al., 1993). This is often a daunting mission, but when successfully fulfilled, it is truly an art form of high order.

1.8 Three-Isotope Diagrams

A three-isotope diagram is simply a cartesian display of one isotope ratio as abscissa and another as ordinate, as schematically illustrated in Figure 1.2. It is a convenient way to visualize isotopic compositional variations among a suite of measurements, particularly to judge how variations in one isotope ratio may be related to those in another ratio. Three-isotope diagrams are used in all branches of geochemistry in which isotope ratios are important, although they are used perhaps more frequently in noble gas than in other studies simply because noble gas isotopic compositions tend to be much more variable than those of other elements; in any case, there are many examples in this book. To the practitioner, the properties of these diagrams are

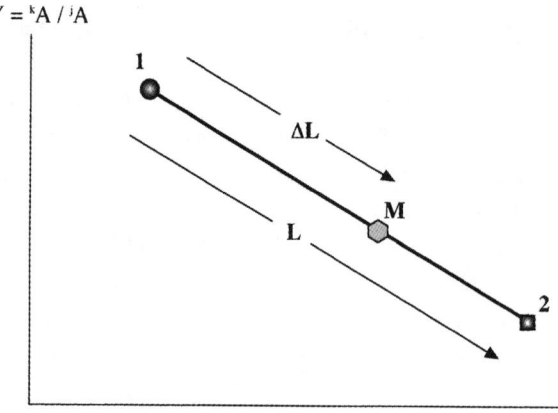

Figure 1.2 Schematic illustration of a three-isotope diagram. The abscissa is any isotope ratio $^iA/^jA$, the ordinate is another ratio $^kA/^jA$ involving a third isotope in the numerator and the same (reference) isotope in the denominator. The points labeled 1 and 2 represent any two components of distinct composition. The locus of all mixtures of components 1 and 2 is the straight line connecting these points. Point M represents any mixture composition; the ratio of the line segment lengths ΔL and L, measured from 1 to M and from 1 to 2, respectively, is the fractional contribution that component 2 contributes to the reference isotope (illustrated for 65%).

about as familiar as the alphabet, but nonspecialists sometimes complain that they are more confusing and obfuscatory than enlightening. A brief general exposition of the properties of these diagrams may thus be useful.

As in Section 1.4, a component is any uniform reservoir with well-defined composition. A component reservoir with at least three isotopes, and therefore two isotope ratios, thus defines a point on a three-isotope diagram (e.g., as labeled 1 on Figure 1.2). A second reservoir with a distinct composition then defines a distinct point (e.g., 2 on Figure 1.2). For any two isotopes ^iA and ^jA, and their ratio $R \equiv {^iA}/{^jA}$, mixing of components 1 and 2 produces an intermediate ratio R, which is given by

$$R = f_1 R_1 + f_2 R_2 \quad \text{where } f_1 + f_2 = 1 \tag{1.2}$$

Here R_1 and R_2 are the compositions of reservoirs 1 and 2, respectively; f_1 is the fraction of the reference isotope ^jA, which is contributed by component 1, and likewise for f_2. This equation is applicable to both the abscissa ratio X and the ordinate ratio Y:

$$X = (1-f)X_1 + fX_2 \quad \text{and} \quad Y = (1-f)Y_1 + fY_2 \tag{1.3}$$

where we have now explicitly substituted $f_1 = 1 - f_2$ and then dropped the subscript, so that f is understood to be the fraction of the reference isotope contributed by component 2. A critical feature of three-isotope diagrams then emerges from Equation (1.3): X and Y are both linear functions of the same parameter f; hence each is a linear function of the other. In other words, mixing two components in various proportions yields a continuum of compositions whose locus on a three-isotope diagram is a *straight line*. The lever rule applies:

$$f = (X - X_1)/(X_2 - X_1) = (Y - Y_1)/(Y_2 - Y_1) = \Delta L/L \tag{1.4}$$

where L is the length of the diagonal line segment from component 1 to component 2 and ΔL is the distance from component 1 to the point representing the mixture (Figure 1.2). Qualitatively, the closer the mixture composition is to component 2, the greater is the contribution of component 2 to the mixture (for the reference isotope). Quantitatively, the fraction f is the ratio of the distance from the mix to component 1 to the distance from component 2 to component 1, as measured along either coordinate axis or along the diagonal. To emphasize this relationship, one may also place tick marks, representing fractional increments from component 2, along the diagonal or along either axis; if the fractional increments are equal (not shown in Figure 1.2), they are equally spaced.

The arithmetic is not restricted to isotopes, and is applicable to any two ratios characterizing distinct reservoirs. Three-element diagrams are often used in geochemistry, for example. More generally, if A, B, and C are three conservative quantities characterizing some type of reservoir, then a plot of C/B versus A/B has the properties of a three-isotope diagram: mixing generates straight lines, and the linear lever rule applies.

It is important, however, that the reference isotope (the denominator) be the same for both coordinates. C/D versus A/B is not a three-isotope diagram, nor is A/B versus B/C, at least as the term is customarily used. Mixing of two components is still defined by only one free parameter so that the locus of mixture compositions is still a line connecting the end-member compositions, but in general the line is not straight; a lever rule still applies, but it is not linear, and one may still place tick marks along the mixing line but even for equal increments the ticks are not uniformly spaced. The exception is the special case in which the denominator on one axis is a constant times the denominator on the other; for example, a U-Pb concordia diagram (^{206}Pb/^{238}U versus ^{207}Pb/^{235}U) does have the linear properties of a three-isotope diagram. Similarly, plots involving normalized ratios (divided by constants) or delta values also have the linear properties, since these variants involve only scale changes and/or shifts in origin.

Some generalizations are also well known. If a third component 3 were introduced, the field of compositions accessible by mixing becomes the triangle whose vertices are the three end-member compositions (because mixing of 1 and 2 generates a line between those two points, and admixture of 3 generates another line connecting 3 with all points on the line between 1 and 2). More components generalize to the convex polygon connecting all end-member compositions. The logic can also be extended in dimensions. Three isotope ratios (four isotopes, one being in the denominator in all three ratios) define a three-dimensional manifold; mixing of two components still generates a straight line, mixing of three components defines a plane, and so on. The obvious generalizations apply to still more ratios and more dimensions, a situation readily accommodated by the formal mathematics and by computers even though they are essentially impossible to display intelligibly by a graph on a piece of paper. The number of dimensions can also be generalized downward to one; the lever rule, for example, is often used to resolve components (e.g., radiogenic and primordial He) mixing to produce an intermediate isotope ratio.

The most common use of a three-isotope diagram is to assist in the visual evaluation of how many components contribute to a given suite of isotopic data. A cluster of data points consistent, within errors, with a single point is deemed consistent with a single component, for example, whereas a finite spread requires at least two components. Particularly important is the case in which ratios vary beyond their errors and the issue is whether the variations can be understood in terms of only two components, in which case the variations will define a straight line trend, or whether more than two components are required. The human eye is very good at perceiving lines, and there are relatively straightforward statistical techniques for assessing whether data are consistent with linear covariation. If two components are sufficient, statistical fitting of the line parameters constrains the end-member component compositions, which must lie on that line on either side of the range of data. As will be evident throughout this book, such a situation arises frequently.

Mixing Line and Correlation Line: In concluding this section, it may be useful to call attention to some variations in interpretation and nomenclature. Some authors will use the term *correlation diagram* for a display similar to Figure 1.2, and *correlation line* for a linear array, particularly when the intent is to call attention to some fundamental geochemical property shared by the samples and to disdain mere mixing as accidental or inconsequential. Often, the correlation line is fundamentally a mixing line anyway. A prominent case is the isochron diagram used in geochronological studies, in which the ordinate numerator isotope is a radiogenic daughter (e.g., ^{87}Sr), the abscissa numerator isotope is its parent (e.g., ^{87}Rb), and the common denominator is a nonradiogenic isotope of the daughter element (e.g., ^{86}Sr). If the sample suite satisfies the usual assumptions – closed system behavior with all samples starting with the same initial daughter element composition at the same time – a linear array, or an isochron, will result. Concern may then be expressed about whether an observed linear array is an isochron or a mixing line. This is better posed as whether a linear array is really an isochron or *only* a mixing line, since a true isochron is a special case of a two-component mixing line. One of the two components is the initial composition of the daughter element (a point on the ordinate); the other is a component containing just the parent and daughter isotopes, plotting at infinity because it contains none of the denominator isotope, but infinity in a certain direction because the ratio of daughter to parent isotope is fixed by the age.

In noble gas studies, it may often be the case that a correlation line is "only" a mixing line reflecting variable admixture of a single sample source component with atmospheric contamination. For examples, samples obtained from a single site or narrowly confined area are commonly characterized by two prominent noble gas components (i.e., air noble gas and a noble gas characteristic to a specific sampling site). In such case, a three-isotope plot (Figure 1.2) gives a mixing line, which represents only a local feature and has little to do with a global characteristics. An example is discussed in the case for mantle neon (Section 6.7).

Chapter 2

Physical Chemistry

2.1 Introduction

The noble gases are also well known as the inert gases, reflecting their characteristic lack of chemical interaction with other elements. In the extreme case, a substance whose atoms fail to interact with other substances except by elastic collisions would always be an ideal monatomic gas. In general, the noble gases approach this extreme more closely than other elements. Nevertheless, of course, the noble gases do not fail completely to undergo interactions, and such interactions as do occur are responsible for governing their geochemical distributions.

A number of the basic parameters characterizing the noble gases as elements are presented in Table 2.1. This chapter will treat those aspects of noble gas interactions with other substances that are important geochemically. A much broader and more extensive treatment of the fundamental physical and chemical characteristics of the noble gases can be found in Cook (1961).

It is now well known that despite their name the noble gases (at least Rn, Xe, and Kr) do, in fact, participate in interactions normally considered chemical, notably with F but also with other elements, and the Xe-F bond strength is a substantial 30 kcal/mole. Noble gas chemistry is accordingly a subject of considerable theoretical interest. Nevertheless, it is extremely unlikely that conditions resulting in the formation of noble gas compounds would be encountered outside the laboratory, so noble gas chemistry will not be important in geochemistry and will not be discussed here. Treatments of noble gas chemistry are presented by Hyman (1963), Classen (1966), Dean (1985), and Pyykkö (1997).

In general, noble gases interact with other elements (and themselves) chiefly through van der Waals forces, also known as dispersion forces, which arise in elec-

Table 2.1. *Physical and chemical properties of the noble gases*

	Units	He	Ne	Ar	Kr	Xe	Rn
Atomic number	—	2	10	18	36	54	86
Atomic weight	amu	4.00260	20.179	39.948	83.80	131.30	—
Triple point	K	—	24.6	83.8	116.0	161.3	202
Normal boiling point	K	4.2	27.1	87.3	119.8	165.0	211
Critical point	K	5.3	44.5	151.9	209.4	289.7	378
Triple point pressure	torr	—	324	516	548	612	—
Heat of fusion at triple point	cal/mole	—	80	281	391	549	—
Heat of vaporization at NBP	cal/mole	19	414	1558	2158	3020	—
Van der Waals const. a	l^2-atm	0.03412	0.2107	1.345	2.318	4.194	—
Van der Waals const. b	l	0.0237	0.01709	0.03219	0.03978	0.05105	—
Atomic radius[a] (crystal)	10^{-8} cm	1.78	1.60	1.92	1.98	2.18	—
Atomic radius[b] (Van der Waals)	10^{-8} cm	—	1.55	1.88	2.00	2.18	—
Ionization energy	eV	24.48	21.56	15.76	14.00	12.13	10.75
Ionization cross-section (80 eV electrons)	10^{-17} cm^2	3.1	6.8	36	51	75	—
Polarizability	10^{-24} cm^3	0.201	0.390	1.62	2.46	3.99	—

[a] Cook (1961). Radii in crystal are calculated from lattice parameters of low-temperature noble gas crystals (FCC).
[b] Pyykkö (1997). Also see Zhang & Xu (1995) for the discussion of noble gas atomic radius.

tronic polarization effects. Although noble gas atoms are spherically symmetric and have no permanent dipole or high-multipole moments, they can interact even with similar atoms because fluctuations in the instantaneous dipole moment of one atom can induce moments in other atoms. The principal term in such resonant fluctuations is the dipole-dipole interaction, whose potential V has an inverse sixth-power dependence on interatomic separation r:

$$V(r) = -C/r^6 \tag{2.1}$$

An exact calculation of the constant C is, in general, not possible; a well-known approximation is that of London (1930), who treated atoms as harmonic oscillators of single frequency v and obtained

$$C = 3/2\, \alpha_1 \alpha_2 h v_1 v_2 / (v_1 + v_2) \tag{2.2}$$

for the interaction of atoms designated by subscripts 1 and 2. The characteristic frequencies v must be determined by empirical fit to this formula, but usually hv is close to the ionization energy. The major variations in C are due to the electronic polarizability α (the ratio of induced dipole moment to inducing field).

At sufficiently close approach, overlap of electronic wave functions leads to a repulsive potential with a sharp spatial dependence, usually modeled as e^{-ar} or r^{-n} with n about 12. Together the repulsive and attractive forces create a potential well with a minimum at interatomic spacing of a few angstroms.

Interaction of the noble gas atom with condensed matter is considerably more complicated and is usually approximate simply by summing or integrating potentials pairwise. Such treatments are necessarily crude; nevertheless, they allow an appraisal of the general features of an interaction and often provide realistic numerical values as well. Young and Crowell (1962), for example, review theoretical treatments of noble gas adsorption along these lines; predicted potentials for adsorption on various forms of carbon, to consider one example, range from a few hundred calories per mole for He to a few kilocalories per mole for Xe, in reasonable agreement with observed heats of adsorption.

Although the details of van der Waals interactions in a given situation may be complex, a generalization worth noting is that the strength of the interaction will usually be roughly proportional to polarizability [cf. Equation (2.2)]. Noble gas polarizabilities (Table 2.1) increase regularly with the number of electrons. This feature accounts for the general tendency of a heavier noble gas to be somewhat less noble than a lighter noble gas. Other parameters, which often are or may be important in interactions of noble gases with condensed matter or otherwise influence their distribution, are atomic size and ionization potential (Table 2.1). They also follow a regular progression according to atomic number. There are a number of ways to define *noble gas atom size*, depending on how noble gas atoms are bound in a condensed phase. These include (i) the extent of the electron cloud, (ii) interatomic length in a noble gas crystal at low temperature, (iii) van der Waals radius, and (iv) those based on relevant physical properties such as gas-viscosity. Among them, the van der Waals radius would be the most appropriate representative of the "atomic size" in common condensed materials of geochemical interest. Further discussions of this subject can be seen in a paper by Zhang and Xu (1995).

Finally, it is worth noting the recent development in the study of noble gas properties at very high pressures. It has been well established that the heavier noble gases (Kr, Ar, Xe) are solid at high pressures at room temperature (e.g. Jephcoat & Besedin, 1996). Jephcoat (1998) compared melting temperatures at pressure and calculated density-pressure curves for argon, krypton, and xenon at 2500 K, with the known internal temperature and density of the Earth (Figure 2.1).

He found that both the melting curves (Figure 2.1a) and the densities (Figure 2.1b) for xenon and krypton are well above the estimated temperature and density of the lower mantle. One might then suppose that Xe and Kr are in solid form in the lower mantle. If this is indeed the case, implications on noble gas degassing from the mantle and hence on the evolution of the atmosphere would be far-reaching (cf. Section 6.9). However, the formation of solid Xe or Kr seems to be unlikely because of its

2.2 Adsorption

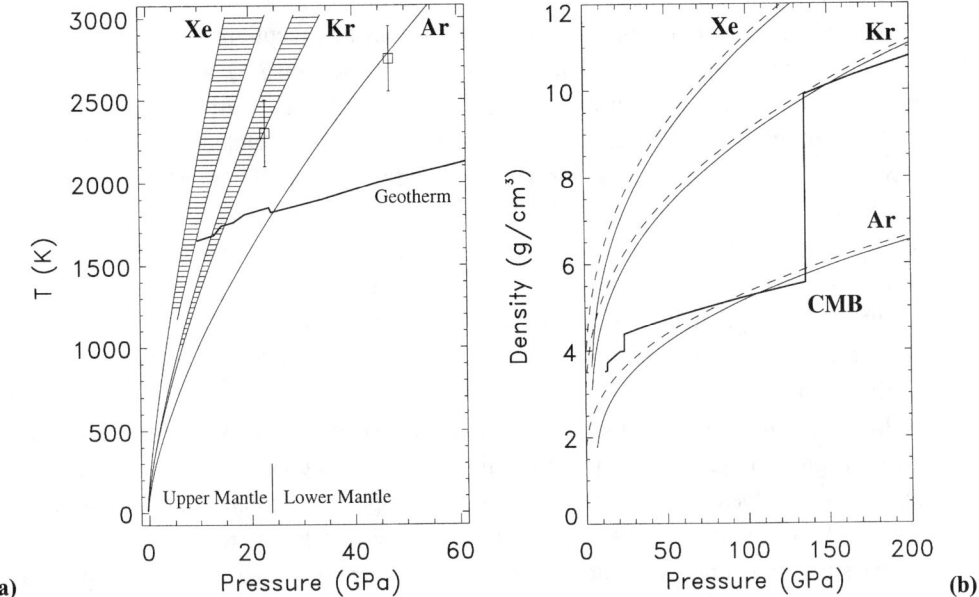

Figure 2.1 (a) Expected melting curves of the heavy noble gases based on data from high-pressure experiments and theoretical models. Bounds on the melting curves of solid Kr and Xe are shown by shaded regions. The two symbols with error bars indicates the P-T region where overlap with the melting temperature of solid iron has been observed; solid Kr is stable in an iron melt above ~23 Gpa and 2300 K, and solid Ar is stable above ~47 Gpa and 2700 K. Note that the temperature of the lower mantle falls below the melting temperatures of the heavy noble gases. (b) Pressure-density data on the heavy noble gases. Solid lines are densities at 2500 K calculated with the Mie-Gruneisen thermal model and the Debye approximation. Dashed lines are densities calculated from the experimental data at room temperature. A bold line indicates the density profile of the Earth obtained from a seismological model. Reproduced from Jephcoat (1998).

extremely low concentration in the mantle, unless noble gases are highly concentrated in some local regions.

2.2 Adsorption

Adsorption is a common mechanism by which the noble gases may interact with other materials, and it is frequently cited as a possibly relevant factor in noble gas geochemistry. Adsorptive phenomena in general are a convenient way of studying intermolecular interactions and the properties of surfaces. The subject has accordingly been afforded extensive theoretical development and empirical investigation. Although many empirical studies have included noble gases, data describing adsorptive interactions between noble gases and materials of geochemical importance are rather scarce. Nevertheless, adsorption is a relatively nonspecific interaction, and much can be inferred on the basis of generalizations. Noble gas adsorption tends to be a rather weak interaction at temperatures of principal geochemical interest. Even so, it can be of substantial importance, and there may be exceptions to this rule. There are many parallels between adsorption and solution (Section 2.3), and in many cases it can be difficult to distinguish between the two in practice (Section 2.4).

The term *adsorption* designates the situation in which the molecules of a gas (the sorbate) are concentrated at the surface of a solid (the sorbent) with which the gas is in contact. The concentration results from an attractive potential experienced by the sorbate at the sorbent surface and is usually viewed as a temporary residence of sorbate molecules "on" the sorbent surface. The gas-solid interaction is the usual picture for adsorption, but it is not the only one possible: the sorbate could be a dilute constituent in a condensed phase (e.g., gases dissolved in water), and/or the sorbent could be a liquid.

In many treatments, a distinction is made between *physical adsorption* and *chemisorption*. Physical adsorption usually designates the case in which the attractive force involved is the van der Waals or "dispersion" force arising in mutual electric dipole induction. This interaction is relatively weak (a potential of a few kilocalories per mole or less, limited by the strong electronic repulsion which prohibits closer approach) but is also ubiquitous and relatively nonspecific, depending principally on atomic number and radius rather than detailed electronic configuration. Chemisorption designates cases of electron sharing or transfer normally considered chemical interaction. The energies involved are higher (tens of kilocalories per mole), and the interaction much more sensitive to the specific identities of sorbent and sorbate. The distinction between physical adsorption and chemisorption is not sharp: a continuum of energies is possible. It is sometimes said that physical adsorption is reversible, whereas chemisorption is irreversible. This is not literally true, of course. Rather, at the higher energies involved in chemisorption, desorption lifetimes can be long compared to realistic laboratory timescales. Clearly, it is expected that noble gas adsorption will be in the physical adsorption rather than the chemical adsorption regime.

In the following discussion, it will be understood that in algebraic expressions and equations all physical quantities are designated in a consistent set of units (e.g., pressure in dyne/cm^2 and energy in ergs); in this case quantities of elements will be dimensionless (number of atoms). Numerical citations in text and tables, however, will be in more familiar laboratory units (e.g., pressure in atm or torr or Pa, energy in cal, and gas quantities in cm^3 STP).

In equilibrium, the quantity N of a given sorbate, which is absorbed on a given sorbent, depends on its partial pressure (fugacity) P in the gas phase and the temperature T. A basic phenomenological description is specification of the functional dependence between N, P, and T. Both experimental observations and theoretical or thermodynamic descriptions are often the case in univariant functional descriptions: the relation between N and P at constant T (an isotherm), between N and T at constant P (an isobar), or between P and T at constant N (an isostere).

An important special case of an isotherm is a linear dependence of N on P (Henry's law):

$$N = \mathcal{H}P \tag{2.3}$$

The Henry constant $\mathcal{H}$ is a function of T but not P. (In some theoretical treatments, the Henry constant is the ratio of fugacity to quantity adsorbed, i.e., the inverse of the sense used here.) It is generally expected that adsorption will be governed by Henry's law at sufficiently low pressures. It is possible to construct theoretical models for adsorption in which an isotherm does not reduce to Henry's law, Equation (2.3), even in the limit $P \to 0$, but it is not clear that such situations obtain in practice and doubtful that they are important in noble gas geochemistry.

The noble gases are all monatomic as gases, have unreactive closed shell electronic configurations, and can be expected to be present at only very low partial pressures in any natural situation of geochemical interest. We will accordingly assume that Henry's law is applicable in all cases of noble gas geochemistry; semiquantitative justifications of this generalization can be made (see following discussion), and in any event there is certainly no evidence to the contrary.

A simple but instructive kinetic theory model due to Langmuir (1918) is often used as a starting point for an atomic scale view of adsorption. It is assumed that the sorbent surface, of total area A, is comprised of N_s sites each capable of binding only one sorbate atom (or molecule). If, at a given instant, a fraction of θ of these sites are occupied by sorbate atoms, the mean life against desorption is τ, and τ is independent of θ (i.e., no interaction among sorbate atoms), the flux ϕ of desorbing atoms is

$$\phi = \theta N_s / A\tau \tag{2.4}$$

If the gas phase is ideal, the flux of impacting atoms on the surface is $P(2\pi mkT)^{-1/2}$, where m is atomic mass and k is Boltzmann's constant. If α is the probability that an atom incident of an unoccupied site will be adsorbed, the flux of adsorbing atoms is

$$\phi = \alpha P(1-\theta)(2\pi mkT)^{-1/2} \tag{2.5}$$

In equilibrium, the two fluxes are equal; hence,

$$N = N_s\theta = N_s bP/(1+bP) \tag{2.6}$$

for

$$b = \alpha\tau a(2\pi mkT)^{-1/2} \tag{2.7}$$

where $a = A/N_s$ is the area of an individual adsorption site. Equation (2.6) is generally designated the Langmuir isotherm.

At sufficiently low pressures ($\theta \ll 1$), the Langmuir isotherm reduces to Henry's law with

$$\mathcal{H} = \alpha A \tau (2\pi m KT)^{-1/2} \qquad (2.8)$$

At sufficiently high pressures ($\theta \to 1$), the adsorbed quantity N approaches the saturation limit N_s.

This model is not only incomplete (e.g., it makes no prediction of α or τ), it is clearly oversimplified, and it is well known that the Langmuir isotherm is not a very good description of most sorbent-sorbate interactions, particularly at high pressures (charcoal is an important exception). It is, nevertheless, quite valuable as a starting point for more sophisticated generalizations and as a framework by which to form a qualitative evaluation of numerical data.

Thus, for example, Equation (2.8) provides a basis for estimating the mean time τ that an adsorbed atom resides on the sorbent surface before it is desorbed. [Assumption of fixed attachment sites is not necessary in the derivation of Equation (2.8), as suggested by the absence of explicit dependence on N_s or a.] As will be seen later, values of geochemical importance are of the order $\mathcal{H} = 1 \, \text{cm}^3 \, \text{STP} \, \text{g}^{-1} \, \text{atm}^{-1}$ and $A = 10^5 \, \text{cm}^2/\text{g}$. For these figures (for Xe at 0°C and $\alpha = 1$), the lifetime τ is 2×10^{-9} s. For comparison, the time required for a gas atom to travel across an atomic distance scale (cf. Table 2.1) at thermal speeds is of the order of 2×10^{-12} s. In this example, the atom is thus adsorbed for a "long" time. We can also infer that unless τ is substantially greater than a thermal flight time across the few angstroms of surface attractive potential, the concept of adsorption will have little meaning.

The Langmuir model also provides a convenient basis for estimating when Henry's law or saturation effects can be expected. If an individual attachment site has an area $a \approx 2 \times 10^{-15} \, \text{cm}^2$, the order of atomic cross-sectional area (cf. Table 2.1), then $N_s \approx 5 \times 10^{14}$ atoms/cm^2 = 2×10^{-5} cm^3 STP/cm^2. Surface concentrations approaching this order of magnitude can be expected to exhibit saturation behavior. Conversely, much lower concentrations indicate $\theta \ll 1$ and lead to the expectation of Henry's law behavior. Possible adsorption effects important in noble gas geochemistry always involve much lower concentration than this illustrative value, which is one reason why Henry's law violation is not expected.

The Langmuir model permits an explicit evaluation of surface areas from experimental isotherm data. Rearrangement of Equation (2.6) gives

$$P/N = (1 + bP)/N_s b \qquad (2.9)$$

The slope of this linear relationship between P and P/N is $1/N_s$, which permits calculation of A for an assumed value of $a = A/N_s$. (An approach to saturation is needed to extend beyond the Henry's law regime $P/N = $ constant.)

Because the Langmuir isotherm is not an adequate description of most systems, Equation (2.9) is not used much for area measurement. A number of other isotherm formulations utilize adsorption in surface area measurements, however (cf. Young & Crowell, 1962, for example). The best known and most widely used is the BET (Brunauer, Emmett & Teller, 1938) theory, a generalization of the Langmuir model to multilayer absorption. Assuming that for the second and succeeding molecular layers the "adsorption" potential is the same as in bulk condensation, the isotherm analogous to Equation (2.6) is

$$N = N_s cx(1-x)^{-1}[1+(c-1)x]^{-1} \tag{2.10}$$

where $x = P/P_0$ (P_0 is the vapor pressure of the sorbate) and c is a dimensionless constant peculiar to the sorbate-sorbent pair (and a function of T). Rearranging to a form analogous to Equation (2.9) gives

$$x/N(1-x) = [1+(c-1)x]/N_s c \tag{2.11}$$

This is a convenient linear relationship against which experimental data can be compared (cf. Figure 2.2.). If experimental data conform to this prediction, the slope and intercept of this linear relation yield both c and N_s, and thus A.

There is substantial theoretical objections to the BET model. Nevertheless, experimental data often conform to Equation (2.11) admirably (Figure 2.2), and the BET model is used extensively to determine areas of microscopically complicated surfaces.

Many materials of geochemical importance, notably clays and shales, turn out to have a large BET area (Table 2.2). Specific areas are often of the order of $10\,\text{m}^2/\text{g}$. For comparison, perfect spheres (of density $3\,\text{g/cm}^3$) would need radius $0.1\,\mu\text{m}$ to have specific areas of $10\,\text{m}^2/\text{g}$. The high areas correspond to a large degree of microscopic surface involution and pores and channels which greatly enhance adsorption capabilities.

One additional isotherm which has been used in parameterizing noble gas adsorption on natural materials (Figure 2.3) is the Freundlich isotherm:

$$N = \text{constant} \times P^{1/n} \tag{2.12}$$

where $n \geq 1$ is a constant (the special case $N = 1$ is Henry's law). It is possible to derive this isotherm theoretically for special models, but it is usually regarded as a semiempirical form to which data can be fit. It does not reduce to Henry's law as $P \to 0$, nor does it predict saturation (Langmuir isotherm) or bulk condensation (BET isotherm) at sufficiently high P, so its application as an interpolation/extrapolation formula is generally considered restricted to an intermediate pressure range.

None of the isotherm models already considered make explicit predictions for absolute values of the relevant parameters nor of their temperature dependence. Useful information on temperature dependence can be obtained by applying thermodynamics, usually via a heat of adsorption. A common parameter is the isosteric heat of adsorption:

$$\Delta H = (RT^2/P)(\partial P/\partial T)_\theta \tag{2.13}$$

which is also the calorimetric heat evolved in a reversible isothermal adsorption. We will restrict further consideration only to cases in which Henry's law applies. In such a case, Equation (2.13) becomes

$$\Delta H = -RT^2 d(\ln \mathcal{H})/dT \tag{2.14}$$

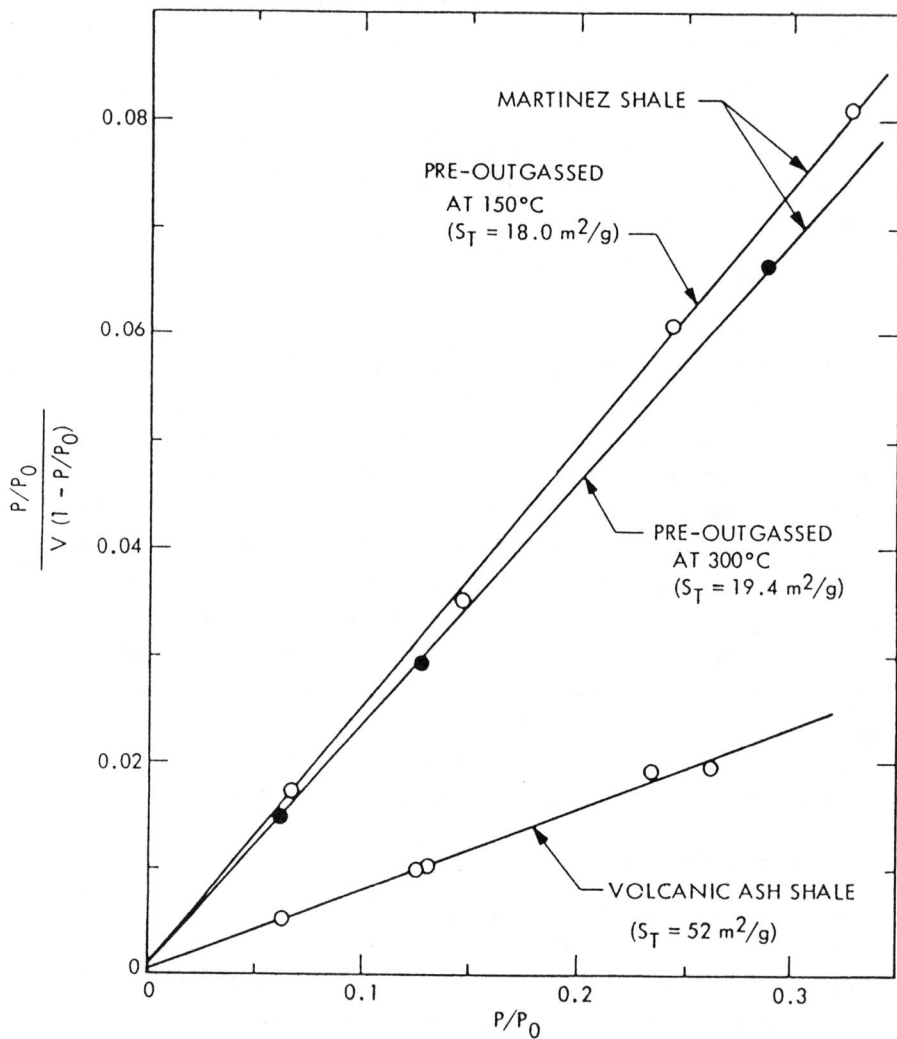

Figure 2.2 BET plot for adsorption of N_2 on shale samples at −195°C. Linear data array indicate conformity to Equation (2.11), by which indicated specific surface areas were calculated. Reproduced from Fanale and Cannon (1971).

If ΔH is known, Equation (2.14) can be integrated to obtain the temperature dependence of $\mathcal{H}$. In general ΔH is not constant because gas and adsorbed phases have different heat capacities, but often in practice the variation in ΔH over a limited temperature range is not severe.

Explicit predictions for adsorption parameters can be obtained by application of statistical mechanics to detailed models for the adsorbed phase. Many such models have been developed (cf. Young & Crowell, 1962; Ross & Olivier, 1964), but a model suitable for a given application cannot in general be chosen a priori. A simple example, used by Podosek, Bernatowicz, and Kramer (1981b) for fitting very-low-pressure (Henry's law region) Xe and Kr adsorption on shales (Figure 2.4) is

2.2 Adsorption

Table 2.2. *Selected data for Xe and Kr adsorption*

Sorbent	Temperature (°C)	Henry constant (cm^3 STP g^{-1} atm^{-1}) Xe	Henry constant Kr	Heat of adsorption[f] (kcal/mole) Xe	Heat of adsorption Kr	Specific area[a] (m^2/g)
Martinez shale[b]	0	3.3	0.27	5.5	3.9	18
	25	1.3	0.15	—	—	—
"Volcanic ash shale"[b]	0	1.2	0.24	5.9	—	52
	25	0.5	—	—	—	—
"Volcanic ash shale"[c]	0	5.6	0.43	4.8	4.0	—
	25	2.5	0.21	—	—	—
	−77	220	10	—	—	—
Kibushi shale[c]	0	1.2	0.43	5.3	3.1	—
	25	0.5	1.1	—	—	—
	−77	69	24	—	—	—
Wyoming montmorillonite[d]	−77	38	7.6	—	—	36
Julian limonite[d]	−77	76	3.0	—	—	16
Vacaville basalt[d]	−77	30	7.6	—	—	4
Columbia River basalt[c]	0	0.7	0.28	4.5	3.3	—
	−77	24	3.9	—	—	—
Uncompaghre quartzite[c]	0	5×10^{-5}	3×10^{-4}	6.7	4.1	—
Allende meteorite[e]	−160	10^4	200	—	—	3
Activated charcoal[c]	0	5835	292	7.5	4.9	—

[a] BET areas (cf. Figure 2.2).
[b] Fanale & Cannon (1971); Henry constants estimated from graphical data presentation and heats estimated from Henry constants and Equation (2.12).
[c] Podosek et al. (1981).
[d] Fanale et al. (1978).
[e] Fanale & Cannon (1972).
[f] All heats are negative.

$$\mathcal{H} = Ah(2\pi mkT)^{-1/2}(kT)^{-1}e^{-\Delta E/RT}(1-e^{-h\nu/kT})^{-1} \quad (2.15)$$

This formula represents the model in which the absorbed phase has perfect lateral freedom (two-dimensional ideal gas) on the sorbent surface and vibrates harmonically (fundamental frequency ν) normal to the surface; the sorbent surface is assumed to have a uniform sorbate potential $E < 0$ (including the zero-point oscillator energy) relative to the gas. By Equation (2.8), the lifetime corresponding to Equation (2.15) in the special case $\alpha = 1$ and $h\nu/kT \gg 1$ is

$$\tau = (h/kT)e^{-\Delta E/RT} \quad (2.16)$$

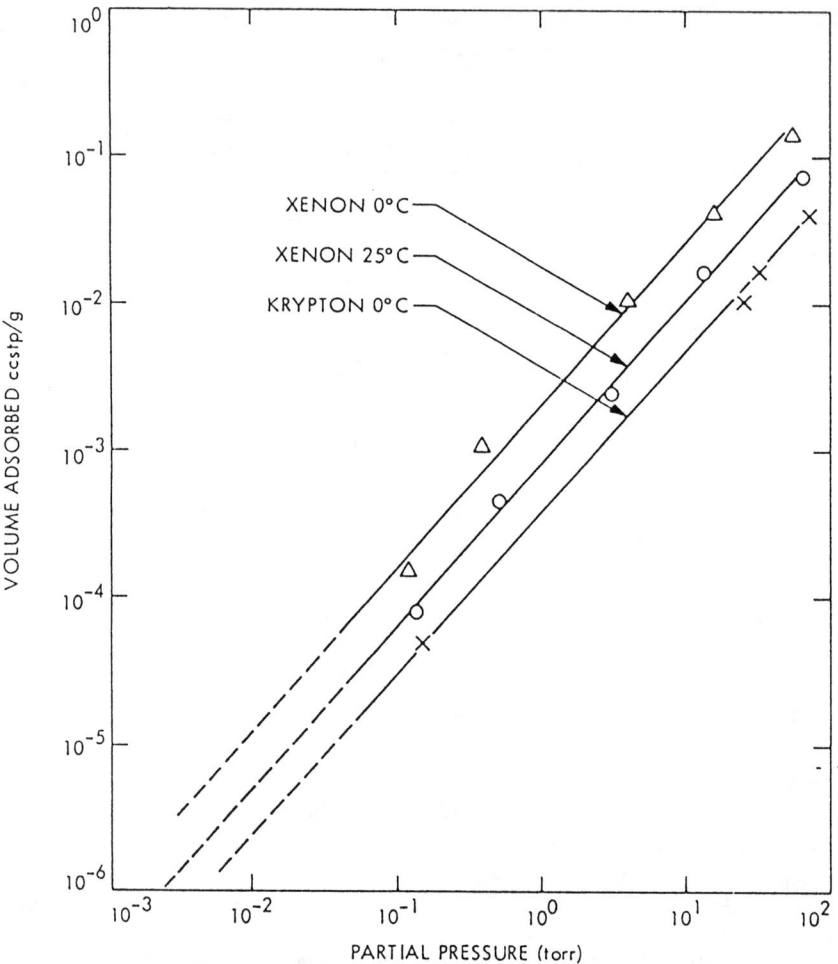

Figure 2.3 Freundlich isotherm [Equation (2.12)] plots for Xe and Kr adsorption on volcanic ash shale. Reproduced from Fanale and Cannon (1971).

Representative adsorption data of geochemical interest are given in Table 2.2. Only Xe and Kr data are listed; the lighter gases are much poorer sorbates, sufficiently so that it is possible to question whether adsorption has any geochemical relevance for them. Note that it is very difficult to measure even Xe and Kr adsorption on the materials and at the T and P of greatest geochemical interest. In general, measurements are made at high P or low T or both so that considerable extrapolation by models of questionable reliability is involved. The accuracy of the surface temperature Henry constants in Table 2.2 is thus probably somewhere between a factor of two and an order of magnitude. For only one sample, the "volcanic ash shale," is there an opportunity to compare different methods. The Fanale and Cannon (1971) measurements (of Figure 2.3) were made at the designated temperatures but at pressures of around 1 torr, a factor of 10^3 or more higher than the atmospheric pressures of primary concern. The Podosek et al. (1981b) measurements were made at characteristic pres-

2.2 Adsorption

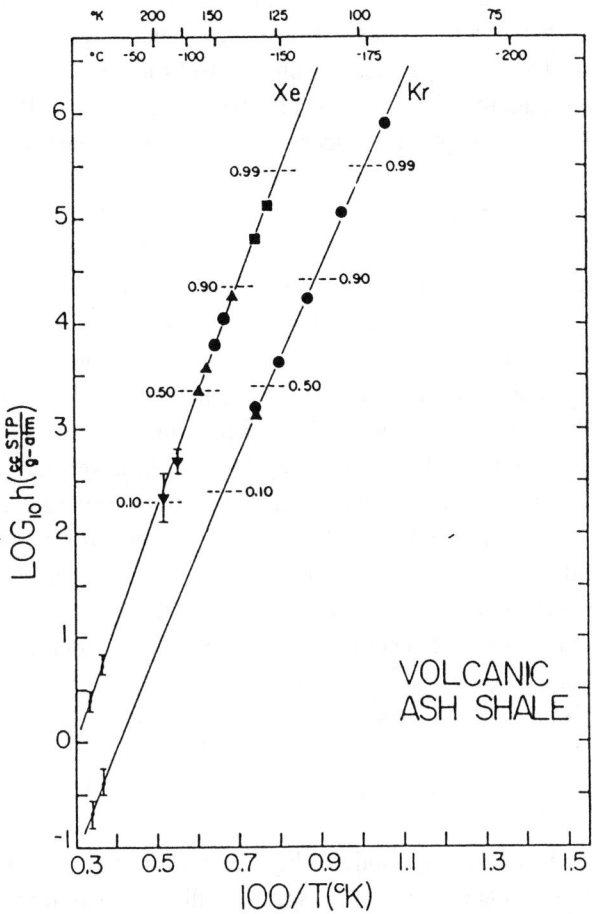

Figure 2.4 Henry constants for Xe and Kr adsorption on "volcanic ash shale" (cf. Figures 2.2 and 2.3). The solid lines are fits according to Equation (2.13), small symbols with bars (lower left) indicate extrapolation to 0°C and 25°C. Reproduced from Podosek et al. (1981b).

sures of 10^{-3} torr and less but require temperature extrapolation of a factor of 10^{-3} or more in $\mathcal{H}$. The agreement (Table 2.2) seems about as good as can be expected.

Adsorption is sometimes invoked not only as a factor in geochemistry but also as a laboratory nuisance. Analyzed samples are often found to have a superficial or "loosely bound" component ascribed to air contamination, which is frequently described as "adsorbed" on the sample. Without further qualification, this makes little sense. An air contamination effect certainly exists, as can be inferred clearly when an "intrinsic" sample gas is isotopically distinct (cf. Section 2.4), but whether adsorption is responsible or even involved is questionable. In all such noble gas analyses, a necessary step is storage in laboratory vacuum before gas extraction. By definition, adsorbed gas is desorbed under vacuum. The relevant factor is the timescale required to reach equilibrium, and in general desorption timescales are short. If Equation (2.15) is even approximately correct, an adsorption energy even as high as 6 kcal/mole (cf. Table 2.2) gives a lifetime against desorption $\tau < 10^{-8}$ s at 25°C. A lifetime of order 10^5 s at 25°C, minimum for laboratory vacuum exposure, requires 24 kcal/mole, a reasonable energy for chemisorption but seemingly well out of reach for a noble gas.

The data in Table 2.2 are for natural samples. It is notoriously difficult to prepare and maintain a "clean" solid surface, since any freshly created surface quickly becomes "contaminated" with adsorbed species even with a rather good laboratory vacuum. Any naturally occurring solid material must be considered to have a surface extensively populated by adsorbed atoms and molecules rather than a "pristine" surface. Such surfaces are what are examined in most laboratory experiments (e.g., those reported in Table 2.2), and, of course, just such surfaces are geochemically relevant for noble gas adsorption. It is interesting to note, however, that in other situations, noble gas adsorption can be rather a stronger effect. Thus, for example, Bernatowicz et al. (1983) examined Xe adsorption on a vacuum-crushed lunar rock and concluded that a small part of the freshly created surface had an adsorption potential as high as 14 kcal/mole but that in a few days at 10^{-8} torr this surface was rendered inaccessible to Xe by other chemical species that were better competitors for the sorbent surfaces.

2.3 Solution

When a condensed phase (the solvent), solid or liquid, equilibrates with a gas phase (the solute), some concentration of the gaseous species will be dispersed in the solid or liquid (i.e., some gas will be dissolved). Solution is the most general way in which a noble gas will interact with other materials. Note, however, that the term *solution* implies a more or less uniform microscopic-scale admixture of solvent and solute molecules or complexes of molecules; this assumption is presumably reasonable for liquid solvents but perhaps not for solids and is difficult to test experimentally.

There is a wealth of both theoretical treatment and empirical data for the phenomenon of solution, including empirical data for noble gas solution. As with adsorption, however, data for solution in geochemically important materials are sparse. A prominent exception, the only one, is water, extensive data for which are presented in Chapter 4. Most of the available data for other materials of principal geochemical interest are summarized in Table 2.3.

As in the case for adsorption (see Section 2.2), in equilibrium, the quantity N of a given solute which is dissolved in a given solvent depends on its gas phase partial pressure (fugacity) P and on the temperature T, and a basic phenomenological description of the equilibrium is specification of the functional relationship between N, P, and T. At sufficiently low pressures, it is expected that the pressure dependence is linear (Henry's law):

2.3 Solution

$$N = \mathcal{H}P \tag{2.17}$$

where the Henry constant $\mathcal{H}$ is a function of T but not P. (In some theoretical treatments, the Henry constant is the ratio P/N, that is, the inverse of the given earlier definition.) In this and subsequent algebraic expressions, it will be understood that physical quantities are in a consistent set of mechanical units (e.g., pressure in dyne/cm^2, and energy in erg), quantities of elements are dimensionless (number of atoms), and numerical citations in text and tables are in more convenient units (e.g., pressure in atm and gas quantities in cm^3 STP).

The preceding basic description (and much of the subsequent treatment that follows) is intentionally similar to that given for adsorption in Section 2.2. This stresses the point that, in terms of macroscopic thermodynamics, there are great similarities between the two phenomena.

Henry's law behavior can be expected as long as the solute concentration is low, in the sense of not approaching a saturation limit or becoming more than a trace constituent of the solution. For liquid solutes, we thus expect that Henry's law will be valid as long as the mole fraction of solute is low. Experimental verification of Henry's law behavior for noble gases in silicate melts has been reported by a number of authors (e.g., Carroll et al., 1994, for a review and references therein). This should be the case in any situation of geochemical interest. In a material of density 3 g/cm^3 and average atomic weight 24, for example, an atomic fraction of 10^{-6} corresponds to a gas concentration of 3×10^{-3} cm^3 STP/g, much higher than any concentration likely to be encountered in nature. For solid solutes, however, gas atoms are likely to be preferentially "dissolved" in special accommodation sites such as vacancies or other lattice defects. These sites are likely to be nonuniform in their interactions with solute atoms and the number of solvent atoms. In such cases, the range of validity of Henry's law solution might be very limited, perhaps even nonexistent, and solubility might be highly variable from sample to sample of the same solvent and sensitive to the particular history of a given specimen. Nevertheless, few theoretical treatments or little empirical data indicate that nonlinearity actually is important for noble gases in nature, and natural levels of dissolved gases are typically so low (atom fraction no more than $\sim 10^{-8}$) that nonlinearity would not be expected. In the rest of this section, we will accordingly assume that Henry's law is adequate for the description of noble gas solution effects in geochemistry and will not pursue a description of any more complicated pressure dependence.

It is known that there are five basic anionic structural units in common silicate melts (Virgo, Mysen & Kushiro, 1980). These are SiO_2 (three-dimensional network), $Si_2O_5^{2-}$ (sheet), $Si_2O_6^{4-}$ (chains), $Si_2O_7^{6-}$ (dimers), and SiO_4^{4-} (monomers). From experimental studies of noble gas solubility in binary CaO, SiO_2 melts at 1100 bar and 1600°C, Shibata, Takahashi, and Matsuda (1996) concluded that among these basic structural units, the three-dimensional network plays a major role in noble gas accommodation. This conclusion was derived from their observation that solubility decreases with increase of Ca: CaO produces nonbridging oxygen, which results in

Table 2.3. *Noble gas solubility in melts and solids*

Solvent	Temperature (°C)	Henry Constant for Solution (10^{-5} cm^3 STP g^{-1} atm^{-1})					Heat of Solution (kcal/mole)					Reference
		He	Ne	Ar	Kr	Xe	He	Ne	Ar	Kr	Xe	

Melts
P = 1 atm

Solvent	Temp (°C)	He	Ne	Ar	Kr	Xe	He	Ne	Ar	Kr	Xe	Ref
Enstatite	1,500	12	7	2	(1.3)	(0.6)						1
Basalt	1,200			2.5	1.3	0.9						2
Tholeiitic basalt	1,200		19	2.3	0.8	0.9						3
Tholeiitic basalt	1,250–1,000		25	5.9	3.0	1.7	7.1	5	5.7	19.5		4
Tholeiitic basalt	1,350	64	35	8.7	6.3	2.7	4.9	1.3	6.8	9.4	2.9	5
Alkali olivine basalt	1,299		22	4.6	1.6							3
Alkali olivine basalt	1,350	57	26	6.4	4.4	1.5	0.54	12.3	14.2	19	14.4	5
Basaltic andesite	1,200		26	9	2.1			4.2	16.7	19.3		3
Basanite	1,350	70	42	12.2	9.3	3.5	1.5	5.1	4.5	3.7	4.4	5
Ugandite	1,350	48	21	4.5	3.0	1.0						5
Andesite	1,350			15.4	11.4	8.3						5
Basalt (synthetic)	1,300		2.8	4.7	1.3	1.1						6
	–1,332		–7.7	–16.7	–3.3	–3.0						
Albite	1,000		150	30	18 (750°C)							7

P = 1100 bar

Solvent	Temp (°C)	He	Ne	Ar	Kr	Xe						Ref
CaO—SiO$_2$	1,600	8.64	6.54	1.09	0.524	0.181						8
	1,600	15.6	13.7	2.98	1.66	0.626						
	1,600	35.8	27.6	8.92	5.67	2.48						

	T (°C)	He	Ne	Ar	Kr	Xe						Ref
$P = 15$ kbar												
Basalt	1,600			4.7		3.1						9
Granite	1,600			39.1		2.2						9
Sanidine	1,600			19.5		1.7						9
Albite	1,600			28.8		2						9
Diopside	1,600			0.6		3.4						9
Anorthite	1,600			2.57		3.4						9
$P = 50–120$ kbar												
Anorthite				5.6								10
Silica				~5.0								10
Solid												
$P = 1$ kbar												
Anorthite	1,300	0.016	0.39	0.8	0.8							6
	−1,332	−3.5	−10.7	−41.3	−119							6
Diopside	1,300	0.018	0.35	0.8	2.7							6
	−1332	−2.1	−7.0	−15.4	−125							6
Forsterite	1,300	6.2	16.7	32.6	32.9							6
	−1,332	−7.8	−30.7	−65.2	−143							6
Spinel	1,300	0.52	1.1	1.5	1.3							6
	−1,332	−2.0	−5.6	−4.8	−4							6
P up to 1.2 kbar												
Silica glass	700			~50					−4.76			11
Fresh water	0	950	1,270	5,360	11,100	22,600	−0.93	−1.87	−4.05	−5.08	−6.23	12

Table 2.3 References: 1, Kirsten (1968): He, Ne, Ar values measured, Kr, Xe values by extrapolation according to equation (2.19); 2, Fisher (1970); 3, Hayatsu & Waboso (1985); 4, Jambon et al. (1986); 5, Lux (1987); 6, Broadhurst et al. (1992); 7, Roselieb et al. (1992); 8, Shibata et al. (1998); 9, White et al. (1989): solubility values are normalized to 1300°C from the data given in White et al. (1995) by Carroll & Stolper (1993); 10, Chamorro-Perez et al. (1996); 11, Carroll & Stolper (1991): solubility value is calculated from the data given in Carroll & Stolper (1991) for activity coefficient of Ar = 1; 12 – Tables 4.1 and 4.3 and Equation (2.17).

breaking the SiO$_2$ network and, thus, reduces the solubility. Shibata et al. (1996) further suggested that with polymerization, there will be larger holes in the network, thereby accommodating more noble gases or resulting in higher solubility. They therefore concluded that noble gas solubility in silicate melts would primarily be defined by degree of polymerization in silicate melts. Lux (1987) observed that increasing content of MgO as well as CaO decreases noble gas solubility in silicate melts. This suggests that Mg also acts by breaking the SiO$_2$ network or reducing solubility.

Noble gas solubility also depends on physical properties of the melts. Lux (1987) observed that noble gas solubility shows a good correlation with density of the melts [cf. Figure 2.5(d)]. Caroll and Stolper (1991, 1993), however, emphasized that solubility is better correlated with "ionic porosity" than with density, where the ionic porosity was defined as IP $\equiv 100\{1 - (V_{ca}/V_m)\}$, where V_{ca} indicates the total volume of spherical cations plus anions in 1 g of melt, and V_m, the melt specific volume (cm^3/g). As shown in Figure 2.5, there is a good correlation for basaltic silicate melts between the Ar solubility (mole fraction of dissolved Ar at 1 bar gas pressure at 1000°C–1400°C) and the ionic porosity. However, considerable deviation from the linear trend is observed in the case of CaO-MgO-Al$_2$O$_3$-SiO$_2$ melts. Shibata et al. (1996) attributed this deviation from the linear trend to the breaking of the SiO$_2$ network due to Ca and Mg ions. Silicate melts are formed by polymer of silicon atoms combining oxygen atoms. The degree of polymerization in silicate melts and glass is represented by NBO/Si (Nonbridging Oxygen per Silicon) ratio, which is the number of nonbridging oxygens per silicon (or tetrahedrally coordinated cation) in the melt (Brawer and White, 1975). Noting that noble gases dissolving in silicate melts are always coupling with bridging oxygen rather than nonbridging oxygen, Shibata, Takahashi, and Matsuda (1998) proposed the parameter NBO/Si as a more satisfactory measure of noble gas solubility in silicate melts. In Figure 2.5, correlations between noble gas solubility in silicate melts and various parameters such as molar volume, density, ionic porosity and NBO/Si are compared, which shows the widest range of linearity for the parameter NBO/Si. The linear relation would give a useful means to estimate noble gas solubility in silicate melts if we know the composition.

The Henry constant (per unit mass of solvent) can be expressed

$$\mathcal{H} = (\rho k T)^{-1} C_l / C_g \tag{2.18}$$

where ρ is solvent density, C_g is the concentration of gas phase solute, and C_l is the equilibrium concentration of solute in the solution. Equal concentrations in gas and solution thus correspond to $\mathcal{H} = (\rho k T)^{-1}$, for example $\mathcal{H} = 1 \, \text{cm}^3 \, \text{STP} \, \text{g}^{-1} \, \text{atm}^{-1}$ for $\rho = 1 \, \text{g/cm}^3$ at $T = 0°C$ and $\mathcal{H} = 0.06 \, \text{cm}^3 \, \text{STP} \, \text{g}^{-1} \, \text{atm}^{-1}$ for $\rho = 3 \, \text{g/cm}^3$ at $T = 1200°C$. Comparing with the data in Table 2.3, we see that noble gas solubility is relatively low even in water and quite low in likely magmas, in the absolute sense that the atoms are more dilute in the solution than in the gas phase. Even the silicate melt solubilities are not negligibly low in other important contexts, however. For the tabulated values, equilibrium with noble gas pressures at air values will produce gas concentrations of the general order of magnitude observed in igneous rocks.

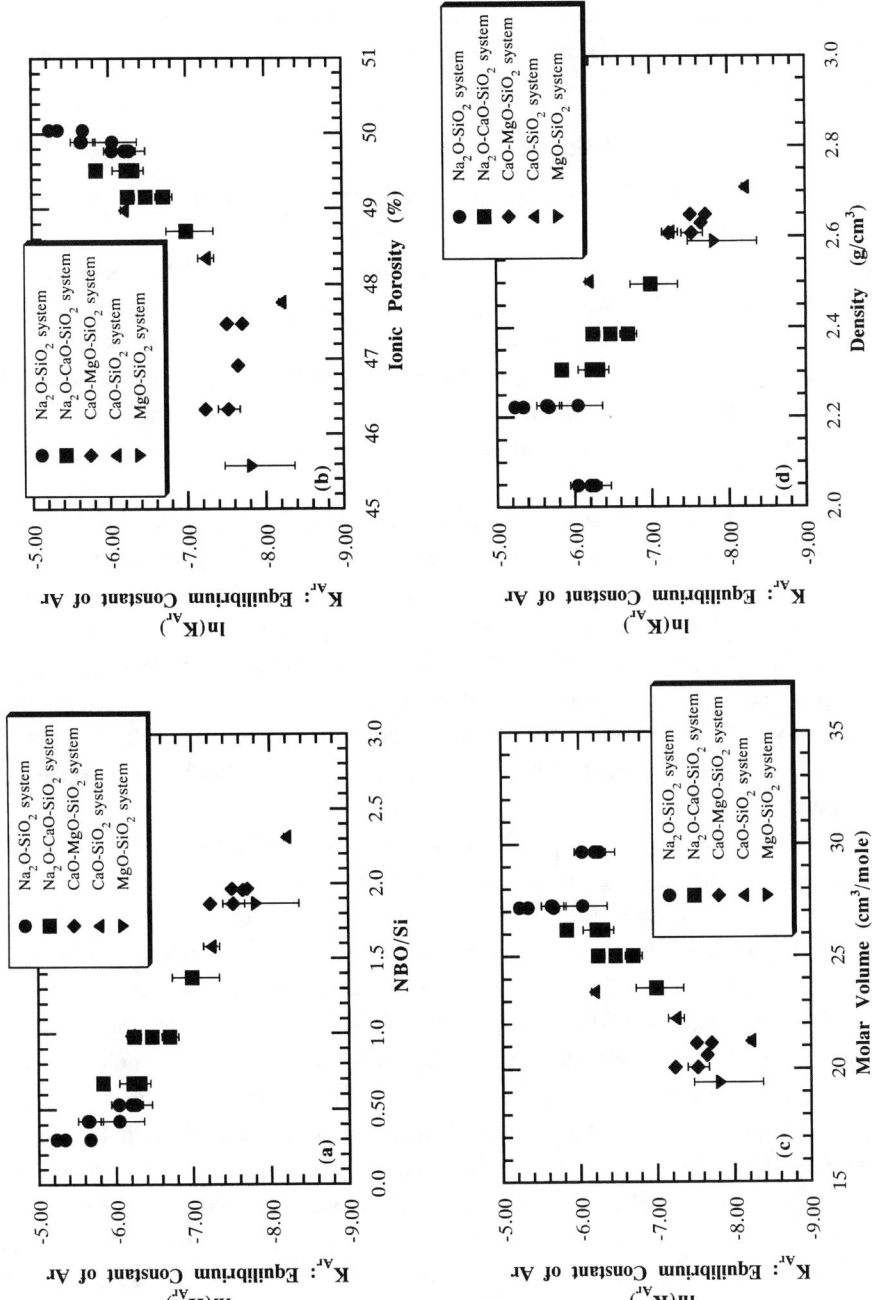

Figure 2.5 Correlation between argon solubility (in natural logarithm of the Henry's law constant) in silicate melts and various parameters: NBO/Si (no bridging oxygen to silicon ratio), molar volume, ionic porosity, and density. Reproduced from Shibata et al. (1998).

47

The temperature dependence of solubility is related to a heat of solution in the same way as previously described for adsorption (Section 2.2). If Henry's law is obeyed, the heat of solution is

$$\Delta H = RT^2 d(\ln \mathcal{H})/dT \qquad (2.19)$$

If the heat of solution is known, Equation (2.19) can be integrated to express the temperature dependence of the Henry constant $\mathcal{H}$ (Figure 2.6). The heat will vary somewhat with temperature, however, because of different heat capacities of gaseous and dissolved phase, and this variation must be taken into account in extrapolation over a large temperature range. In contrast to the case for adsorption (Table 2.2), note that heats of solution (Table 2.3) are generally positive (greater solubility at higher temperatures). Also, heats of solution are of a greater magnitude than heats of adsorption so that the temperature dependence is characteristically stronger than for adsorption.

Pressure dependence of argon solubility in silicate melts were measured by White, Brearley, and Montana (1989). They measured argon solubility in various silicate melts from 5 to 25 kbar with argon as pressure medium (Table 2.3) and found that the Henry's law holds to good approximation for the pressure range. Although experimentally determined noble gas solubility in silicate melts at extremely high pressures pertinent in the lower mantle is still scanty, noble gas solubility at high pressures is likely to be lower than that at the atmospheric pressure (e.g., Carroll et al., 1994; Shibata et al., 1996). In Table 2.3, we compiled noble gas solubility data in various materials relevant to earth science.

Chamorro-Perez et al. (1996, 1998) reported unusual behavior of argon solubility in silicate melts at very high pressures. They measured argon solubility in anorthite and silicate melts up to about 10 GPa (100 kbar) with a laser-heated diamond anvil cell (DAC). They observed that in both cases the Henry's law approximately holds up to about 5 GPa, but the solubility leveled off (anorthite) or even drastically decreased (silica) above 5 GPa. Assuming that noble gases are accommodated in holes in silicate melts, they suggested that increasing pressure makes the holes smaller, resulting in fewer sites being available to accommodate noble gas and thus decreasing the solubility. They further suggested that, with a decrease of the hole size, the spectrum of size distribution of the holes becomes sharper; once the peak size in the spectrum becomes smaller than the radius of a molecule of noble gas, there is an abrupt reduction in the number of accomodation sites and a sharp decrease in solubility. In Chamorro-Pérez's experiment, argon was used as the pressure medium. Hence, in this experimental arrangement, the amount of argon in the melts was above 0.1 weight percent. In contrast, the concentration of argon in the mantle-derived materials, which would reflect argon concentration in the mantle, is generally less than 1 ppb. Chamorro-Pérez et al. attribute the abrupt decrease in solubility to the availability of fewer "holes" that can accommodate argon atoms, and for a large amount of argon dissolved in the melts this may be the case. However, argon atoms are extremely diluted in the mantle, and there may be more than enough accommodation sites even at the depths where the holes are much smaller in size and number. Therefore, it is important to examine whether or not the abrupt decrease of argon solubility in silicate melts at high pressures is indeed observed for much lower concentration of argon.

It is interesting to consider a simple model for noble gas solution in liquids. From the Boltzmann distribution theorem, the ratios of equilibrium concentrations of solute molecules in two phases (here we consider liquid and gas phases) can be written as

$$C_l/C_g = \exp(-E/kT)$$

where k denotes Boltzmann constant and E is the energy needed to transfer a molecule from the gas phase to dissolve in solvent. Therefore, from Equation (2.18), we have

$$\mathcal{H} = (\rho kT)^{-1} \exp(-E/kT) \tag{2.20}$$

It is then convenient to consider the solubility process of introducing the solute atom into the solvent as consisting of two steps and to consider E as consisting of two parts E_1 and E_2 corresponding to each step (e.g., Pierotti, 1963). Step I is the creation of a cavity in the solvent to accommodate the solute atom. Step II is the introduction of the solute atom into the cavity. The latter process is to transfer a solvent molecule from the gas phase into the solvent.

The energy (E_1) to create a cavity can be approximately equated to $4\pi r^2 \sigma$, where σ represents the surface tensions between the fluid and a perfect rigid wall of the cavity and r denotes the radius of the cavity (e.g., Blander et al., 1959). Therefore, if the cavity creation is the dominant controlling factor in noble gas solubility, which is likely to be the case in common silicate melts, the Henry's law constant can be approximately given by

$$\mathcal{H} = (\rho kT)^{-1} \exp(-4\pi r^2 \sigma/kT) \tag{2.21}$$

where r corresponds to the effective atomic radius of a specific noble gas. Because of a positive value of σ, Equation (2.21) suggests that Henry's law constant would give a positive temperature dependence. Lux (1987) reported that except for He, which showed little temperature dependence of the solubility, solubilities of heavier noble gases (Ne, Ar, Kr, Xe) in lancite-basalt, tholeiite basalt, and alkali-olivine basalt melts, all showed mild positive temperature dependence (Figure 2.6). The similar positive temperature dependence of noble gases in a basalt (MORB) melt was also reported by Jambon, Weber, and Braun (1986) and Hayatsu and Waboso (1985). Lux (1987) also observed that for basalt melts, there was a rough linear relation between $\ln(\mathcal{H})$ and r^2 in accordance with Equation (2.21). However, it is to be noted that more rigorous calculation of the energy to create a cavity in melts (e.g., Reiss et al., 1960) shows that the energy of the cavity creation is a complicated function of an atomic radius, and the preceding simple relation should be understood only as a crude approximation, although the general rule suggested from Equation (2.21) that the

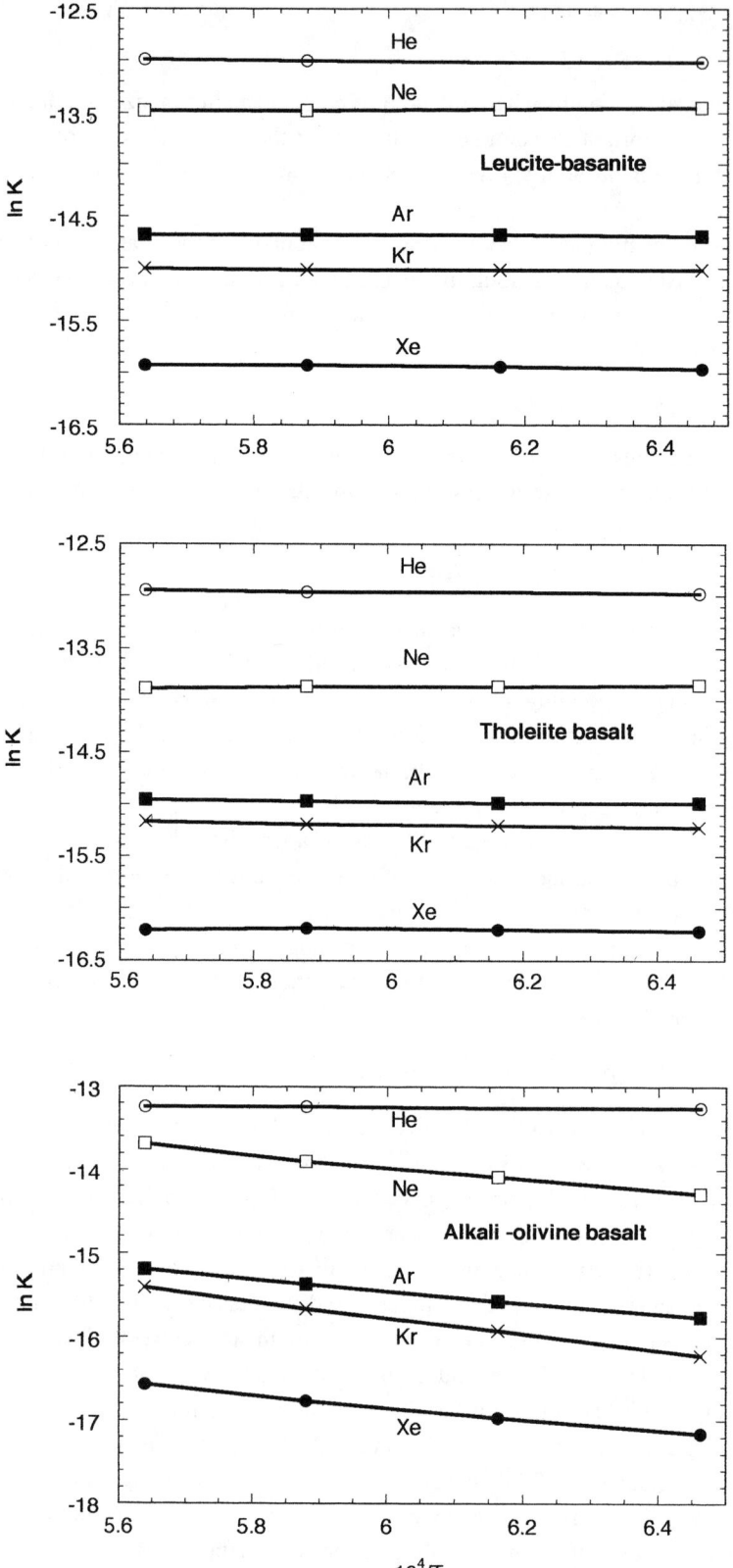

lighter noble gases (hence smaller size) are more soluble in basaltic melts than the heavier ones would still be valid.

In general, transferring a molecule from a gas and putting it in a more restricted site in a condensed phase involves the evolution of heat (Carroll & Stolper, 1991). Therefore, the energy (E_2) associated with the second process is negative in this case. Positive interaction (attraction) between a molecule and solvent gives a negative E_2. Hence if the latter energies (negative E_2) becomes more important than the cavity creating energy (E_1), negative temperature dependence of solubility would be expected. Noble gas solubility in water is a typical example. Because of high polarizability of water, interaction between water molecules and a noble gas atom becomes a dominant factor in gas solubility. This effect is more pronounced in the case of the heavier noble gases because the heavier noble gases have the higher polarizability. This is reflected in the noble gas solubility in water; the heavier noble gases are more soluble than the lighter ones in spite of the larger atomic sizes (see Figure 4.1).

Noble gas solubility in metallic iron is also important in earth science because of the possibility of noble gas storage in the metallic core. A little experimental work has been done on helium solubility in metals at atmospheric pressure such as in solid nickel (von den Driesch & Jung, 1980). Wolfer et al. (1989) made a theoretical estimation of helium solubility in solid and liquid nickel. This theoretical estimate of helium solubility in nickel was about eight orders of magnitude smaller than the measured values. They therefore questioned whether the measured values represented equilibrium solubilities. If noble gas solubility in iron were to be similar to that of nickel, even the larger value (i.e., the experimentally determined one), would give much less noble gas concentration in the core than in the mantle, provided that noble gas partition took place under the condition similar to those in the preceding experiments (i.e., at the atmospheric pressure and slightly above the melting temperature). However, if the core-mantle segregation took place deeper in the earth's interior and at much higher temperatures, we must consider these effects on the noble gas solubility. Stevenson (1985) made a theoretical estimation of noble gas solubility in metal at extremely high pressure. For this, he calculated the energy to immerse noble gas atoms in metal (immersion energy). He found that the immersion energy first increases with increasing ambient pressure, but above a few hundred gigapascals, they reach a maximum and then decrease with further increase of the pressure. This result suggests an interesting possibility that at an extremely high pressure exceeding the pressure at the center of the Earth, xenon would become more solvable in metal. However, the turnover pressure at which the immersion energy starts to decrease exceeds the highest pressure in the Earth's interior, and the drastic change of Xe solubility with pressure is unlikely to occur in the terrestrial planets.

◀

Figure 2.6 Temperature dependence of the Henry's law constants (K) for noble gas solubility in three silicate melts. Reproduced from Lux (1987).

Noble gas solubility has also been measured in solids: olivine crystals (Hiyagon & Ozima, 1986), silicate glass (Carroll & Stolper, 1991), and anorthite, diopside, forsterite, and spinel crystals (Broadhurst et al., 1992). In a marked contrast to the case of noble gas solubility in the melts, the experimentally determined solubility values showed large variation. Also, the observed characteristic features of the solubility such as the relation between the solubility and the atomic size and its temperature dependence are quite varied among the investigators. For example, Ar solubilities obtained by Carroll and Stolper (1991) and Broadhurst et al. (1992) are more than an order of magnitude larger than those obtained by Hiyagon and Ozima (1986): 4×10^{-4} cm³ STP/g atm (Carroll & Stolper, 1991); $\sim 10^{-4}$ cm³ STP/g atm (Broadhurst et al., 1992); and $\sim 3 \times 10^{-7}$ cm³ STP/g atm (Hiyagon & Ozima, 1986). Also, Broadhurst et al. observed a positive correlation between the solubility and noble gas atomic size, whereas Hiyagon and Ozima (1986) observed a negative correlation. The discrepancy among the experiments may partly be attributed to adsorption effect. For example, the positive correlation of noble gas solubility with the atomic size can be understood if the solubility measured by Broadhurst et al. (1992) resulted from adsorption rather than dissolution, although Broadhurst et al. (1992) concluded that their experimental results cannot be due to adsorption because retentivity of the incorporated noble gases in the minerals were very high; most of noble gases were degassed only around the melting temperature. At any rate, it is obvious that the problem is still open to future investigation. In view of its fundamental importance in noble gas geochemistry such as in the determination of noble gas partition coefficient (cf. Section 2.4), it is urgent to carry out well-controlled laboratory experiments to resolve this problem.

2.4 Crystal-Melt Partitioning

The crystal-melt partition coefficient $K_D = C_s/C_l$, where C_s is concentration in a solid and C_l is concentration in coexisting liquid, is a key parameter in trace element studies of igneous systems. A noble gas crystal-melt partition coefficient is the ratio of the gas solubilities considered here. As seen in Table 2.3, solubilities have now been reported for a variety of melt compositions, but solubility data are still very scarce for solids in general.

There are a number of trace elements commonly described as incompatible because their partition coefficients (for typical mantle mineralogy) are low, often as low as 10^{-2} or 10^{-3} or less. Prominent examples are K, Rb, U, Th, and the rare earth metals. The principal reason for their incompatibility is evidently their large ionic radii. Concentrations of these elements in mantle-derived rocks are relatively low, and the generalization emerges that they have been expelled from the mantle (at least the upper mantle) and concentrated in the crust, especially continental crust.

It is generally believed that the noble gases should behave like incompatible elements, only more so. Their atomic radii are also relatively large (Table 2.1). More-

2.4 Crystal-Melt Partitioning

over, they have only very weak interactions with other elements; in igneous conditions they will not be ionized or share electrons in covalent bonds. It is not expected that they could substitute for regular lattice atoms except by accident (i.e., if there is a structural vacancy). In analogy with the lithophile incompatibles, their abundances are low in mantle-derived rocks, and they are greatly concentrated in a "crust" (the atmosphere). On the other hand, the mantle clearly contains at least some juvenile gas that has never been part of the atmosphere, and there is substantial uncertainty about what significance is to be attached to low concentrations in mantle materials and even exactly how low these concentrations actually are (see Chapters 6 and 7). There is precious little actual observational data by which to judge this expectation of very low crystal-melt partition coefficients for the noble gases. This lack hampers theoretical treatments of mantle degassing and atmospheric evolution.

Estimation of a noble gas partition coefficient can be made by measuring noble gas concentration both in rock-constituting minerals and in the matrix phase, which was supposed to be in equilibrium. Marty and Lussiez (1993) measured He and Ar concentration in olivine and glass separates from a MORB picritic basalt and estimated values of crystal-melt partition coefficients of 0.008 for He and of 0.003 for Ar. One difficulty in obtaining a reliable partition coefficient with the use of natural samples is in ensuring whether the melt (glass) and the crystal were in thermodynamic equilibrium. Hiyagon (1994) questioned the results by Marty and Lussiez on the grounds that the He/Ar ratio observed both in the olivine and the glass were almost identical; if the olivine and the melt were in equilibrium, the He/Ar ratio in the former must be higher by about a factor of 2.7 (= 0.008/0.003) than in the latter in accordance with the difference in their solubilities. Valbracht et al. (1994) also attempted to obtain the partition coefficient of noble gases between olivine and glass on Hawaiian basalts. However, they found differences in Kr and Xe isotopic compositions between the olivine and the glass, which indicates that the crystals and the melts were not in thermodynamic equilibrium.

Hiyagon and Ozima (1986) employed a laboratory approach of measuring crystal-melt partition coefficients. They measured noble gas concentration in olivine crystals and basalt melts, which were synthesized at 1370–1300°C under an atmospheric pressure, and also at 1360–1050°C under high pressure (0.2–1.5 GPa), of noble gas mixture. From these experimental results, they obtained ranges for noble gas partition coefficients: $K_{He} = 0.07$, $K_{Ne} = 0.006–0.08$, $K_{Ar} = 0.05–0.15$, $K_{Xe} = 0.3$. These partition coefficients are much larger than the values obtained by Marty and Lussiez (1993) and also these of common incompatible elements such as U (~0.002) or K (0.0002 – 0.008) between olivine and basalt melt (e.g. Henderson, 1982).

A common difficulty in these partition experiments, with the use of either natural or synthesized samples, is in achieving perfect separation of the melt from the crystal phase for determining the noble gas content. Even a very small amount of glass (melt) contamination in crystal would increase the partition coefficient considerably, since noble gasses are much more enriched in glass. To circumvent this difficulty, Broadhurst et al. (1990, 1992) prepared natural minerals and synthetic silicate melts that

Table 2.4. *Crystal/matrix partition coefficients of noble gases in basaltic melts*

	He	Ne	Ar	Kr	Xe	Reference
Olivine	0.01–0.25	0.01–0.15	0.03–0.2	0.03–0.2	0.05–0.5	1
	0.008^a		0.003^a			2
	0.05^a					3
	0.08–0.8	0.02–0.8	0.08–1.5	0.09–2		4
			0.04–0.16			5
	0.076–0.18^a	0.019–0.13^a	0.013–0.048^a	0.016–0.17^a	0.029–0.37^a	6
Clinopyroxene			0.0029–0.1			5
Plagioclase			0.008–0.036			5
Diopside		0.013–0.37	0.15–0.84	0.31–5.9	3.2–47	7
Spinel		0.034–0.088	0.04–1.3	0.63–2	0.61–1.8	7
Forsterite		2.7	6.7	45	90	7

^{a} Measurements were made on natural samples. Others are laboratory experiments.

Table 2.4 References: 1, Hiyagon & Ozima (1986); 2, Marty & Lussize (1993); 3, Trull & Kurz (1993); 4, Shibata, Takahashi & Ozima (1994); 5, Matsuda et al. (1993); 6, Valbracht et al. (1994); 7, Broadhurst et al. (1992).

were separately equilibrated with 1 bar noble gases at 1300°C and 1320°C. From the noble gas concentration in the melt and mineral phases, they deduced noble gas partition coefficients. Partition coefficients, especially of Xe, thus determined were much larger than those obtained by other authors. As discussed in Section 2.3, the solubilities estimated for the crystal might be affected by adsorbed noble gas components due to the extremely small grain size of the sample, which was purposely made to ensure equilibration of noble gas distribution in the sample. It may be possible that the unexpectedly high noble gas solubility obtained by Broadhurst et al. (1992) is due to the adsorbed noble gas.

In Table 2.4, we compiled experimental data on noble gas partition coefficients in mineral-melt systems. Very large spreads in the partition coefficient values may primarily be attributed to the experimental difficulties. Because there is little published data, which nevertheless show large spreads, it is not possible to conclude definite values for partition coefficients. However, the experimental data at least urge us to consider seriously the possibility that noble gases are not as incompatible as what we have hitherto assumed, and could be less incompatible than geochemically interesting common trace elements such as potassium, rubidium, lead, uranium, or thorium. If this were the case, most of the mantle degassing models published so far would probably need fundamental revision, since higher incompatibility of noble gases than relevant trace elements such as K or U is a basic premise in these models (cf. Section 6.9).

Matsuda et al. (1993) studied noble gas partition between iron melt and silicate melt at 1550°C for pressures ranging from 5 to 100 kbar. Noble gas partition coefficients thus determined showed a systematic decrease with increasing pressure (Figure 2.7), from about 5×10^{-2} at 5 kbar to about 5×10^{-4} at 100 kbar. The results may

2.5 Trapping and Implanting

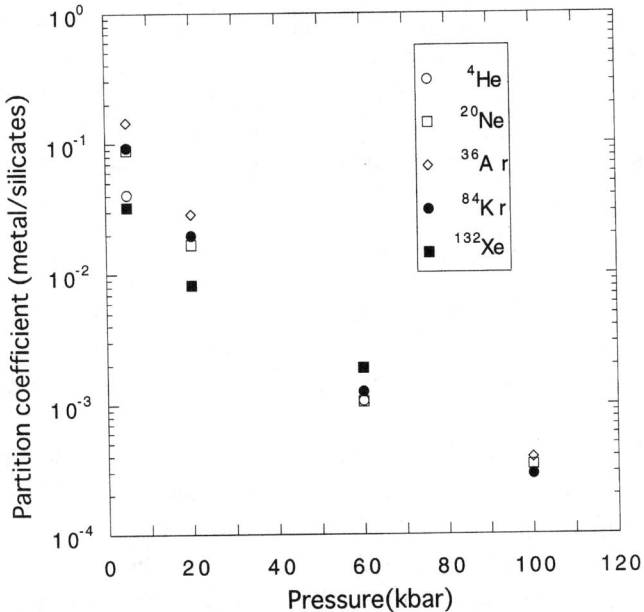

Figure 2.7 Partition coefficients of noble gases between iron melt and silicate melt are plotted against pressure. Reproduced from Matsuda et al. (1993).

suggest that noble gases are unlikely to be accommodated into the iron core. However, at pressures much higher than the experimental ones, the behavior of noble gases, especially the heavier ones, may be quite different as suggested by experiments by Jephcoat and Besedin (1996) (Section 2.1). Hence, it remains to be seen whether noble gases are accommodated in the iron core.

2.5 Trapping and Implanting

The term *trapped*, as applied to noble gases, has come to have a well-defined denotation as designating noble gases that were not produced in situ (Section 1.4). Its original connotation, however, is that gases somehow got into a sample by a process or processes, which are unnecessary and often impossible to specify, and have subsequently been unable to get back out of it. The term is still used with this connotation.

Strictly speaking, there are only two ways in which gases can be in a sample. If gas atoms are mixed with host atoms on a microscopic scale, the gas is dissolved; if gas atoms are on the surfaces that bound the sample, the gas is adsorbed. There are quite a few complications and variations on these basic themes, however, as will be discussed. From a macroscopic viewpoint, and from the viewpoint of laboratory practice, it is sometimes difficult to distinguish between solution and adsorption, particularly for samples inferred to have extensive and complicated internal surfaces (embayments, pores, cracks, etc.). It can be and has been argued, for example, that in some experiments purporting to measure noble gas solubility the relevant phenomenon is actually adsorption, and vice versa. When the structure of the sample in question is sufficiently complicated (e.g., in clays, activated charcoal, or poorly crystalline or amorphous materials), it is not clear that a distinction can or should be made between a trace element's presence in (solution) or on (adsorption) its host. From a macroscopic viewpoint, however, it is also unnecessary to make the distinction in such cases. As pointed out in Sections 2.2 and 2.3, the thermodynamic descriptions of solution and adsorption are quite similar. The Henry constants, which describe both phenomena (Tables 2.2 and 2.3), are examples of partition coefficients and have a meaning independent of the microscopic interpretation of how the gas atoms interact with the host.

An additional connotation of the term *trapped* gases deserves explicit note here. The gases must be observed experimentally before they can be identified as trapped gases. The experimental procedure always involves loading the samples into a laboratory vacuum system before analysis of gases subsequently (hours or months later) released by heating or other means; usually, the samples are also heated to approximately 100°C in vacuo to remove "air contamination" or "loosely bound" gases or other volatiles that would interfere with the analysis. Solution and adsorption are both equilibrium effects for which the gases contained in the sample are functionally dependent on external partial pressure, and when this pressure is effectively zero, as in the laboratory vacuum, so too are the equilibrium gas contents. The same remarks apply to the immediately prior history of the samples: field residence and laboratory storage in which the external medium is air and equilibrium gas contents should be dictated by the partial pressures in air.

Trapped gases are thus those gases contained within samples, which, for kinetic reasons, are not in equilibrium with their present or immediately prior external pressures: at some epoch in the past they were incorporated in the samples, in equilibrium with that past environment or otherwise, and have survived their later history, usually on a geological timescale, without equilibration in response to a changing environment. If the gases are dissolved, there is no particular problem with maintaining this disequilibrium even over geological timescales. At ambient temperatures on the surface of the Earth (and in interplanetary space), noble gas diffusion coefficients in many materials are quite low, and the timescales for equilibration by diffusive redistribution are quite long (see Section 2.7). Indeed, it can be argued that noble gas atoms can be so efficiently encaged in a crystalline lattice that they move only few lattice spacings in the age of the solar system. Adsorption timescales, in contrast, are quite short, and adsorption/desorption are often considered essentially instantaneous on a laboratory timescale.

Trapped noble gases are thus usually considered to be of the dissolved rather than the adsorbed variety. It must be remembered that the relevant definition is operational, however: Gases that appear in the mass spectrometer in the usual analysis were trapped no matter what their original microscale distribution in the sample. Zaikowski and Schaeffer (1979), for example, report that noble gases dissolved in phyllosilicates may equilibrate with air – both lose and gain gas – in laboratory (and museum) conditions and timescales. Conversely, Podosek et al. (1981b) suggest that, in materials of sufficiently convoluted surfaces, adsorbed gases may not escape the sample in the laboratory vacuum. Whether the term *trapped* should be applied to gases thus contained in natural samples is a semantic issue we will not pursue here.

Although trapped gases are either dissolved or adsorbed, there are nevertheless a variety of "ways" in which they may be trapped. One such possibility is that gases may be contained in small regions, fluid or crystalline, wholly enclosed in a crystal of another material. A good example is Mid Oceanic Ridge Basalts (MORB), in which most of indigenous mantle-derived noble gases reside in CO_2 inclusions in its glassy margin. Among MORBs, "popping rock" is notable for its abundant CO_2 inclusions and thus contains comparatively high mantle-derived noble gas (cf. Section 6.2). The gases in this case are not really *in* the enclosing material, but this circumstance may not be evident in the laboratory. Gases present on surfaces or in nearby materials, or even in a gas phase, might be "fixed" (dissolved?) in transient high pressures and temperatures resulting from high-velocity impact shock (e.g., Bogard, Horz & Johnson, 1987; Wiens & Pepin, 1988), or even in static compression at only moderately high pressures (cf. Honda et al., 1979).

It is often suggested that noble gases can be trapped during the growth of crystals, especially during condensation from vapor but also during crystallization from a liquid. Gases may be present on or near growth surfaces because of solution, adsorption, energetic implantation, or some other mechanism, and then covered over – trapped – by the continuing growth. The term *occlusion* seems to be used for a variety of meanings in the literature, but it frequently suggests this process.

A number of researchers have studied the extent to which noble gases are trapped during the laboratory synthesis of solids, which are very difficult to describe in terms of any simple model. These include direct (and decidedly nonequilibrium) condensation from vapor (Honda et al., 1979; Kothari et al., 1979; Matsuda, Fukunaga & Ito, 1991; Matsumoto et al., 1996) and also precipitation from solution and from an electric discharge (Frick, Mack & Chan, 1979). In general, an adequate description of the processes involved or the conditions of formation is not possible; even a description of the resultant materials is difficult. Nevertheless, noble gases are trapped. The results can be described in terms of partition coefficients, but in most cases the variation of trapped gas content with ambient gas pressure was not examined, so the "partition coefficients" are only ratios of trapped gas to pressure in a single experiment. The partition coefficients are in some cases quite high, and in some cases produce quite strong elemental fractionation.

Ion implantation is another common noble gas trapping mechanism in nature. For example lunar soils are implanted by solar emission particles that essentially consist of lighter noble gases. A few laboratory experiments have been carried out with the hope that experiments may be relevant to incorporation of noble gases in early solar

Table 2.5. *Mean penetration depths (µm) of noble gas atoms implanted in minerals at 20 KeV (after Futagami et al., 1993)*

	^{4}He	^{20}Ne	^{40}Ar
Olivine	0.147	0.028	0.016
Ilmenite	0.124	0.026	0.015

system materials (e.g., Bernatowicz & Fahey, 1986; Bernatowicz & Hagee, 1987; Futagami et al., 1993). Bernatowicz & Hagee (1987) and Bernatowicz & Fahey (1986) observed significant mass-dependent isotopic fractionation in Xe and Kr, which were implanted into the glass wall of an experimental apparatus under a closed system condition at very low energies (50–500 eV). The isotopic fractionation was explained as being the result of balancing between implantation into the glass wall of an experimental vessel (a Modified Bayard-Alpert ionization gauge) and sputtering from the wall; both processes are mass-dependent (Bernatowicz & Fahey, 1986; Bernatowicz & Hagee, 1987). Hence, the fractionation appears to be essentially due to the specific closed system experimental condition. Since noble gas implantation in minerals in nature is more likely to take place in an open system, it is difficult to judge how far the experimental results can be extrapolated to infer isotopic fractionation effect in lunar soils or in meteorite grains.

Futagami et al. (1993) implanted He and Ne on olivine and ilmenite target plates with a high irradiation dose (up to 10^{16} dose cm^{-2}) at 20–200 KeV and found that the amount of implanted noble gases were linearly proportional to the irradiation dose (also see Section 7.8). Noble gas implantation depends on the mass of an impinging particle: lighter noble gases are more deeply implanted into target material. Futagami et al. (1993) calculated the mean depth of ion penetration depth in both olivine and ilmenite at 20 KeV (Table 2.5). This suggests that lighter noble gases would be more abundantly implanted into target material. However, larger diffusivity of lighter noble gases would counteract to retain lighter noble gases relative to heavier ones in target material. This may explain the large deficiency of He and slight enrichment of heavier noble gases in lunar soils relative to SW noble gas abundance (cf. Section 7.3). From experimental results on noble gas diffusion (Section 7.7 and Figure 7.14), Futagami et al. also suggested that ilmenite has higher noble gas retentivity than olivines. The latter result is in good accordance with observations on noble gas retention characteristics in lunar soils; hence, lunar ilmenites can be regarded to be the best sample to study implanted SW noble gases.

Mechanical shock applied in ambient noble gas atmospere also emplaces substantial amount of noble gases into target material. Wiens and Pepin (1988) studied the effects of mechanical shock (20–60 GPa) applied to rock specimens (basalt disk and powder: 63–125 µm in size) in ambient air. They found that emplacement efficiencies (number densities in shocked samples/number densities in ambient gases)

2.5 Trapping and Implanting

Table 2.6. *Concentrations for incorporation of noble gases by various processes at gas partial pressures in air*

Sample	Gas Concentration (cm^3 STP/g)^a				
	^{4}He (10^{-9})	^{20}Ne (10^{-9})	^{36}Ar (10^{-9})	^{84}Kr (10^{-12})	^{130}Xr (10^{-12})
Solution^b					
Water (0°C)	50	210	1,690	72,100	802
Serpentine (340°C)		6.1	14	490	4.2
Enstatite melt					
(1500°C)	0.63	1.2	6.3	8.5	0.021
Adsorption^c					
Martinez shale					
(0°C)				175,500	11,685
Kibushi shale					
(0°C)				275,500	4,250
Shock^d					
Basalt disk					
(3 atm, 40 GPa)	521	164	560	12,900	56
Basalt powder					
(2 atm, 20 GPa)	2,680	204	3,300	73,300	312
Crystallization (noble gas pressures follow)^e					
Fe$_3$O$_4$ (500°C)	<114	140	132	113	93.1
Fe$_3$O$_4$ (1,200°C)	<15.5	7.4	11.9	17.7	33.8
Others^f					
Soot		2.4	750	91,800	6,030
Kerogen					
(discharge)	1.2	10	1,100	18,400	1,600

^{a} Noble gas incorporation is made at noble gas partial pressures in air except for Adsorption.
^{b} Table 2.3.
^{c} Table 2.2.
^{d} Wiens & Pepin (1988).
^{e} Matsumoto et al. (1996); Fe$_3$O$_4$ was synthesized by oxidizing Fe with water vapor. Noble gas pressures = 8.3×10^{-4}, 7.5×10^{-4}, 2.7×10^{-4}, 4.7×10^{-4}, 2.2×10^{-4} atm for He, Ne, Ar, Kr, and Xe, respectively.
^{f} Data from Frick et al. (1979).

are constant over the range of 0.045–3.0 atm total ambient gas pressure and roughly constant from 35 to 50 Gpa shock pressure, with efficiency falling off at lower and higher pressure. The emplacement efficiencies for a disk sample ranges from 2.0 to 6.7% and from 40 to 50% for powderd sample (Table 2.6). Abundances of emplaced noble gases were linearly proportional to the ambient gas pressure for constant shock pressure. However, they found neither elemental nor isotopic fractionation with Ar, Kr, Xe, or N$_2$. The experimental results would provide empirical grounds to relate the trapped noble gases in shergottite (meteorites that are generally assumed to have been ejected from the Martian surface) to the Martian atmosphere.

A comparison of the various ways in which noble gases can be put into various materials is made in Table 2.6. The tabulated values are the gas concentrations that correspond to the partition coefficients in Tables 2.2–2.4 and partial pressures in air.

2.6 Clathrates and Fullerene C_{60}

Clathrates are generally described as "cage" compounds in which one substance, the host, forms a lattice with large structural "holes" in which molecules of a second substance, the guest, reside. Often the presence of the guest is essential in that the host will not form that crystal structure without the guest. The structure of a unit cell may be quite complicated and involve many molecules, and so the proportion of guest to host is variable; however, the ratio of guest to host will still be one of relatively small integers (i.e., the guest is not a trace element).

In the ideal case, guest and host will occur in definite stoichiometric proportions; thus, at least from a thermodynamic perspective, the clathrate is a specific substance (i.e., a chemical compound). In normal usage the distinction between such a clathrate and a chemical compound is presumably that in a clathrate the forces involved are van der Waals forces rather than transfer or sharing of electrons.

If the gas phase activity of the host is controlled by the presence of a pure condensed phase, solid or liquid, the equilibrium between host and guest in a stoichiometric clathrate can be described in terms of the gas phase pressure of the guest. This is, in effect, a vapor pressure for the guest. At higher pressures the guest will condense to form clathrate, and at lower pressures the clathrate will decompose. Temperature variation of this pressure will follow the Clapeyron equation which, with the usual assumptions (ideal gas behavior of the vapor and negligible volume of the condensed phase), reduces to the Clausius-Clapeyron equation:

$$d(\ln P)/dT = \Delta H/RT^2 \tag{2.22}$$

[cf. Equations (2.14) and (2.19)]. Here ΔH is the heat of evaporation of the guest, specifically the enthalpy change for the process in which clathrate decomposes a gaseous guest and condensed-phase host. Over a limited temperature range in which ΔH is approximately constant, this can be integrated to

$$P = P_0 e^{-\Delta H/RT} \tag{2.23}$$

where P_0 is a constant of integrations. This can be used as an approximation to vapor pressure as a function of temperature.

More generally, clathrates often tend to be nonstoichiometric. The clathrate is a distinct phase but its composition is not definite (i.e., the degree to which the structural holes are filled depends on the gas pressure of the guest). In such cases, the clathrate can presumably be viewed as an extreme case of selective solution or adsorption, according to whether the holes are viewed as interior to the host structure or as its surface (cf. Sections 2.2 and 2.3). In such cases, however, the solute/sorbate (guest) is not a trace constituent, and Henry's law will not apply.

Noble gases are known to form clathrates with a number of hosts, notably organic compounds such as hydroquinone. An important special case, the only one of even potential significance in geochemistry, is the ice-noble gas clathrate. For ideal stoichiometric behavior, the parameters for vapor pressure, according to Equation (2.23) calculated from fits by Sill and Wilkening (1978) to data of Barrer and Edge (1967), are given in Table 2.7. The heats are higher than the heats of vaporization of the pure noble gases, and the decomposition temperatures are higher than the normal boiling points (Table 2.1). In the presence of sufficient H_2O, pure noble gases would condense as ice clathrates before they would condense as pure liquids. The heats of clathrate formation are lower, however, than the corresponding heats for aqueous solution (Table 2.3).

Noble gas clathrates will not now form on the Earth, as can be seen from the air pressure decomposition temperatures in Table 2.7. They might, however, form in cooler regions of the primitive solar nebula (see Lunine & Stevenson, 1985). Sill and Wilkening (1978) note that for pressures in a plausible model nebula, pure ice clathrates of Ar, Kr, and Xe could form at 40, 45, and 62 K, respectively.

As with other compounds, solution effects can elevate the condensation temperatures of clathrate guest species. Sill and Wilkening calculated that in a gas of solar composition the major clathrate, and the first to form, will be ice-methane, and that noble gases can substitute for the methane at temperatures higher than decomposition temperatures for noble gas clathrates. They calculate, for example, that in a total nebular pressure of 2×10^{-3} atm (high in comparison with most model pressures currently considered of about 10^{-4} atm), ice-methane clathrate at 80 K will have dissolved 99% of the available Xe (and substantially smaller amounts of the other noble gases).

The Sill and Wilkening proposal that clathrates formed in the cold outer parts of the solar system and then transported to the inner solar system (e.g., in comets) might help account for the atmospheres of the terrestrial planets. They contend that infall of 1 ppm of ice-methane clathrate with noble gases dissolved as just described could account for the present inventories of Ar, Kr, and Xe in the terrestrial atmosphere.

The cagelike structure to accommodate noble gases is also seen in fullerenes C_{60}, the recently discovered giant carbon molecule (cf. Saunders et al., 1996, for a review). The remarkable feature of fullerene is the presence of a closed central cavity, which is large enough to accommodate atoms (Figure 2.8). Saunders et al. (1993) showed that heating fullerenes in 1 atm of He at 600°C for one hour resulted in the trapping of He in the fullerene. The trapped amount was about 6.7×10^{-10} cm^3 STP per 14 μg of fullerene. Similar experiments with neon (in 3 atm of neon at 600°C for one hour) yielded 7.4×10^{-9} cm^3 STP per 22 mg of fullerene. From the stepwise degassing

Table 2.7. *Decomposition vapor pressure data for ice-noble gas clathrates*

	Ne^b	Ar	Kr	Xe
P_0 (10^4 atm)[a]	2.82	2.09	1.77	6.11
ΔH (kcal/mole)[a]	0.92	2.96	3.90	5.78
Temperature (K) for $P = 1$ atm[c]	45	150	201	264
Temperature (K) for $P = $ air pressure[c,d]	22	102	84	107

[a] Parameters for eq. (2.23), equivalent to fits by Sill & Wilkening (1978) to experimental data of Barrer & Edge (1967).
[b] Data by extrapolation of properties of heavier gases.
[c] From P_0 and ΔH in first two rows.
[d] For 1 atm air pressure with composition given in Table 1.2.

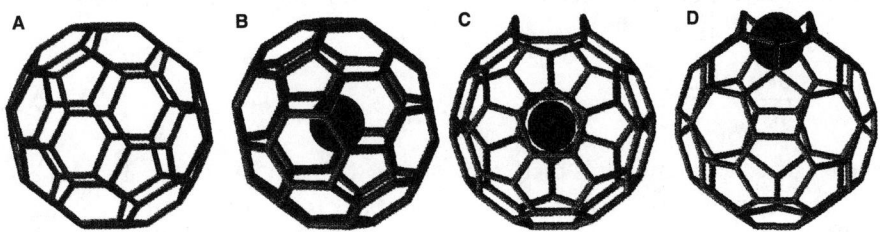

Figure 2.8 (A) C_{60} fullerene. (B) C_{60} with an atom inside. (C) C_{60} with an atom inside and with a bond broken (open window). (D) Same molecule as in (C) but with the atom (noble gas) moving out through the window. Reproduced from Saunders et al. (1996).

experiment, Saunders et al. found that helium and neon were released over a temperature range from 600 and 900°C. The approximate rates of release are consistent with an activation energy of 70–80 kcal/mol for a simple first-order process (Saunders et al., 1996). From the relatively high and also similar temperature release both for helium and neon, they concluded that helium and neon were trapped inside the C_{60} cage. However, since the estimated activation energy for the helium release was considerably smaller than the energy necessary to push a helium atom through an intact ring (about 200 kcal/mol), Saunders et al. suggested that breaking a C-C bond, which requires much less than 200 kcal/mol would open a "window" through which helium or neon atom sneaked into the cage. This is illustrated in Figure 2.8.

Fullerenes are known to occur in nature, for example in the Sudbury impact crater and in a meteorite (cf. Section 5.3). In the former case, it is also known that the fullerenes contained a significant amount of He with high $^3He/^4He$ ratio. However, the amount of noble gases trapped in fullerene in air would be too small to be of any significance in common geological settings. Assuming that the amount of noble gas trapped in fullerene is proportional to the ambient noble gas pressure, we estimate from these experimental results that at the helium partial pressure in air, fullerene would contain about 4×10^{-16} cm^3 STP/g-C_{60} of ^3He, which is less than 1% of the dissolved ^3He in sea water (Table 2.6).

2.7 Diffusion

Diffusion of noble gases into and out of magmas, minerals, and rocks is a pervasive process that alters not only noble gas concentrations but also the ratio of one element to another and, to a lesser extent, the isotopic composition of a given element. An appreciation of the basic processes involved is thus fundamental to deriving any useful conclusions from observations of elemental compositions of noble gases in natural samples. Since the subject of diffusion is one of the most extensively developed areas of mathematical physics, this section only briefly describes relevant data for the noble gases.

Quantitative treatments of diffusion almost invariably begin with the diffusion equation

$$\frac{\partial C}{\partial t} = D\nabla^2 C \qquad (2.24)$$

where C is concentration, t is time, and D is the diffusion coefficient. This equation assumes an isotropic medium in which diffusive flux is proportional to concentration gradient. An additional term must be included if the concentration is nonconservative (i.e., if there are sources such as production by radioactive decay or sinks). Solutions of the diffusion equation for a variety of initial and boundary conditions are available in standard texts such as Carslaw and Jaeger (1959) or Crank (1975); an illustrative example is given in the next section.

It is also generally assumed that the diffusion coefficient D is independent of concentration and has a temperature dependence given by an Arrhenius equation:

$$D = D_0 e^{-E/RT} \qquad (2.25)$$

where D_0 is a constant and E is the activation energy. Selected data for noble gas diffusion coefficients are given in Table 2.8 and Figure 2.9.

Solutions of the diffusion equation inevitably involve the dimensionless parameter Dt/a^2 in such a way that diffusive redistribution becomes significant as this parameter approaches unity. Here a is some characteristic dimension of the diffusive region. In the case where the medium is homogeneous, without a microscopic substructure (e.g., a glass or a liquid), a is the macroscopic dimension. In rocks that consist of numerous mineral grains, the relevant a is usually the individual grain size rather than the macroscopic dimension.

The importance of the characteristic size a as well as the diffusion coefficient D in controlling the rate of diffusion is evident. Subdivision of mineral grains into smaller grains by exsolution or alteration can thus markedly increase the speed of diffusion. As an illustrative example, K-feldspar often exsolves into perthite, a fine-grained mixture of albite and orthoclase or microcline, and this perthitization has been invoked to help explain the frequently poor retention of radiogenic ^{40}Ar in K-feldspar (e.g., Foland, 1974).

Diffusive loss of gases from crystalline rock is often viewed as a two-stage process: first volume diffusion from grain interiors to grain boundary surfaces and then grain boundary or surface diffusion for diffusion out of the macroscopic sample. Each process has its characteristic diffusion coefficient and activation energy; in general, volume diffusion has a lower diffusion coefficient and higher activation energy than grain boundary or surface diffusion. As in the foregoing example, it is usually considered that volume diffusion from grain interiors is the rate-limiting process, and that once gas reaches grain exteriors its diffusion along grain boundaries is relatively rapid. In some cases, however, the converse may be true. Ozima and Takigami (1980), for example, studied the loss of radiogenic ^{40}Ar by stepwise heating and concluded that even though volume diffusion is the limiting process in granites, grain boundary

Table 2.8. Selected data for noble gas diffusion in minerals

Gas	Sample	Temperature (°C) and Pressure	E (kJ/mol)	D_0 (cm^2/s)	Remark
^{3}He	Olivine	150–650	105	1.995×10^{-4}	Trull et al. (1991), ^{3}He: Cosmogenic
	Quartz	150–650	105.8	1.585	Trull et al. (1991), ^{3}He: Cosmogenic
	Diamond	1200–1700	~150	6.1×10^{-11}	Wiens et al. (1994), ^{3}He: Cosmogenic
^{4}He	Diamond	1200–2050	1134	1.821	Honda et al. (1987)
	Diamond	1700–2000	1480	6.2×10^{-7a}	Ozima (1989)
	Diamond	650–1550	103	6.3×10^{-14}	Zashu & Hiyagon (1995)
	Hornblend	180–850	120	4.6×10^{-2}	Lippolt & Weigel (1988)
	Hornblend	180–850	131	3.63×10^{-3}	Lippolt & Weigel (1988)
	Sanidine	180–850	94	4.57×10^{-4}	Lippolt & Weigel (1988)
	Ilmenite	1000	46.2	4.97×10^{-11a}	Futagami et al. (1993), He: Implanted with 20 eV
	Olivine	700–1400	420	1.259×10^5	Trull & Kurz (1993)
	Pyroxene	700–1400	290	125.9	Trull & Kurz (1993)
Ne	Vitreous albite	200–500	95	2.22×10^{-2}	Roselieb et al. (1992)
	Ilmenite	1000	63	1.95×10^{-11a}	Futagami et al. (1993), He: Implanted with 20 eV
	Obsidian	200–600	73.5–84.4	$(0.55–7) \times 10^{-3}$	Matsuda et al. (1989)
Ar	Diamond	1700–2000	1919	8.1×10^{-6a}	Ozima (1989)
	Rhyolite	400–900	144.9	6.86×10^{-3}	Carroll & Stolper (1991)
	Orthoclase	400–900	121.8	6.64×10^{-3}	Carroll & Stolper (1991)
	Albite	400–900	137.7	1.62×10^{-3}	Carroll & Stolper (1991)
	Ilmenite	1000	92.4	1.0×10^{-10}	Futagami et al. (1993), ^{3}He: Implanted with 20 eV
	Silica	1200 2500 bar	99	3.1×10^{-6}	Roselieb et al. (1995)
Kr	Silica	700–950 215–3150 bar	115.9	3.16×10^{-7}	Carroll et al. (1993)
	Silica	1200 2500 bar	165	1.6×10^{-5}	Roselieb et al. (1995)
Xe	Silica	1200 2500 bar	293	6.3×10^{-2}	Roselieb et al. (1995)

[a] Calculated from data given in figures.

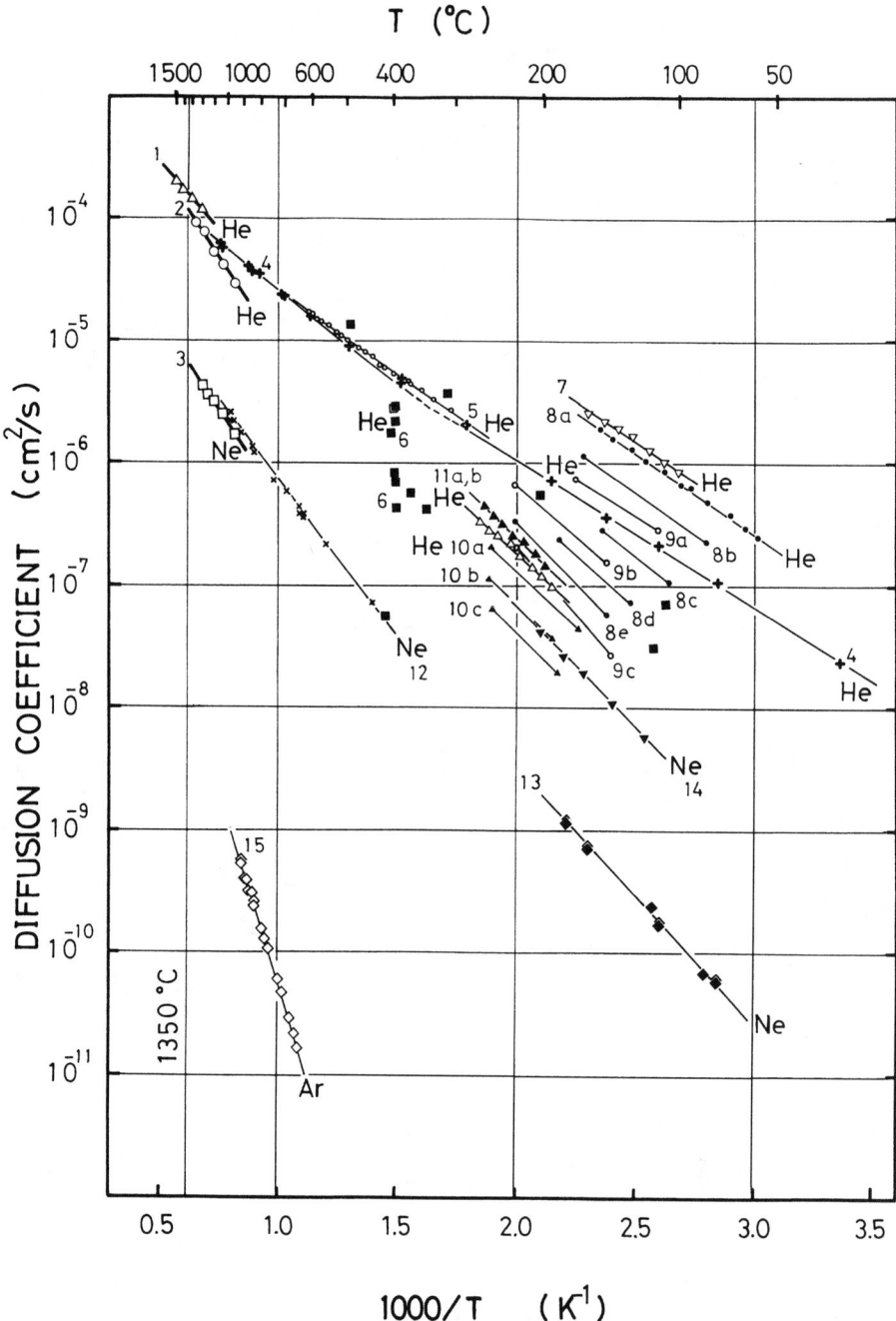

Figure 2.9 Arrhenius plot showing temperature variation of noble gas diffusion coefficients. Samples 1–3 are glass melts; 4, 5, and 14 are vitreous silica; 6 is commercial glass; 7 and 14 are B_2O_3; 8–10 are mixtures of alkali oxides with B_2O_3, SiO_2, and Al_2O_3; 11 and 12 are obsidians; 13 and 15 are SiO_2. Reproduced from Hiyagon (1981).

diffusion is the limiting process in oceanic basalts. This conclusion is consistent with the view that K is distributed throughout relatively small grains in granites.

Diffusion of noble gases in geological materials is generally very slow (cf. Figure 2.9) at relatively low temperatures, so laboratory diffusion experiments are usually conducted at fairly high temperatures and extrapolated to lower temperatures by means of Equation (2.25). Alternatively, a geologic estimate of diffusion coefficient can be made (e.g., by measurement of radiogenic ^{40}Ar in rocks whose age is determined by independent geochronometers). Various authors (e.g., Wescott, 1966) have pointed out that the geologic estimates are generally higher than the laboratory measurement extrapolations to the relatively low temperatures of principal interest (≈200°C). This discrepancy might be attributable to faster diffusive loss in nature because of grain size reduction by weathering or exsolution. Much of the discrepancy may also be due to failure of Equation (2.25) when extrapolated over too wide a temperature range. This is particularly plausible when significant structural changes occur in the relevant minerals at temperatures spanned by the extrapolation. Wescott (1966), for example, studied Ar diffusion in biotite and concluded that two different activation energies were involved: 35 kcal/mole below 800°C and 49 kcal/mole above 800°C. However, House, Farley, and Kohn (1999) reported that empirical diffusivity of He of some apatites at relatively low temperatures (67–97°C) is consistent with the diffusivity extrapolated from high-temperature laboratory diffusion experiments.

A remarkable example of the multiactivation process was reported by Zashu and Hiyagon (1995) in the case of He diffusion from carbonado diamond (Figure 2.10). In the Arrehenius plot, two distinct linear segments are evident; the slope of the linear trend corresponds to the activation energy (i.e., $E = 103$ kJ/mol for <1500°C and about 2090 kJ/mol for the higher temperature range). Since the latter activation energy (enormously high!) is close to that of graphitization of diamond, Zashu and Hiyagon (1995) concluded that even though the low-temperature degassing represents a volume diffusion, the degassing at the higher temperatures was more likely to be due to graphitization. An order of magnitude differences in the activation energy of He diffusion from diamonds among different authors (Table 2.8) is very likely to reflect the difference in the temperature range at which the experiments were carried out. Zashu and Hiyagon (1995) also observed that, of the total degassed He, 99.2% was degassed at the higher temperature range. This suggests that a major fraction of He trapped in diamonds is characterized by the higher activation E; thus, He in diamond may essentially be retained even at mantle temperature for geological time.

There is now a large amount of noble gas diffusion data obtained for rocks or minerals. However, very few studies have been done on noble gas diffusion in silicate melts. The latter bears a central importance in understanding the noble gas evolution in the mantle. Figure 2.11 shows one such scarce example (Lux, 1987), where the diffusion coefficients obtained for a tholeiite basalt melt at 1350°C are plotted as a function of noble gas radius. Diffusion obeys more or less the same linear relationship with r^2 as does the solubility. As Lux (1987) noted, it is remarkable that the

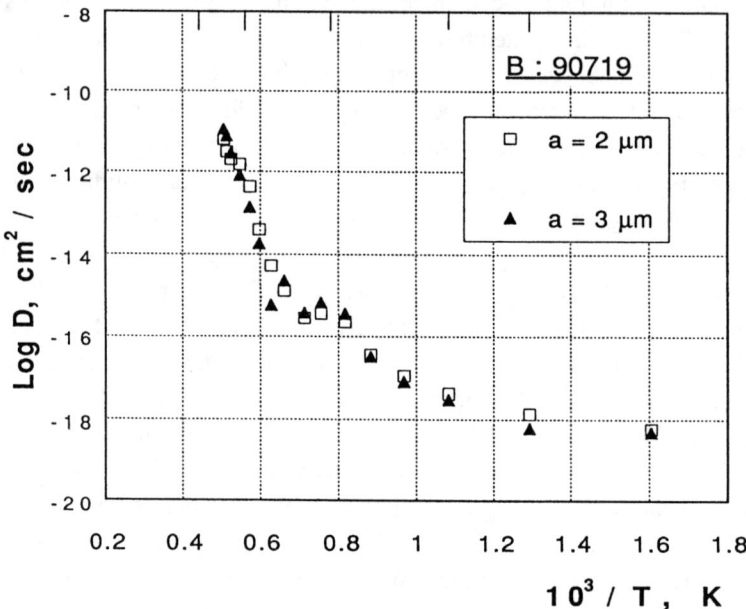

Figure 2.10 Arrhenius plot showing temperature variation of He diffusion coefficient in carbonado (diamond). *a* indicates the size (in micrometers) of the pulverized powder. Note that the diffusion is characterized by two distinct activation energies. Reproduced from Zashu and Hiyagon (1995).

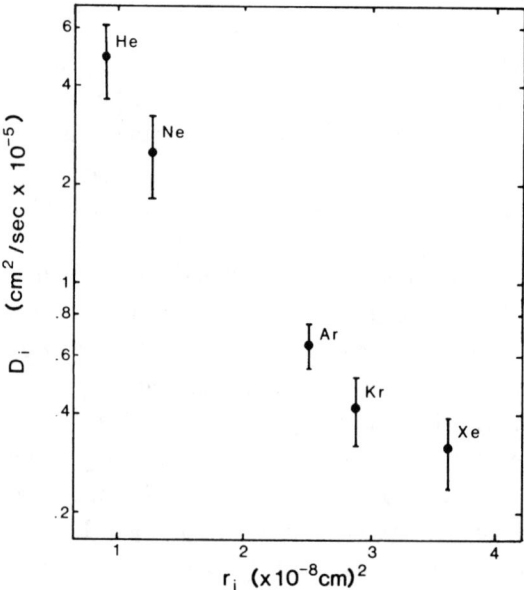

Figure 2.11 Diffusion coefficients (D_i) for He, Ne, Ar, Kr, and Xe in a tholeiite basalt melt at 1350°C as a function of the radius of the gas atom. Reproduced from Lux (1987).

measured diffusion coefficients are much larger than would be expected from extrapolation of D values measured on silicate glass at lower temperatures (cf. Figure 2.9).

2.8 Stepwise Degassing

Stepwise degassing is a very powerful and widely used technique in noble gas analysis. The approach is simple: piecemeal extraction. The sample is heated at a given temperature for a given time, and the evolved gases are collected and analyzed; then the sample is heated to a higher temperature, gases are again collected and analyzed, and so on. Typical parameters are 100°C temperature steps and half-hour to hour heatings, but these vary considerably.

Often in terrestrial samples, and almost always in extraterrestrial samples, the total noble gas complement is a superposition of components with different origins, sited in physically different locations in the samples. In general, then, they will have different release patterns in stepwise heating (i.e., gases collected in different temperature steps will have different proportions of the various components). This partial separation of components is the great utility of stepwise heating: the constraints on composition and quantity that can be established thereby and, indeed, the demonstration of the very existence of distinguishable components. Examples abound. Much of what has been learned (cf. Section 7.3) about noble gas components in the early solar system, for example, is based on stepwise heating analysis of meteorites.

Note that the components separated in stepwise heating are implicitly understood to be components of a single element. Different elements may separate in stepwise heating even if they are in the same place to begin with because they can have greatly different diffusion coefficients. Different isotopes of the same element will be released together, however, so that a single component in the sample remains a single component in stepwise heating. (As is discussed in Section 2.9, this is not quite true, but isotopic fractionation will be much smaller than elemental fractionation and follow a regular pattern in mass, and observers are usually alert to the possibility.)

Component separation is often particularly prominent when one or more of the components present is in situ. There is no requirement that different components reside in different minerals; indeed, it is often not known at all *where* the components reside. An in situ component will be sited wherever its parent element is, and this siting might be rather different from that of some other in situ component or a trapped component, as illustrated in Fig. 2.12. Radiogenic gases are common in situ components, and stepwise heating is frequently applied to resolving them from other components.

There is a family of quite similar geochronological methodologies that are particularly well suited for demonstrating the features of stepwise heating: the ^{129}I-^{129}Xe method (Jeffrey & Reynolds, 1961), the ^{40}Ar-^{39}Ar method (Sigurgeirsson, 1962; Merrihue & Turner, 1966), the ^{244}Pu-^{136}Xe method (Podosek, 1970; 1972), and the fission

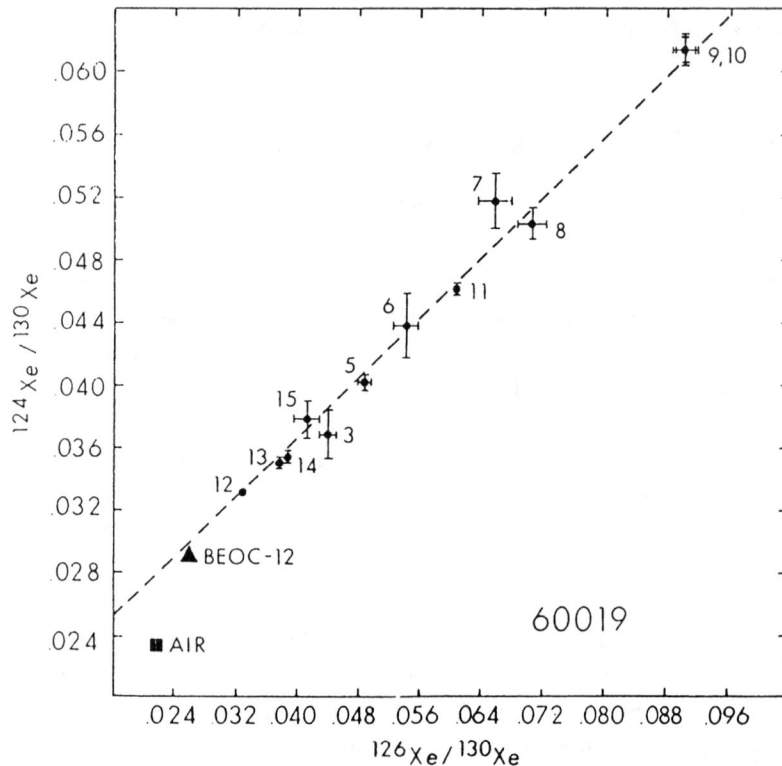

Figure 2.12 Correlated isotopic variations observed in stepwise heating of lunar breccia 60019. The linear array suggests that only two components, in varying proportions, are present: one is trapped solar wind, presumably near BEOC-12, the other is in situ cosmic ray-induced spallation, relatively rich in ^{124}Xe and ^{126}Xe. The numeral 3 indicates 300°C extraction, whereas the numeral 4 represents a 400°C extraction, and so on. Reproduced from Bernatowicz et al. (1978).

Xe method (Teitsma & Clarke, 1978; Shukolyukov et al., 1994). All these methods are based on natural radionuclides (^{129}I, ^{40}K, ^{244}Pu, ^{238}U), which decay to noble gas daughters (^{129}Xe, ^{40}Ar, fission Xe, fission Xe), and the first essential step in the method is a neutron irradiation that produces a different isotope of the daughter element (^{128}Xe, ^{39}Ar) or a distinct fission Xe component, from a different isotope of the parent element (^{127}I, ^{39}K, ^{235}U, ^{235}U) (there is no extant isotope of Pu, so in the third case it is merely hoped that U is now sited as Pu was when ^{244}Pu was extant). If the natural daughter has accumulated undisturbed since a given time in the past, it will now be found in fixed ratio to its parent and thus to the artificial daughter, and the value of this ratio carries the desired geochronological information. Moreover, the natural daughter and the artificial daughter should both be in the same physical locations (wherever the parent element is) and thus constitute a single component. (A qualification: production of ^{39}K involves a finite recoil range, and the methodology is compromised if the phases that host the parent are very small; cf. Huneke & Smith, 1976. Fission Xe fragments have even longer recoil ranges, but fission is involved in both

2.8 Stepwise Degassing

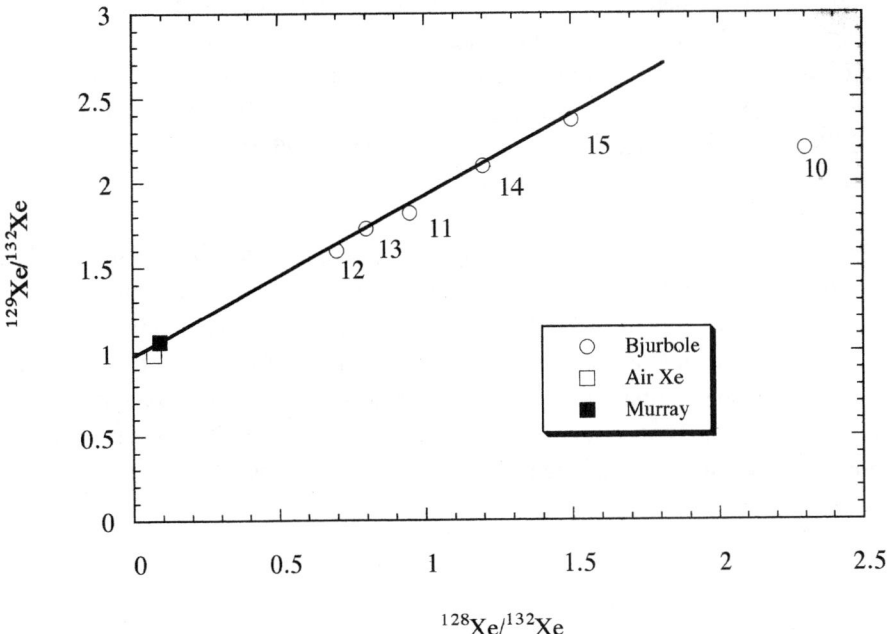

Figure 2.13 Correlated isotopic variations observed in stepwise heating of (neutron-irradiated) meteorite Bjurböle (cf. Figure 2.12). Trapped Xe is at lower left, and variations are caused by addition of in situ ^{129}Xe (from decay of ^{129}I) and ^{128}Xe (from neutron irradiation of ^{127}I). The linear variation indicates a constant ratio of the added ^{129}Xe and ^{128}Xe. The 1000°C datum (and lower temperature data, not shown) deviate from this linear relationship in the direction indicating partial loss of radiogenic ^{129}Xe. Reproduced from Drozd and Podosek (1976).

the natural decay and the activation, so there is no significant separation.) It is noteworthy that these methods eliminate problems due to sample heterogeneity and can establish a relative chronology on the basis of an isotopic ratio (natural daughter to artificial daughter) rather than a typically less accurate elemental ratio (natural daughter to parent). More important, however, is the second essential step: analysis by stepwise heating to test the basic assumption (i.e., to determine whether such a well-defined component actually exists, and if so, to measure its composition). Of these methods, the most familiar and widely applied is the ^{40}Ar-^{39}Ar method.

The application of the basic methodology is illustrated in Figure 2.13. The in situ I-derived component is a mixture of natural daughter ^{129}Xe and activation daughter ^{128}Xe; the spread in observed compositions reflects variable proportion of this component to trapped Xe. The linear correlation at high temperatures indicates a well-defined ratio of I-derived ^{129}Xe-^{128}Xe (and thus ^{129}I to ^{127}I), whose value is the correlation slope. It is noteworthy that the lower temperature data deviate from the line in the direction usually interpreted as partial loss of radiogenic ^{129}Xe from the least retentive sites. The most important feature of the methodology is that the desired chronological information can be obtained from the high-temperature data in spite of the low-temperature loss. The same considerations apply to the ^{40}Ar-^{39}Ar method, for which the corresponding diagram is ^{40}Ar/^{36}Ar versus ^{39}Ar/^{36}Ar, but in practice such a diagram is seldom used. The principal objective of examining this correlation is separation of trapped and in situ components. Trapped ^{40}Ar is usually negligible in extraterrestrial samples and assumed to be 296 times ^{36}Ar (the air ratio) in terrestrial samples. The effects of partial ^{40}Ar loss are usually more serious, however.

To proceed with a quantitative illustration, it is necessary to adopt some specific model for gas release. It is usually assumed that Fick's law applies, and the problem can be treated as volume diffusion (Section 2.7). As a first approximation, it is often assumed that the sample is composed of uniform spheres of radius a. If it is further assumed that at some initial time $t = 0$ the gas concentration is uniform, and that the concentration of gas is always nil at the surface of the sphere. Then, after diffusion to time t, the fraction f of the original gas that remains in the sphere is (Carslaw & Jaeger, 1959)

$$f(x) = \frac{6}{\pi^2} \sum_{n=1}^{\infty} \frac{1}{n^2} \exp(-n^2 x) \qquad (2.26)$$

The extent of diffusion is conveniently parameterized by the dimensionless variable

$$x(t) = \frac{\pi^2}{a^2} \int_0^t D(t')dt' \qquad (2.27)$$

which scales both the time and sphere size.

To model the effects of partial loss, we will suppose that some rock forms at time $t = 0$ and accumulates uniform radiogenic ^{40}Ar (designated component ^{40}Ar$_1$) until some time t, when it experiences a partial degassing event. Diffusive loss in this event is specified by some value x_1 [Equation (2.27)], so the ^{40}Ar remaining is ^{40}Ar$_1 f(x_1)$. Subsequently, diffusion is arrested, and the rock again accumulates uniform radiogenic ^{40}Ar, that can be designated a second component, ^{40}Ar$_2$, until the present. The rock is then irradiated, producing a ^{39}Ar component, and then heated in the laboratory to an extent that can be described by an (unsubscripted) value x [Equation (2.27)]. At any stage x in the laboratory, heating the Ar remaining in the rock is then given by

$$^{40}\text{Ar} = {}^{40}\text{Ar}_1 f(x_1 + x) + {}^{40}\text{Ar}_2 f(x) \qquad (2.28)$$

$$^{39}\text{Ar} = {}^{39}\text{Ar}_0 f(x) \qquad (2.29)$$

where ^{39}Ar$_0$ is produced from all the K in the rock. The instantaneous Ar composition released by any stage x is obtained by differentiation:

$$\frac{(^{40}\text{Ar})}{(^{39}\text{Ar})} = \frac{(^{40}\text{Ar}_1)}{(^{39}\text{Ar}_0)} \frac{g(x_1 + x)}{g(x)} + \frac{(^{40}\text{Ar}_2)}{(^{39}\text{Ar}_0)} \qquad (2.30)$$

where

2.8 Stepwise Degassing

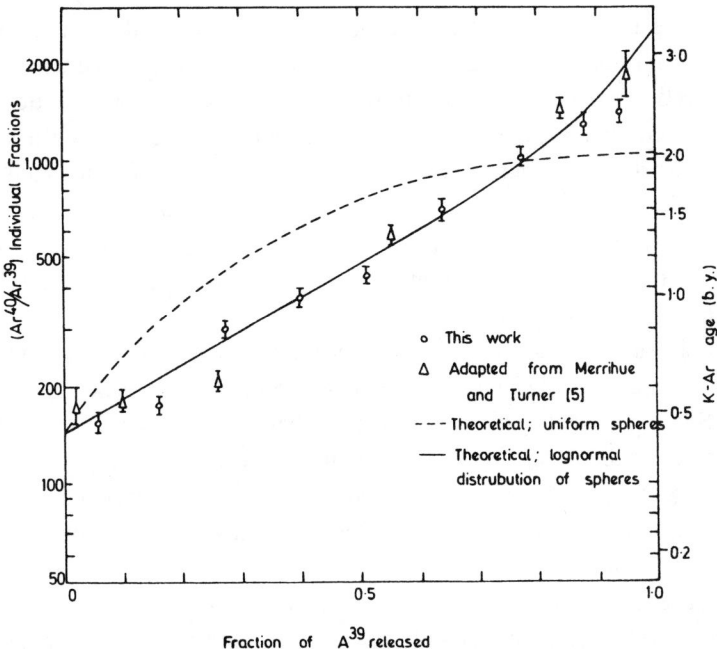

Figure 2.14 Release pattern obtained in a ^{40}Ar-^{39}Ar analysis of the Bruderheim meteorite. The theoretical curves are for a primary age of 4.5 Ga and a 500-Ma thermal event in which 90% loss of radiogenic ^{40}Ar occurred. Reproduced from Turner et al. (1966).

$$g(x) = \frac{df}{dx} = \frac{-6}{\pi^2} \sum_{n=1}^{\infty} \exp(-n^2 x) \tag{2.31}$$

At the beginning of the experiment, $x = 0$, the measured ^{40}Ar/^{39}Ar ratio is ^{40}Ar$_2$/^{39}Ar$_0$, from which the time t_1 of the episodic loss can be determined. Near the end of the release, for large x, the measured ^{40}Ar/^{39}Ar ratio asymptotically approaches the sum of ^{40}Ar$_2$/^{39}Ar$_0$ and e^{-x_1} ^{40}Ar$_1$/^{39}Ar$_0$; if x_1 is small, this sum will be close to the nominal value that would have been obtained had there been no loss ($x_1 = 0$) at t_1, (i.e., the high temperature ratio gives a close estimate of the 'true' age in spite of the partial loss at t_1).

Turner, Miller, and Grastry (1966) were the first to apply this kind of model analysis to ^{40}Ar-^{39}Ar dating. In general, this approach is reasonably successful in accounting for the observations. Turner et al. were able to show, for example, that data for the Bruderheim meteorite (assuming a log-normal grain size distribution) fit a model for an 0.5-Ga event, which caused extensive but not complete loss of ^{40}Ar accumulated in a much older primary age (Figure 2.14).

In actual experiments, of course, things are not quite so simple. In the first place, one does not observe the instantaneous release but only its integration in discrete temperature steps. Problems also arise because of the recoil effect mentioned earlier and various interferences and other uncertainties arising in the neutron irradiation.

However, a more serious problem would be the assumption of the uniform distribution of noble gas in a sample that was taken for granted as the initial condition. Because of diffusion loss, natural samples would hardly preserve the uniform noble gas concentration but are more likely to show a diffusion-loss-concentration profile; the less concentration appears toward the outer boundary. Considering the nonuniform initial distribution of noble gases, Albarède (1978) extended the mathematical treatment of the Turner's ^{40}Ar-^{39}Ar analysis. In this regard, it is important to note that diffusion experiment on natural sample must take account of the possible nonuniform initial distribution; a false assumption of a uniform distribution would result in the underestimation of diffusion coefficient.

In much the same way as in the stepwise heating discussed earlier, stepwise application of acid treatment can be used to separate noble gas components. The technique (named CSSE for closed system step etching) has been developed at ETH-Zurich (e.g. Signer, Baur & Wieler, 1993). In applying this technique, samples are exposed to acid vapor (HNO_3 or HF), which is introduced by heating an acid container attached to a gas-extraction system, for a few minutes to weeks depending on the mineral. After each step of the acid-leaching, gases evolved from a sample were analyzed with a mass spectrometer, which is attached to the gas extraction line. Stepwise acid treatment progressively dissolves a sample from the surface to the interior. Therefore, noble gases liberated during each step of the acid-leaching would reflect depth-profile of noble gas distribution in the sample. With the CSSE technique, the ETH group succeeded in resolving solar emission noble gases into components characterized with different energies: The initial step liberated mostly a component of lower energies such as SW of which implantation depth is a few hundred nanometers, whereas further steps yielded a more energetic component implanted into the deeper part of mineral grains (cf. Table 2.5). From the stepwise acid treatment of lunar soils, they could first identify a distinct component SEP (solar energetic particle) of which energy ranges from a few KeV/amu (amu: *a*tomic *m*ass *u*nit) to 0.1 MeV/amu (Wieler, Baur & Signer, 1986; Wieler & Baur, 1994). Figure 2.15 shows the result of the CSSE method applied to plagioclase separates from the lunar soils (Signer et al., 1993). Ne isotopic compositions of the unetched samples show rough linear trend in a neon three-isotope plot, suggesting the mixing of two independent Ne components. However, the linear trend shows considerable scatter, and it is not possible to identify the end members. After the acid etching (CSSE), the data reveal a nice linear array, from which one can easily identify the cosmogenic Ne component as one of the end member of the mixing (compare a similar experiment by the step-heating method shown in Figure 2.12).

2.9 Isotopic Fractionation

When the equilibrium partitioning or transfer rate of a given element between two reservoirs depends on atomic mass, isotopic fractionation may arise. For elements

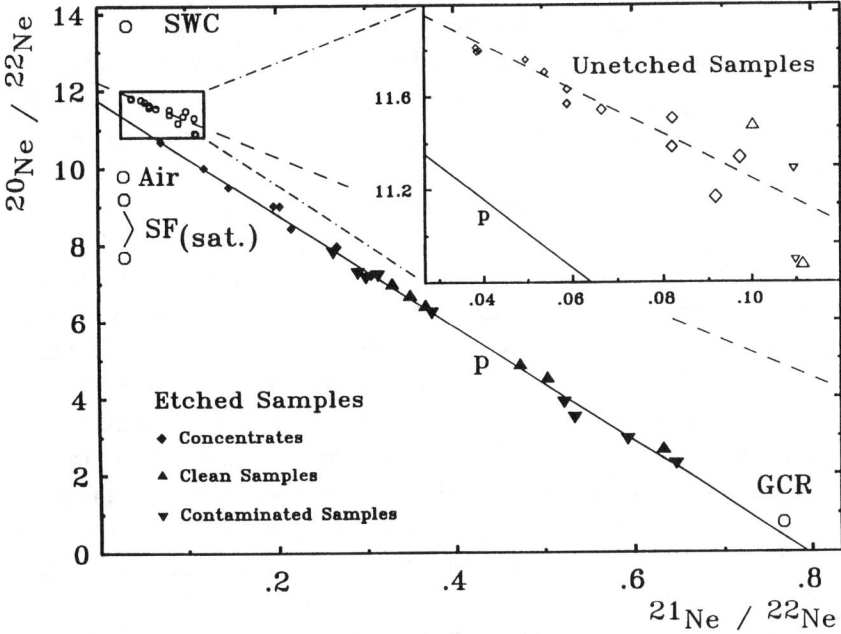

Figure 2.15 Ne three-isotope plot for a grain-size suite of plagioclase separates from lunar high land soil that were treated by the CSSE treatment (see text). The best fitted line through the data from all etched samples (line p) passes close to the data point GCR (galactic cosmic ray) of cosmogenic Ne. On the left side, the path of mass fractionation of SWC (solar wind composition)-Ne intersects line p at a $^{20}Ne/^{22}Ne$ ratio of ~11.3, which is interpreted to represent SEP (solar energetic particle) Ne (cf. Section 2.8). Open symbols: unetched sample. Solid symbols: etched samples. SF: Solar flare Ne. Reproduced from Signer et al. (1993).

such as O, C, H, S, N, and others that make chemical bonds, significant isotopic fractionations are common, occasionally reasonably well understood, and often exploited to provide information about the processes responsible for the fractionation or as geochemical tracers. Noble gas fractionations share some of these features.

Isotopic fractionation in equilibrium partitioning between two reservoirs arises in two ways. The density of quantum states, and thus the partition function, usually varies as some low power of mass (e.g., $m^{3/2}$ for the translational partition function of an ideal gas). Such mass dependence may be nearly but not quite equal in the two reservoirs; hence, the partition coefficient will depend on mass. Fractionation so arising will be relatively small, the isotopic partition coefficient for a mass difference δm differing from unity by some small fraction of $\delta m/m$, and only weakly or not at all dependent on temperature. In a manner of speaking, by this means, the mass enters the Gibbs free energy through the entropy term. The other way is through the energy term, most commonly because of the mass dependence of the zero point vibrational energy ($\frac{1}{2}\hbar v$, where v is proportional to $m^{-1/2}$). This effect contributes to the isotopic partition function of a factor in which the term constant x ($\delta m/mkT$) is exponentiated, where the proportionality constant depends on the difference in bond strengths in the two reservoirs. Although this fractionation effect is very important for elements, O, C, H, N, and S through their chemical bonds, it may be safely neglected for noble gases, which, in general, do not form chemical bonds but rather a weak van der Waals "bond." For the van der Waals bond, Bernatowicz and Podosek (1989) showed that the isotopic fractionation effect is unlikely important at any temperatures of geochemical interests.

Fractionation also arises in the nonequilibrium transfer of material from one reservoir to another if the transfer rate is mass dependent (and the transfer is incomplete). As for nonequilibrium partitioning, we can distinguish two basically different cases according to whether the mass enters as a relatively low power or in an exponential. In the first case, the fractionation is typically small, a fraction of $\delta m/m$, and insensitive to temperature. The most common example is diffusion, in which fractionation is usually modeled by assuming that the diffusion coefficient is proportional to kinetic velocity (i.e., to $m^{-1/2}$). (This relationship is intended to encompass different isotopes of a given element and *cannot* be expected to be correct for different elements.) In the latter case, the exponential is usually the Boltzmann factor $e^{-Q/kT}$, proportional to the probability that a given atom at a given time has an energy at least equal to Q (if $kT \ll Q$). In this case, the fractionation can be large and in general will be sensitively temperature dependent. Probably the most important case in geochemistry is escape from a gravitational field (Q proportional to mg). The problem is a complicated one involving not only the probability that an exosphere atom has enough energy to escape but also the rate at which the exosphere supply is replenished from below; nevertheless, it is clear that light isotopes will escape faster than heavy ones, and the escape rate can be quite sensitive to mass. Escape from the Earth's atmosphere is important for He, for example, and it is noteworthy that even though thermal (Jeans) escape is the dominant loss mechanism for ^{3}He, Jeans escape (cf. Lewis & Prinn, 1984, and references therein) for the heavier isotope is sufficiently slow that it is not the dominant loss term for ^{4}He.

It is well known that even when single-stage fractionation is small, fractionation effects may be cumulative in appropriate circumstances. Diffusion provides a convenient example as illustrated in Figure 2.16. The fractional mass difference between ^{134}Xe and ^{132}Xe in $\delta m/m = 15‰$. For a mass dependence according to $m^{-1/2}$, the diffusion coefficient for ^{134}Xe will be approximately $½(\delta m/m) = 8‰$ less than for ^{132}Xe. The first gas to diffuse out of material with an initially uniform concentration of both isotopes will be depleted (relative to the initial composition) in the heavier isotope by $¼(\delta m/m) = 4‰$. Because of the continuing preferential loss of ^{132}Xe, the ^{134}Xe/^{132}Xe in the residual gas becomes larger the greater is the depletion of the original reservoir. At a certain point, the escaping gas has the composition of the original reservoir, and beyond that point both the residuum and the escaping gas are enriched in the heavier isotope (Figure 2.16). Conversely, comparable effects of light isotope enrichment occur for gas diffusing into the reservoir of interest. Similar cumulative effects can occur in any process involving fractionation; the equivalent and more

2.9 Isotopic Fractionation

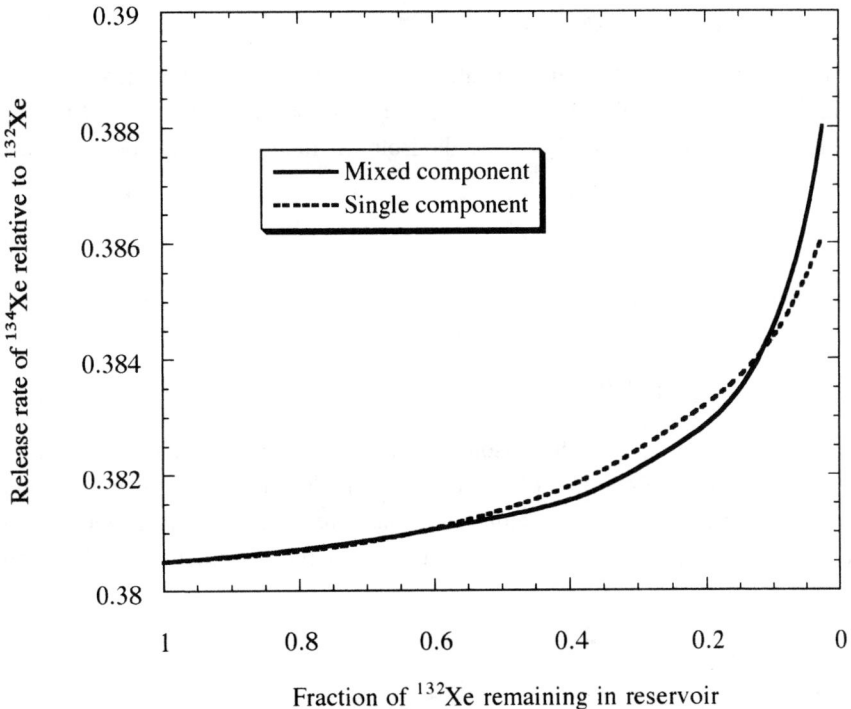

Figure 2.16 Illustration of isotopic fractionation effects in diffusion. The model is that ^{132}Xe and ^{134}Xe are initially uniformly distributed throughout spheres in the ratio ^{134}Xe/^{132}Xe = 0.382 and then allowed to escape by diffusion with the boundary condition that the concentration vanishes on the surface. The figure shows the instantaneous composition of the released gas at various stages, assuming that the diffusion coefficients varies as $m^{-1/2}$. The single-component locus is for all spheres having the same radius; the mixed-component locus is for distribution of sizes. Reproduced from Funk, Podosek, and Rowe (1967).

familiar effect for equilibrium partitioning between different phases is Rayleigh distillation, for example. A characteristic feature of such cases is that (unless there is reflex or its equivalent) the extreme effects occur only in a very small part of the total quantities involved.

Since the Rayleigh distillation is likely to be the most common fractionation process in noble gas geochemistry, we will give some quantitative discussion of this process next. Suppose atoms (or isotopes) 'i' and 'j' are escaping at a rate k_i and k_j from a reservoir which contains n_i and n_j numbers of the atoms, we have

$$\frac{dn_i}{dt} = -k_i n_i$$
$$\frac{dn_j}{dt} = -k_j n_j \quad (2.32)$$

Integrating these equations, we have

$$(n_j/n_i)/(N_j/N_i) = \{n_i/N_i\}^{k_j/k_i - 1} \tag{2.33}$$

where n (present) and N (initial) indicates the amount of the component i and j. Denoting an isotopic (or elemental) ratios $r_{ji} = n_j/n_i$ (present) and $R_{ji} = N_j/N_i$ (initial), and defining a depletion factor $F_i = n_i/N_i$, the isotopic fractionation can be expressed as

$$(r_{ji}/R_{ji}) = \{F_i\}^{k_j/k_i - 1} \tag{2.34}$$

If the escape velocity of an atom is inversely proportional to a square root of its mass m, $k \propto (m)^{-1/2}$, and from Equation (2.34) we have

$$(r_{ji}/R_{ji}) = \{F_i\}^{\sqrt{(m_i/m_j)} - 1} \tag{2.35}$$

If we deal with more than three isotopes such as in the case of Kr or Xe, it is convenient to modify Equation (2.35) to accommodate all isotopes in a single equation. This can be done by setting up equations similar to Equation (2.34) for all the isotopes, $k, l, m, \ldots$, and dividing the equations by each other. Then we have

$$(r_{jk}/R_{jk}) = \{F_i\}^{\sqrt{(m_i/m_k)} - \sqrt{(m_i/m_j)}} \tag{2.36}$$
$$\{F_j\} \equiv \{F_i\}^{\sqrt{(m_i/m_j)}}$$

where k and j stand for any pair of species among $j, k, l, \ldots$, and i was chosen as a reference. From Equation (2.36), it is easy to see that logarithms of ratios of isotopic ratios of any pairs of i, j, k, should form a line passing through the origin in a $\log\{r_{kj}/R_{kj}\}$ versus $\{(m_i/m_k) - (m_i/m_j)\}^{1/2}$ diagram, if isotopes i, j, k are related to each other by Rayleigh distillation. The slope of the line then corresponds to a depletion factor F_j. A remarkable example of this Rayleigh distillation type fractionation can be seen betwen the terrestrial atmospheric noble gases and the solar noble gases (Section 7.5).

Since isotopic fractionation is mass dependent, there will be a "smooth" relationship between effects observed in various isotopic ratios. In most cases of interest, the magnitude of fractionation and the fractional mass range are both small in an absolute sense, and to first order the degree of fractionation is approximately linear in mass: δ_m is proportional to δm, where δ_m is a fractional measure (e.g., in per mil) of the fractionation effect between isotopes of mass m and m_0 [Equation (1.1)] and $\delta m = m - m_0$. In elements of three or more isotopes, this is the signature by which fractionation is distinguished from specific isotope effects. If two or more isotopic ratios are observed to vary in this fashion appropriate for fractionation, it is usually considered that this is not an accident and that fractionation has indeed occurred (rather than two or more independent effects that depend on the nuclear identity of the isotopes in question).

Among the noble gases, the element whose isotopes span the greatest relative mass range is He, and it is thus expected that He should exhibit the greatest fractionation effects in natural occurrence. This is presumably true, but it is usually difficult to tell. Because He has only two isotopes, the smoothness test for mass dependence cannot be applied, and there are so many possible and potentially large specific isotope effects that unambiguous assignment of some modest isotopic variation to fractionation is usually impossible. Ne has three isotopes, so that a smoothness test can be applied. In practice, however, ^{21}Ne is difficult to measure accurately because its abundance is so low, and a specific isotope effect at ^{21}Ne is not uncommon (Sections 5.4 and 5.5), so that it is not too easy to judge competing hypotheses for variation of ^{20}Ne/^{22}Ne – fractionation or specific isotope effect (cf. Section 6.7) – simply by appeal to Ne compositional variation alone. In contrast, Ar also has only three isotopes, and specific effects at ^{40}Ar are so common that ^{36}Ar and ^{38}Ar must often be treated as a two-isotope system decoupled from ^{40}Ar. Even so, however, plausible specific isotope effects at ^{36}Ar and ^{38}Ar are so uncommon and so small that, except in very unusual samples (Section 6.2 and 6.7), any observed variations in ^{38}Ar/^{36}Ar ratio are usually *assumed* to be due to fractionation. The smoothness test really comes into its own for Kr and Xe, which have so many isotopes that fractionation is obvious, even when it is not anticipated or understood and even when there actually *are* complicating specific isotope effects at some isotopes.

A characteristic of noble gas chemistry is that there are many cases where a fractionation signature is so clear that one must conclude that fractionation has occurred even when there is no adequate understanding of how it has occurred. An important example is the apparent very strong fractionation of terrestrial Xe relative to any plausible identified or even hypothesized source elsewhere in the solar system (cf. Section 7.3). Also, large fractionation appears between the atmospheric Ne and Ne in the mantle. Many investigators have pointed out that the elemental composition of the terrestrial noble gas shows the Rayleigh-distillation-like fractionation relative to the solar component, although the origin of the fractionation is not understood (cf. Section 7.3).

In concluding this section, it is pertinent to take note of a special kind of isotopic fractionation: ubiquitous, often quite severe, and arguably the most important source of fractionation that must be taken into consideration in noble gas geochemistry. This fractionation arises in mass spectrometric analysis: contributory effects can and do arise in gas extraction and transport through the vacuum system, in the ion source (especially when a source magnet is used), in beam transmission, and in ion collection and detection (especially when an electron multiplier is used). As noted in Section 1.3, sample data are corrected for instrumental (and procedural) discrimination, which is calibrated by analysis of some standard gas (usually air). This is a roundabout and imperfect near-equivalent to the δ value convention, which is the norm in "stable isotope" geochemistry (O, C, H, S, N, etc.). The reproducibility of instrumental discrimination inferred from repeated calibration analysis is usually quite satisfactory, but seldom is any care taken to try to match operating conditions in samples and calibration analyses. It is thus a matter of faith – undoubtedly quite justified in most cases, but faith nonetheless – that discrimination is the same for sample and calibration. The only unequivocal method for distinguishing natural from laboratory fractionation is double-spike analysis, which to our knowledge has never been applied in noble gas geochemistry.

Chapter 3

Cosmochemistry

For noble gases much more than for other elements, it is impossible to have a coherent appreciation of circumstances on the Earth without a background appreciation of circumstances in the solar system as a whole, or more generally the cosmos at large. This generalization reflects the situation that the original complement of noble gases in the Earth is likely at least partially different from noble gases in other objects in the solar system, and that one cannot adequately understand the unique modifications arising in the Earth's own evolution without understanding the nature of the starting materials. The "cosmochemistry" of the noble gases, based largely but not entirely on the study of meteorites, is a discipline very rich both in data and in competing theories seeking to explain the data. It may also be noted that this is one of the disciplines that has changed dramatically since publication of the first edition of this book. It is not our intention to provide a comprehensive review of noble gas cosmochemistry, which would go considerably beyond our scope; instead, the intention is to focus on aspects most relevant as background to the study of terrestrial noble gases. A number of specific issues and questions in which cosmochemistry is important are raised in Chapter 7.

3.1 Cosmic Abundances

The atoms, specifically their nuclei, that comprise the solar system were made in a variety of environments (cf. Woolum, 1988). H and most of the He were made in the Big Bang. The He/H ratio, and the isotope ratios D/H and ^{3}He/^{4}He, are adequately accounted for in Big Bang models; perhaps a better statement is that these ratios provide important constraints on such models. The Big Bang made just a little heavier nuclides. Most of the relatively scarce elements Li, Be, and B were evidently produced by cosmic ray-induced spallation reactions in the interstellar medium. Some of the He and essentially all the elements heavier than B were synthesized by nuclear reactions occurring in the interiors of a variety of stars.

Current understanding is that the solar system formed from a cold dense interstellar molecular cloud about 4.57 Ga ago. The interstellar medium from which it formed already had a long and complicated nuclear/chemical history. The heavier (than B) elements in particular had to have been made in stellar interiors and then returned to the interstellar medium, in stellar winds or explosions, and some of it must have been in and out of stars multiple times. The particular mixture of diverse nucleosynthetic sources that constitutes our solar system is called *cosmic composition*. The term is, of course, too grandiose; rather than describing the cosmos at large, it refers to the composition of a particular region of the interstellar medium in a galaxy whose composition not only varies as a function of position within the galaxy but also evolves with time.

Although the concept is simple, the determination of actual quantitative values for cosmic composition is a challenging intellectual exercise, involving spectroscopic observations, chemical analyses of meteorites that have never experienced igneous differentiation processes, and applications of regularities in theories of stellar nucleosynthesis. The most widely accepted estimate of cosmic composition is that of Anders and Grevesse (1989); values for selected isotopes are listed in Table 3.1. Uncertainties in *elemental* composition are probably in the range of a few to perhaps a few tens of percent (see Anders and Grevesse, 1989, for a more extensive discussion); in the present context such uncertainties are not important. *Isotopic compositions* are much more tightly constrained, as will be discussed later.

For the noble gases, Table 3.1 also lists equivalent concentrations in cm^3 STP per 0.17 g of Si, which would be cm^3 STP/g in a rock that contained a representative 17%

Table 3.1. *Cosmic abundances[a] of selected isotopes*

Isotope	Atoms per 10^6 Atoms of Si	cm^3 STP per 0.17 g Si "Cosmic Rock"
^{1}H	2.79×10^{10}	
^{4}He	2.72×10^9	3.69×10^5
^{12}C	9.99×10^6	
^{14}N	3.12×10^6	
^{16}O	2.37×10^7	
^{20}Ne	3.20×10^6	434
^{32}S	4.89×10^5	
^{35}Cl	2.86×10^3	
^{36}Ar	8.50×10^4	11.5
^{84}Kr	25.7	3.48×10^{-3}
^{132}Xe	1.24	1.68×10^{-4}

[a] From Anders & Grevesse (1989).

Si. No real rock contains this much noble gas; indeed, no rock *could* contain so much He or even Ne and still be a rock. These values are provided as a convenient basis for evaluating the extent to which a given real rock is depleted in noble gases, relative to cosmic composition (cf. "Cosmic Rock" in Table 3.1) .

Aside from numerical values, the concept of cosmic composition, even in the restricted sense of referring to the solar system rather than the whole cosmos, carries the implicit assumption that there actually *is* a single composition that represents the whole solar system. Given that most of the mass of the solar system is in the sun, this assumption must be at least nearly true, but we tend to have disproportionate interest in the bodies that we live on or have samples of. It was once thought that this assumption was literally true, at least for the sun and rocky bodies (everything inward of Jupiter, and maybe also comets). That is, we once thought that at one time everything in at least the inner solar system had exactly the same composition down to samples so small that Poisson statistics were important. Given that a variety of nucleosynthetic processes are required to produce cosmic composition, and that at least the dust component of the interstellar medium is likely to preserve specific nucleosynthetic compositions in diverse grains, this requires the further assumption of a process that will thoroughly mix all the different nucleosynthetic components (e.g., by totally vaporizing at least the inner solar system). As will be discussed in more detail later, it has been known for some time that this proposition of complete uniformity is manifestly untrue. Some objects in the solar system have never been thoroughly mixed into a reservoir of uniform cosmic composition. The study of such objects has attracted a great deal of effort and attention, and justifiably so, but this should not lead us to lose sight of cosmic composition as a very useful and almost true generalization on a grand scale. It is the *primordial composition* which provides the initial condition for evaluating the modifications involved in the chemical and nuclear processes that have produced the objects that now populate the solar system.

3.2 Solar Noble Gases

In the part of the interstellar medium from which the solar system formed, at least during the process of formation, most of the noble gas inventory must have been present in the gas phase rather than in the dust. At least for the terrestrial planets, most of this gas has been lost or was never incorporated in the first place; relative to cosmic composition, all noble gases are depleted by several orders of magnitude in all the terrestrial planets (including meteorites). The depletions are not uniform, suggesting the possibility of learning something about the depletion process through how it affected the different gases. Moreover, the extreme depletion of noble gas abundances in the terrestrial planets allows the possibility that minor components trapped in interstellar dust, trivial compared to the overall cosmic inventory of noble gases, might be important in the terrestrial planets. In both respects, there is therefore considerable interest in knowing just what the chemical and isotopic compositions of presolar noble gases actually were. The best prospect for obtaining such information is characterization of noble gases in the sun, which is believed to indeed represent the materials from which the solar system was made. Absent any prospect of actually obtaining a sample of the sun, most interest has focused on characterization of the solar wind. The subject has a long history of active investigation, which continues to the present (e.g., Wiens et al., 1999). Table 3.2 shows the compilation of elemental and isotopic abundances in solar wind and in other prominent noble gas components in the solar system.

Table 3.2. *Elemental (atomic) abundance ratios of prominent noble gases in the solar system*

	^{4}He/^{20}Ne	^{20}Ne/^{36}Ar	^{36}Ar/^{84}Kr	^{84}Kr/^{130}Xe	Remark/Reference
Solar wind (SW)[a]	550 ± 50	48.5 ± 7			Al + Pt foil on the moon: SW (1)
			1870	54.5	Lunar ilmenite 71501 (2)
			2120	28.4	Lunar ilmenite 79035 (2)
	650 ± 50	47 ± 3			Fe-Ni phase in chondrites (3)
	612 ± 10	41 ± 3			Fe-Ni phase in chondrite Acfer 111 (4)
(SEP)[b]	460 ± 50	47 ± 10			Spacecraft measurement (5)
Solar (= sun)	850	37	3320	122.5	(6)
Q[c]	112 ± 42	0.049 ± 0.008	92 ± 9	4.6 ± 1.0	(7)
Earth atmosphere	(0.319)	0.524	48.4	183	(8)
Mars atmosphere		0.305	28.67	130.96	(9)
Jupiter atmosphere	6783	2.3			(10)

[a] Energy range: $E < \sim 1$ KeV/amu.
[b] ~ 10 KeV/amu $< E < \sim 0.1$ MeV/amu (see e.g. Wieler et al., 1986). (SEP: solar energetic particles).
[c] Noble gases in the ill-defined carrier "Q" in primitive meteorites. Q is HF/HCl-resistant, but releases noble gases on oxidation with HNO_3 or $HClO_4$. A similar component to Q-noble gases is also found in various classes of meteorites, suggesting a ubiquitous occurrence in early solar nebula (see e.g. Wieler, 1994).

Table 3.2 References: 1, Cerutti (1974); see also Murer et al. (1997) for recent solar wind. 2, Wieler & Baur (1995). Elemental abundances of Ar, Kr, Xe in SW and SEP components are nearly identical. Sample 71501 trapped solar gases in past ~100 Ma; sample 79035, about 1 Ga ago. 3, Murer et al. (1997). Elemental abundances of He, Ne, Ar in SW and SEP are nearly identical. 4, Pedroni & Begemann (1994). 5, Brenemann & Stone (1985) and Stone (1989) and references therein. 6, Anders & Grevesse (1989). 7, Wieler et al. (1992) and Busemann, Baur & Wieler (1997). Mean and standard deviation of CSSE (see Section 2.8) runs of Allende, Murchison, and Dimmitt meteorites. 8, from Table 1.3 and references therein. 9, Pepin (1991). Weight ratios given in the reference were converted to atomic ratios. Data are based on SNC meteorites (see Section 6.9). 10, calculated from the Galileo Probe Mission data by Niemann et al. (1996).

3.2 Solar Noble Gases

Besides its literal dictionary meaning, as used in the preceding paragraph, in noble gas cosmochemistry the term *solar* has long been used to designate a specific *pattern* of noble gas elemental abundance ratios observed in planetary materials (Signer & Suess, 1963; Pepin & Signer, 1965). The defining feature is that the various noble gases occur in cosmic (i.e., solar) proportions (Figure 3.1), even when their absolute abundances vary substantially between various samples. Solar gases are found in many classes of meteorites, typically in only a small fraction of known representatives of any given class, but are most prominent in some classes of differentiated meteorites that are strongly depleted in volatile elements in general and especially so in noble gases. Solar gases are particularly prominent in lunar samples. The moon is very depleted in volatiles, and trapped noble gas concentrations in lunar rocks are so low that no unambiguous identification of any indigenously lunar primordial noble gases has yet been made, yet some lunar soils and breccias contain strikingly high concentrations of noble gases in the solar abundance pattern (Figure 3.1).

There is no doubt about the origin of the gases which are designated solar on the basis of elemental abundance patterns. They are literally solar gases (i.e., trapped gases implanted in materials directly exposed to solar corpuscular emanation – predominantly the relatively low-energy solar wind component – on the surfaces of solar system bodies with insufficient atmosphere and magnetic field to shield their surfaces from the solar wind). This conclusion follows from correlation of solar-type noble gas abundances with other indicators of solar wind exposure (Suess, Wänke & Wlotzka, 1964) as well as their very shallow implantation depths (Eberhardt et al., 1970). Besides lunar samples and macroscopic meteorites, solar noble gases are also found in interplanetary dust particles (IDPs) (e.g., Rajan et al., 1977; Nier & Schlutter, 1990), presumably acquired during their careers as small grains independently orbiting the sun. The solar gases in IDPs may be an important item in the crustal noble gas budget (Section 5.3).

Although solar gases are clearly from the sun, their *elemental* abundance ratios are generally recognized as not being exactly those of the sun (i.e., cosmic abundances); indeed, they are not even constant from one sample to the next (Figure 3.1). Some elemental fractionation likely arises in acceleration of different elements into the solar wind, and certainly some elemental fractionation arises in the degree to which different gases are retained in the target surfaces exposed to the solar wind. The degree of fractionation from lightest to heaviest gases is about an order of magnitude. This is rather large in absolute terms, but it is still small in comparison with the elemental fractionation that defines "planetary" noble gases (Section 3.3), so it is still traditional to consider solar gas to be a well-defined component as long as it is being compared to planetary gas.

The obvious elemental fractionation in solar gas, as retained in meteorite or lunar samples, leads to the question of whether there may be some isotopic fractionation as well. This may be a nontrivial issue because various estimates of solar gas isotopic composition may differ largely by possible fractionation. Probably the best view is that solar noble gases represent the isotopic composition of the sun, except possibly for a superposed mass-dependent fractionation probably of order a permil.

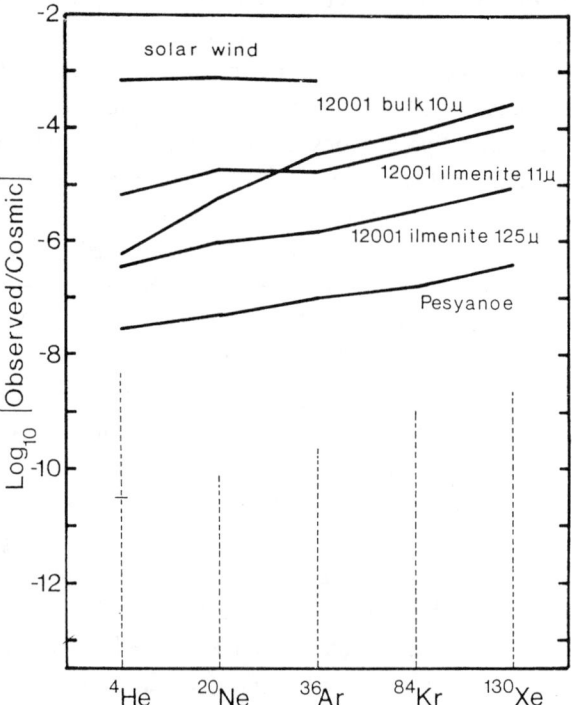

Figure 3.1 Illustration of noble gas abundance patterns. For each specific noble gas isotope, the ordinate is the logarithm (base 10) of the ratio to the "cosmic rock" (cf. Table 3.1) concentration. The ordinate position of the solar wind composition is arbitrary. The vertical dashed lines illustrate roughly the magnitude of typical in situ contributions; at He, the horizontal tick is for ^{4}He, whereas the upward extension is for ^{3}He; trapped gases at lower concentrations would be difficult to identify. Solar wind composition from Table 3.2, lunar ilmenites from Eberhardt et al., 1972 and Pesyanoe from Marti, 1969. All the patterns shown here are of the solar type (cf. Figure 3.2 for the planetary type).

Somewhat ironically, there is one prominent case in which the sun is well understood not to preserve primordial isotopic composition, namely the noble gas He. Main-sequence hydrogen burning in the solar core is producing He, of course, but this is not believed to affect strongly the composition near the surface (Bochsler, Geiss & Maeder, 1990). During pre-main-sequence evolution, however, the sun experienced deuterium burning, in which all the primordial D (^{2}H) was converted to ^{3}He, thereby increasing the ^{3}He/^{4}He ratio. In principle, knowledge of the primordial presolar D/H ratio (and the He/H ratio) would allow calculation of primordial ^{3}He/^{4}He from present solar ^{3}He/^{4}He. D/H in terrestrial planetary materials is highly variable, however, and typically much higher than any plausible primordial value, evidently because of enormous isotopic fractionation of H in cold interstellar clouds (e.g., Messenger & Walker,

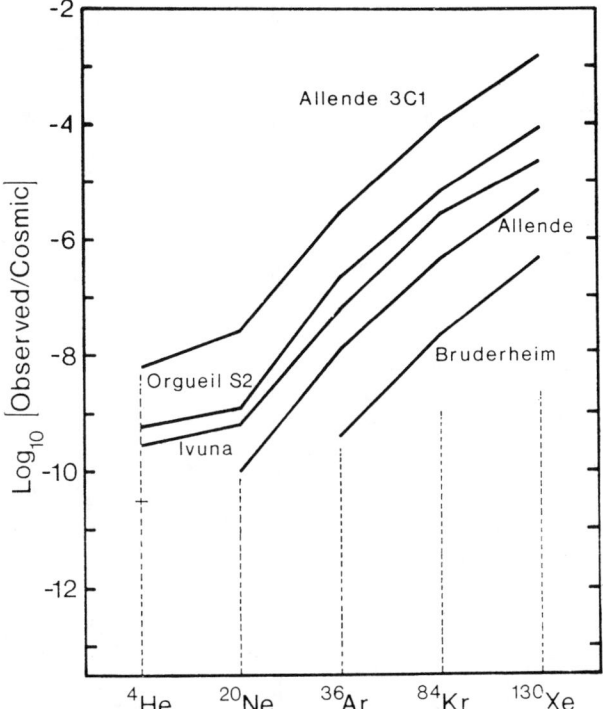

Figure 3.2 Illustration of noble gas abundance patterns (cf. Figure 3.1). All the patterns shown here are the planetary type. Allende from Lewis et al. 1975; Bruderheim from "Berkeley standard Bruderheim" (circulated by J. H. Reynolds); Orgueil and Ivuna from Mazor, Heymann, and Anders (1970).

1997), so that it is hard to make a reliable and accurate independent estimate of cosmic D/H. Instead, the greater interest is typically in the converse calculation: estimation of cosmic D/H from the difference between solar ^{3}He/^{4}He and primordial cosmic ^{3}He/^{4}He (e.g., Geiss & Reeves, 1972). Although this calculation was originally based on meteorite He as a proxy for primordial solar system He, it has become evident that meteoritic He is not necessarily representative of cosmic He (Section 3.4), so that estimation of primordial D/H must be based on other data, such as the composition of He in the atmosphere of Jupiter [see Table 3.3(a)].

3.3 Planetary Noble Gases

Like *solar*, the term *planetary*, as used to describe noble gases, carries a set of connotations well beyond the dictionary definition. *Planetary* is also used to refer to a pattern of noble gas elemental abundances, and the terms *solar* and *planetary* evolved to describe contrasting patterns. The defining feature of planetary noble gases is a much stronger (than in the case of solar gases) elemental fractionation pattern in

Table 3.3(a). Isotopic (He-Kr) compositions of prominent noble gas components in the solar system

	^{3}He/^{4}He	^{20}Ne/^{22}Ne	^{21}Ne/^{22}Ne	^{36}Ar/^{38}Ar (^{40}Ar/^{36}Ar)	^{78}Kr/^{84}Kr	^{80}Kr/^{84}Kr	^{82}Kr/^{84}Kr	^{83}Kr/^{84}Kr	^{86}Kr/^{84}Kr
SW[a]	0.000457	13.8	0.0328	5.6	—	—	0.2028	0.2004	0.3037
SW[b]	0.000426	13.7	0.0333	—	0.006359	0.04075	0.20501	0.20276	0.30115
SEP[c]	0.000217	11.2	0.0295	4.87	—	—	—	—	0.3205
Q[d]	0.000159	10.70	0.0294	5.29	0.00616	0.0394	0.2014	0.2018	0.3101
P1[e]	—	—	—	5.33	0.0066	0.03967	0.2024	0.2021	0.3091
Primordial Ar[f]				($<2.9 \times 10^{-4}$)					
Earth atmosphere[g]	0.0000014	9.80	0.02899	5.32	0.006087	0.039599	0.20217	0.20136	0.30524
Mars atmosphere[h]	—	10.1	—	4.1	0.00637	0.0409	0.2054	0.2034	0.3006
Jupiter atmosphere[i]	0.00012								

[a] He, Ne, and Kr: Ilmenite from lunar soil 71501 by the CSSE method (Benkert et al., 1993; Wieler and Baur, 1994; see also Section 2.8). The soil trapped its solar gases in the last ~100 Ma. Ar: Becker et al. (1996).
[b] He, Ne: Al-foils on Moon (Geiss et al., 1972). Kr: Pepin et al. (1995).
[c] Benkert et al. (1993). Lunar ilmenite 71503.
[d] CSSE method (see Section 2.8) on Murchison and Allende (Wieler et al., 1992).
[e] Stepped pyrolysis on 14 chondrites of seven classes (Huss et al., 1996), renormalized to atmospheric Kr isotopic composition.
[f] The lowest ^{40}Ar/^{36}Ar ratio observed in the Dyalpur ureilite (Göbel, Ott & Begemann, 1978). This may be assumed to be nearly pure primordial Ar as of 4.55 Ga ago in the solar system with only a negligible contributin of ^{40}Ar from ^{40}K decay (Anders and Grevesse, 1989).
[g] Air composition data are from Table 1.3 and the references therein.
[h] Compiled by Pepin (1991) and references therein. Data are based on SNC meteorites (see Section 6.9).
[i] Niemann et al. (1996). Galileo Probe Mission.

Table 3.3(b). *Isotopic (Xe) compositions of prominent noble gas components in the solar system*

	$^{124}Xe/^{132}Xe$	$^{126}Xe/^{132}Xe$	$^{128}Xe/^{132}Xe$	$^{129}Xe/^{132}Xe$	$^{130}Xe/^{132}Xe$	$^{131}Xe/^{132}Xe$	$^{134}Xe/^{132}Xe$	$^{136}Xe/^{132}Xe$
SW[a]	—	—	0.0849	1.054	0.1656	0.8255	0.3660	0.2970
SEP[a]	—	—	—	(0.996)	0.1513	0.812	0.3841	0.3172
Q[c]	0.00462	0.00407	0.08257	1.036	0.1625	0.8200	0.3803	0.3180
P1[d]	0.00462	0.00410	0.0830	1.040	0.1629	0.8227	0.3777	0.3161
Earth atmosphere[d]	0.003537	0.003300	0.07136	0.9832	0.15136	0.7890	0.3879	0.3294
Mars atmosphere[e]	0.003793	0.003282	0.07379	2.5387	0.1584	0.7967	0.4005	0.3551

[a] Lunar soil 71501 by the CSSE method (Benkert et al., 1993; Wieler and Baur, 1994).
[b] CSSE method on Murchison and Allende (Wieler et al., 1992).
[c] Stepped pyrolysis on 14 chondrites of seven classes (Huss et al., 1996), renormalized to atmospheric Xe isotopic composition.
[d] Air composition data are from Table 1.3 and references therein.
[e] Compiled by Pepin (1991) and references therein. Data are based on SNC meteorites (see Section 6.9).

which lighter gases are depleted relative to heavier gases (Figure 3.2), amounting to several orders of magnitude from lightest to heaviest. Planetary gases are typically found in undifferentiated meteorites, notably in carbonaceous chondrites. There is significant variation in relative gas abundances from one meteorite to another, but these are generally small compared to the difference between the average planetary and solar abundance patterns. The term *planetary* was coined in recognition that the pattern in undifferentiated meteorites is more similar to that in air (i.e., in planet Earth) than to the solar pattern. In many ordinary chondrites, absolute noble gas abundances are rather lower than in carbonaceous chondrites, and only for the heaviest gases can it be inferred that there are any gases present at all above in situ production levels; it is generally *assumed* that the lighter gases are present in planetary proportions. Even though this assumption seems reasonable, remember that it is indeed an assumption.

In contrast to the case of solar noble gases, the origin of planetary noble gases in meteorites is not at all well understood, and accounting for the origin of planetary noble gases is a traditional problem that has attracted attention for decades. Although understanding of the overall chemical compositions of the terrestrial planets (and meteorites) is far from complete in detail, there is nevertheless a broad understanding that the terrestrial planets and meteorites are made of the "condensible" elements, those that would be present as solids when material of cosmic composition is subjected to plausible thermodynamic conditions in the solar nebula. More generally, no single set of thermodynamic conditions can account for planetary and meteoritic compositions, but it is widely held that observed or modeled compositions can be accounted for by mixing of materials formed in a range of conditions, with due allowance for plausible fractionation processes based on mechanical or magnetic properties, grain size, and so on (e.g., see Sears and Dodd, 1988). The basic problem is that, under any plausible conditions involving gas-solid equilibrium that could account for the bulk compositions of terrestrial planets and meteorites, the abundances of the noble gases in the solids should be vanishingly (unmeasurably) low, several orders of magnitude lower than is observed for the abundance of planetary gases trapped in meteorites (or in the atmospheres of Earth, Mars, and Venus).

Besides the problem of accounting for the chemical abundances of planetary noble gases, there are characteristic differences in isotopic composition between planetary noble gases in meteorites and the solar gases that presumably represent the nebula from which meteorites formed. For Ar and Kr the differences are modest or perhaps nonexistent or can ultimately be explained in terms of a reasonable degree of mass-dependent isotopic fractionation. For Ne (Figure 3.3) and Xe (Figure 7.6), the

Figure 3.3 Display of various Ne compositions (see Table 3.3(a)). SW is solar wind, SEP is solar energetic particles, Q is the "local" trapped components, and *planetary* is the traditional whole-rock carbonaceous chondrite value. Ne-E represents presolar components nearly pure ^{22}Ne. The abundance of ^{21}Ne is very low in all these components. In many materials exposed to cosmic rays, compositions may be significantly sifted to the right by admixture of the GCR (galactic cosmic ray spallation-induced) component, distinguished chiefly by subequal amounts of all three isotopes. In the inset, *mfl* designates a mass-fractionation trajectory.

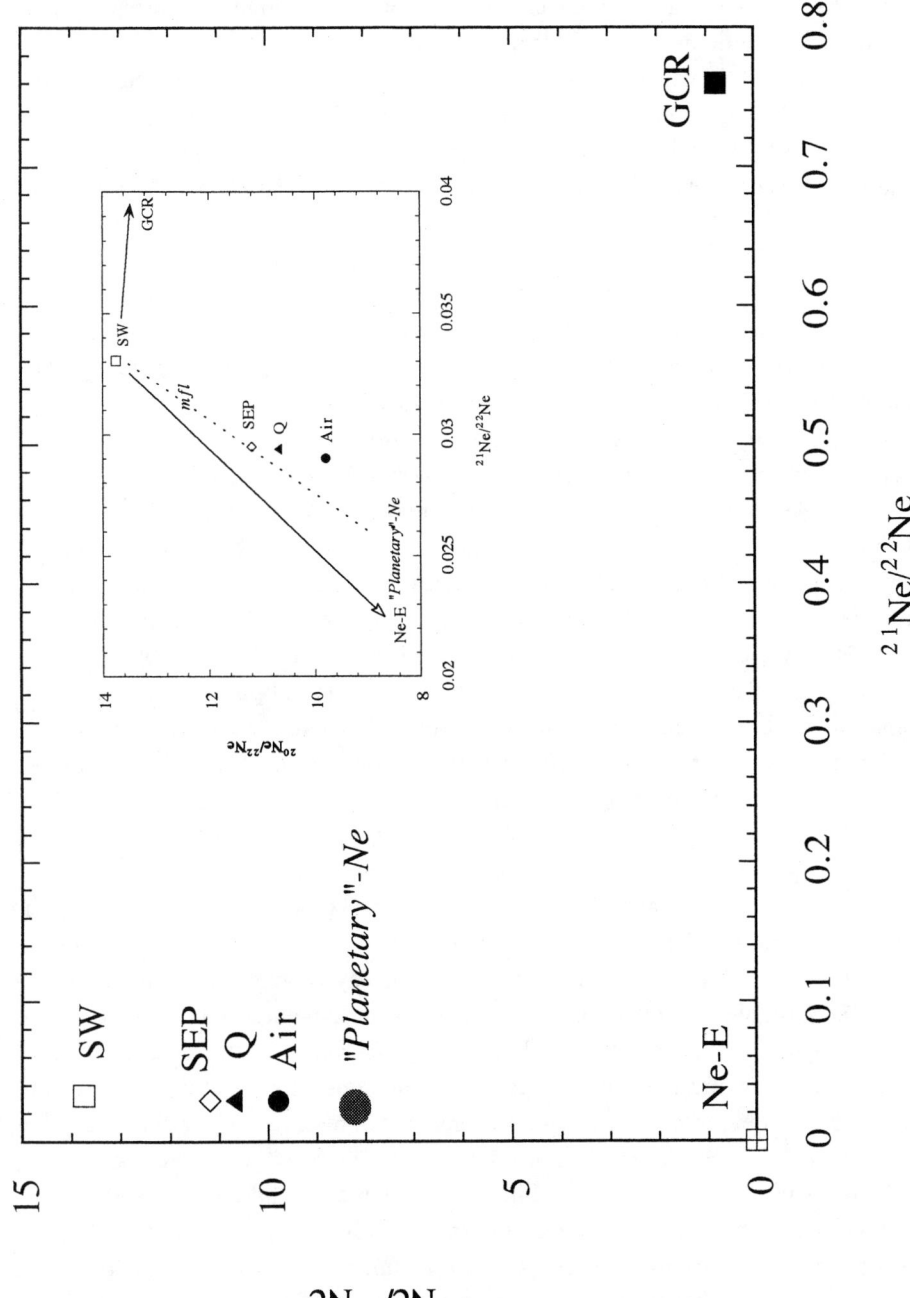

differences between solar and planetary gas isotopic composition are more profound and still lack quantitatively satsifactory explanation.

The framework in which comparison of solar and planetary gas isotopic compositions should be made has changed with the advent of the recognition that some or much of planetary gases are extrasolar components, carried into the solar system in solid grains incorporated into and sometimes preserved in meteorites (Section 3.4). For Ne, for example, there are two well-known extrasolar components of Ne-E(H) and Ne-E(L) (both are nearly pure ^{22}Ne but have different carriers and likely different origin; e.g., see Huss, 1997), essentially pure monisotopic ^{22}Ne, which, mixed with solar Ne (see Figure 3.3) in the right amounts, can account for virtually any Ne composition, particularly if coupled with allowance for cosmic ray-induced spallation production of the scarce isotope ^{21}Ne. For Xe (Figure 7.6), the explanation may lie in admixture of the heavy-isotope-rich extrasolar component Xe-H (e.g., see Anders, 1988). It remains somewhat unclear how satisfactory such interpretations are. Again for Ne, for example, there are more components and more explanations than there are isotope ratios to test them. For Xe, a comprehensive explanation of prominent solar system isotopic compositions in terms of extrasolar components (and fractionation) is algebraically feasible (for example, see Pepin & Phinney, 1978; Pepin, 1991) but involves an uncomfortably large amount of extrasolar component in the sun. It remains true that quantitative accounting of solar and planetary gas isotopic compositions requires special (and ad hoc, in some viewpoints) assumptions and does not emerge naturally from broader models that seek to explain the overall compositions of planetary materials (see Chapter 7).

Although the term *planetary* reflects the similarity of nonsolar trapped gases in meteorites to the noble gases in the Earth's atmosphere, there are some important differences between terrestrial gases and planetary trapped gases in meteorites. One such difference is that the elemental abundance of Xe in air is more than an order of magnitude lower than expected from the meteorite analogy (Figure 7.11); this "missing Xe" effect has given rise to the persistent but unproven hypothesis that the Earth actually has (or had) a much greater supply of Xe than is evident in air (see Section 7.5). There are also significant isotope differences, again most prominently in Ne and Xe. The ^{20}Ne/^{22}Ne ratio in air is lower than in any known reservoir of trapped Ne in the solar system, perhaps because the Earth formed with more extrasolar Ne-E carrier or perhaps because Ne was fractionated in the evolution of the Earth's atmosphere (see Section 7.4). Terrestrial Xe is unique, different from Xe known anywhere else in the solar system. The first-order difference between atmospheric and planetary (or solar) Xe is evidently a very strong mass-dependent isotopic fractionation (Figure 7.6), often (because of its uniqueness) taken to be a result of atmospheric evolution (e.g., Pepin, 1991) rather than a primordial feature. Superimposed on the fractionated Xe are specific-isotope excesses that are plausibly attributable to radiogenic contributions from ^{129}I, ^{244}Pu, and ^{238}U (see Sections 7.6 and 7.7).

As already noted, the origin of planetary noble gases remains mysterious. Equilibrium solubility of gases in solids yields concentrations far too low; adsorption may produce approximately the right pattern and abundances but only at unreasonably low temperatures and still requires appeal to some other mechanism to actually trap the gases. It is also speculated that gases can be somehow occluded and trapped during grain growth, but it is difficult to evaluate such nonequilibrium hypotheses theoretically or by laboratory simulations. Besides such "local" models based on direct gas-solid interaction there are also models in which planetary gas is created as some global component whose characteristics are determined by ionization potential (e.g., Jokipii, 1964) or mass, as in Rayleigh distillation (e.g., Ozima et al., 1998), but in such models there is no obvious astrophysical scenario for creation of the planetary component, and they still require some mechanism to get the gases trapped in the solids. There are many hypotheses, but it is fair to say that none has gained consensus acceptance.

Important as it has been for understanding of noble gases in general, recognition that some of what has traditionally been called planetary noble gases is actually of presolar origin (Section 3.4) has changed the problem of the physical origin of the gases somewhat, but only somewhat. Some of these gases clearly entered the solar system already trapped in solid presolar grains, evidently formed in circumstellar environments, but the mechanism by which the gases got into the grains remains unclear. The problem of origin for the bulk of the gases, the local rather than exotic component now often called the Q (see Section 7.3) or P1 component (Huss, Lewis & Hemkin, 1996), remains as it has for decades. An interesting alternative hypothesis is that Q-P1 gases are also exotic rather than local (e.g., Huss et al., 1996); in other words, they also entered the solar system already trapped in solids so that essentially none of the noble gases in planetary materials (including the Earth) were actually derived from ambient nebular gas during formation of the solar system. This hypothesis merely exports the problem, of course, to the presolar interstellar medium, but it does enlarge the available parameter space (e.g., to include very low temperatures or UV irradiation). In addition, if trapping of planetary (Q or P1) gas occurred over long times, during which the isotopic composition of total noble gases evolved, this hypothesis could help explain isotopic differences between planetary and solar gases.

3.4 Extrasolar Noble Gases

Several aspects of cosmochemistry (and astrophysics) have been dramatically transformed in recent years by the growing characterization and appreciation of extrasolar (or presolar or interstellar) materials preserved within the solar system, prominently including the study of noble gases. It has been discovered that several different kinds of presolar grains are preserved in certain kinds of undifferentiated meteorites (e.g., see Anders & Zinner, 1993; Zinner, 1997). These grains were present in the interstellar medium from which the solar system formed and have survived, evidently essentially intact, through whatever conditions they endured during the formation of the solar system and during the 4.6 Ga since.

Several different phases of presolar grains are known; silicon carbide, diamond, and graphite are particularly prominent and well characterized (e.g., Zinner, 1997). These phases share the characteristics that they are thermally refractory and in many circumstances particularly resistant to chemical reaction. These characteristics may have been critical to their survival through early solar system processing; they are certainly critical in the laboratory because these phases are typically isolated for detailed study by dissolution of nearly all the other phases of their host meteorites. Besides these three phases, a handful of others, also thermally refractory and chemically resistant, have been identified (e.g., Nittler, 1997). As a broad generalization, all the elements (to the extent that they have been analyzed) in all these phases have isotopic compositions that are radically different from average solar system composition. Therefore, they are inferred to be not only interstellar but also circumstellar, having condensed from expanding stellar atmospheres or explosions and incorporating the products of specific nucleosynthetic processes in individual stars, not just the mixture of many nucleosynthetic products, which defines average solar system composition. (The principal qualification to this generalization is that the C in diamonds is essentially normal. It may be that the C in individual diamonds is indeed radically anomalous as well, but the diamonds are so small as to defy analysis as individual grains, so that any analysis averages over many grains. It remains conjectured, however, that the carrier of anomalies in other elements is some other phase, less abundant than the diamonds and not separated from them by laboratory procedures so far employed.)

As is implicit in the preceding paragraph, the noble gases in silicon carbide, diamond, and graphite presolar grains have isotopic compositions that are radically different from solar system normal. This feature led to the discovery of presolar grains. Early in the exploration of noble gas contents in meteorites, it was noted that stepwise heating of carbonaceous chondrites revealed correlated variations in the isotopic composition of Xe (Reynolds & Turner, 1964). At the time, however, there were many instances of isotopic variations in the noble gases, especially Xe, which were not fully understood, and the effects observed by Reynolds and Turner (1964) were not recognized as isotopic anomalies as presently understood (i.e., a manifestation of incomplete mixing of presolar nucleosynthetic components). Subsequently, observation of Ne compositions, also revealed in stepwise heating of carbonaceous chondrites, outside the field that could be accounted for by mixing solar, planetary, and spallation Ne (Fig. 3.3) led Black (1972) to suggest specifically an isotopic anomaly interpretation. Similarly, Manuel, Hennecke, and Sabu (1972), stressing that correlation of light- and heavy-isotope variations in Xe (Figure 3.4) could not plausibly be attributed to an in situ origin such as fission, also advocated an isotopic anomaly interpretation. These interpretations did not attract a great deal of support or attention in the mainstream geochemistry or cosmochemistry communities at the time, however, perhaps because many mainstream geochemists and even cosmochemists regarded noble gas studies as complex, arcane, and rife with isotopic variability. It was not until after isotopic anomalies in the major element O were reported

3.4 Extrasolar Noble Gases

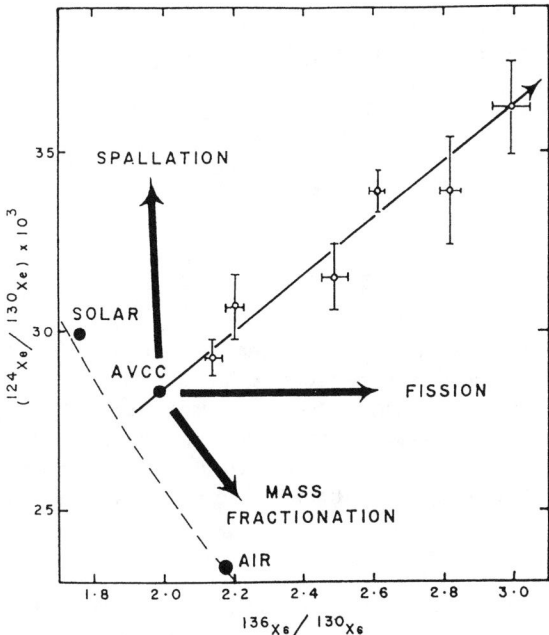

Figure 3.4 Isotopic compositions observed in stepwise heating of carbonaceous chondrites. The linear correlation requires the existence of a Xe component (upper right) that cannot be simply fission, spallation, or a fractionated version of common (solar, AVCC, air) Xe. AVCC is an acronym for average value carbonaceous chondrite, i.e., it is equivalent to planetary. Reproduced from Manuel et al. (1972).

(Clayton, Grossman & Mayeda, 1973) that these and other isotopic variations in noble gases were widely understood to reflect nonhomogenization of presolar nucleosynthetic components.

This recognition spurred extensive laboratory work aimed at identifying and isolating the phases within meteorites that carried the anomalous noble gases, using the noble gas isotopic compositions themselves as a guide to successful or unsuccessful enrichment of their carriers. The history of this development is complicated, but it eventually led to observed noble gas compositions with anomalies, relative to normal solar system composition, measured not in permil or percent but in factors of two or even powers of ten (Figure 3.5), in materials that could be identified as extrasolar and even circumstellar because essentially all their elements were radically anomalous (Lewis, Srinivasan & Anders, 1975; also see Bernatowicz et al., 1987; Anders and Zinner, 1993; Zinner, 1997).

As is the case with trapped planetary gases in meteorites in general, the physical/chemical factors responsible for the presence of noble gases in these circumstellar grains remain largely conjectural. Otherwise, however, although it was the noble gases that led to identification of these extrasolar phases, the noble gases are not excep-

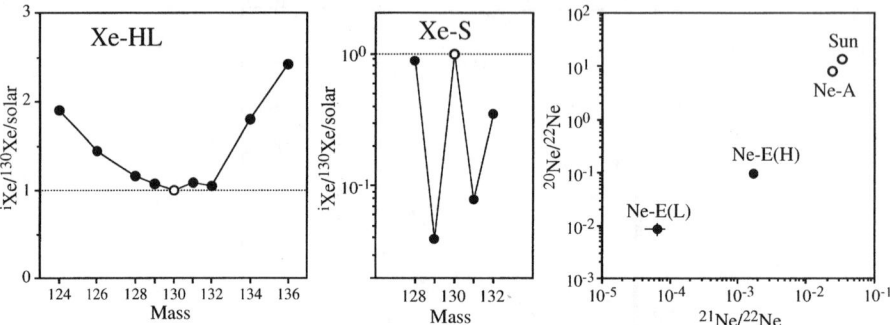

Figure 3.5 Radically anomalous noble gas isotopic compositions in extrasolar materials isolated from undifferentiated meteorites (from Anders & Zinner, 1993). Stepwise heating of whole-rock meteorites liberates slightly more or less of components such as Xe-HL and Ne-E, relative to other noble gas reservoirs in the rock, leading to the modest isotopic variations (e.g., Xe compositions as illustrtated in Figure 3.4, or Ne compositions to the lower-left of the air-spallation-solar wind triangle in Figure 3.3) from which the presence of anomalies was originally inferred.

tional compared to other elements in these materials. Other elements are just as radically anomalous as the noble gases. A better view is that all the elements, including the noble gases, in such phases are reasonably understandable in terms of nucleosynthetic processes occurring in individual stars, superimposed on the material from which the star was originally made. The study of these isotopic structures has had a major impact in providing "ground truth" for astrophysical models of stellar evolution and nucleosynthesis.

In terms of solar system cosmochemistry and terrestrial geochemistry, however, the noble gases in extrasolar materials stand out from the other elements, again because of their scarcity. The abundances of these well-characterized preserved presolar phases in meteorites are low in absolute terms, and the amount that they contribute to the total inventory of most elements is also small. For most elements in such phases, even though they are highly anomalous, their presence or absence would not make a detectable change in bulk isotopic composition. Because the overall abundances of the noble gases are so low, however, the quantities attributable to the presolar phases known to carry noble gases – diamond, silicon carbide, and graphite – are not trivial.

Presolar diamonds (or perhaps some other phase that follows the diamonds through the laboratory separation procedures), for example, are the host phase for Xe-HL, the component so named because of its relatively high abundance of the heaviest and lightest isotopes of Xe (Figures 3.4 and 3.5). Besides numerous other non-noble-gas elements the diamonds also contain other noble gases related to Xe-HL, plus other noble gas components, more nearly solar in isotopic composition, that can be distinguished on the basis of release temperature (Huss & Lewis, 1994). For elements other than noble gases, the diamonds themselves do not contribute enough material to alter bulk isotopic compositions significantly, but Xe-HL may be several percent of the total Xe in meteorites and the associated light gases He and Ne can account for *most* of the total He and Ne (i.e., much or most of the He and Ne in planetary gases, at least in many meteorites, are actually exotic components imported in presolar grain carriers). This is clearly an important boundary condition in consideration of models for the ultimate origin of noble gases in planetary bodies such as the Earth.

It seems likely that other interstellar phases may survive in undifferentiated meteorites, carrying with them isotopically exotic components acquired from specific stars or from the mixture of nucleosynthetic components characteristic of the interstellar medium. There is indeed evidence that this is so for some other elements (e.g., Messenger & Walker, 1997; Podosek & Nichols, 1997), although the phases involved are not well characterized, perhaps because they are not well defined (e.g., kerogen) or perhaps just because they do not share resistance to the reagents used to isolate the known forms of presolar grains in the laboratory. It is unclear whether any significant fraction of the "normal" planetary noble gases is carried by presolar phases not yet characterized (cf. Section 3.3). It is perhaps suggestive that the carrier of normal heavy planetary noble gases is thought to be some form of carbonaceous material, as is also the case for the carrier of isotopically variable H and N considered to have been severely fractionated at very low temperatures in the interstellar medium (e.g., Messenger and Walker, 1997; Robert et al., 2000).

Recognition that nontrivial to major fractions of the noble gases in meteorites are exotic components imported into the solar system in presolar grains widens the plausible latitude available to models for the origin of planetary gases, and especially for how closely the noble gas inventories of the major terrestrial planets, including Earth, might be related to those in meteorites. The latter question is related to the broader issue of how uniformly, or otherwise, presolar solids are distributed on a planetary scale, an issue not yet very well explored.

Chapter 4

Water

4.1 Introduction

As a good first-order generalization, the noble gases found in natural waters are acquired from air and are present in concentrations approximately consistent with air equilibration. Solubility data (Tables 4.1–4.4 and Fig. 4.1) are thus of central importance in evaluating noble gas observations in water. A comprehensive review and data evaluation for the general phenomenon of gas solution in water is given by Wilhelm et al. (1977).

On the whole, noble gases exhibit about the same order of magnitude of solubility in water as do other gases that do not react chemically with the water. Ar, in particular, is approximately as soluble as the major atmospheric gases: its solubility (pure water at 0°C) is 2.26 times that of N_2 and 1.09 times that of O_2. As a group, however, the noble gases exhibit a fairly wide spread in solubilities, with the characteristic features of strongly increasing solubility and temperature dependence of solubility with increasing atomic weight. This signature, combined with the useful feature that (with exceptions discussed later) they are conservative – no sources or sinks in organisms or other material in sea water and unlikely to participate in complex chemical reactions – makes the noble gases useful in a variety of geochemical studies.

A noteworthy feature of such studies is that they frequently make the most stringent demands encountered in noble gas geochemistry for high-precision absolute elemental abundances. Particularly in marine studies the effects of interest are manifested in elemental abundance variations measured in percent. For Ar, accuracy at the required level can be achieved in gas chromatography; for mass spectrometric abundance determinations at the percent level, isotope dilution is usually necessary. The technical problems are also different from those in most other noble gas spec-

4.1 Introduction

Table 4.1. *Solubility of gases in water and in sea water*[a]

Gas	A_1	A_2	A_3	B_1	B_2	B_3
He	−34.6261	43.0285	14.1391	−0.042340	0.022624	−0.0033120
Ne	−39.1971	51.8013	15.7699	−0.124695	0.078374	−0.0127972
N_2	−59.6274	85.7661	24.3696	−0.051580	0.026329	−0.0037257
O_2	−58.3877	85.8079	23.8439	−0.034892	0.015568	−0.0019387
Ar	−55.6578	82.0262	22.5929	−0.036267	0.016241	−0.0020114

[a] Values for use in Eq. (4.4) for T in Kelvin, salinity S in per mil, Bunsen coefficient β in cm³ STP gas in cm³ solution under partial pressure of 1 atm; from Weiss (1970a, 1971a).

Table 4.2. *Moist-air saturation concentration of gases in water and sea water*[a]

Gas	A_1	A_2	A_3	A_4	B_1	B_2	B_3
He	−167.2178	216.3442	139.2032	−22.6202	−0.044781	0.023541	−0.0034266
Ne	−170.6018	225.1946	140.8863	−22.6290	−0.127113	0.079277	−0.0129095
N_2	−177.0212	254.6078	146.3611	−22.0933	−0.054052	0.027266	−0.0038430
O_2	−177.7888	255.5907	146.4813	−22.2040	−0.037362	0.016504	−0.0020564
Ar	−178.1725	251.8139	145.2337	−22.2046	−0.038729	0.017171	−0.0021281

[a] Values for use in Eq. (4.5) for T in Kelvin, salinity S in per mil, c^* in cm³ STP in 1 kg of solution under 1 atm total pressure of water-vapor-saturated air; from Weiss (1970a, 1971a).

Table 4.3. *Solubility of gases in pure water*[a]

Gas[b]	A_0	A_1	A_2
He	5.0746	4,127.8	−627,250
Ne	4.2988	4,871.1	−793,580
O_2	4.0605	5,416.7	−1,026,100
Kr	3.6326	5,664.0	−1,122,400
Xe	2.0917	6,693.5	−1,341,700

[a] Values for use in Eq (4.7) for T in Kelvin, giving Henry's law constant k as the ratio of fugacity (in atm) to solute mole fraction x; from Benson & Krause (1976).
[b] Wilhelm, Battino, & Wilcock (1977) tabulate Rn solubility data.

trometry: ample gas amounts are available, but special care must be taken to ensure complete extraction of gases from the water sample and then to separate the water vapor from the noble gases. Care must also be exercised in sampling, since a small air contamination can product large distortions (cf. Table 4.5).

To highlight the differences between observed and expected gas concentrations, a delta value (usually in percent) representation is often used:

$$\Delta \equiv (c/c^* - 1) \times 100\% \qquad (4.1)$$

Table 4.4. *Relative gas solubilities in sea water and fresh water*[a]

Temperature (°C)	He	Ne	N_2	O_2	Ar	Kr	Xe
0	0.809	0.786	0.749	0.766	0.764	0.72	0.73
25	0.838	0.825	0.785	0.798	0.798	0.77	0.78

[a] Tabulated quantity is c^*_{35}/c^*_0, where c^*_0 is wet air saturation solubility in pure water and c^*_{35} is corresponding value for sea water (35‰ salinity). Data from Weiss (1970a, 1971a) (cf. Table 4.2) except for Kr and Xe, from Wood & Caputi (1966), as tabulated by Kester (1975). Data are for c^* by weight water; values for c^* by volume water will be slightly different.

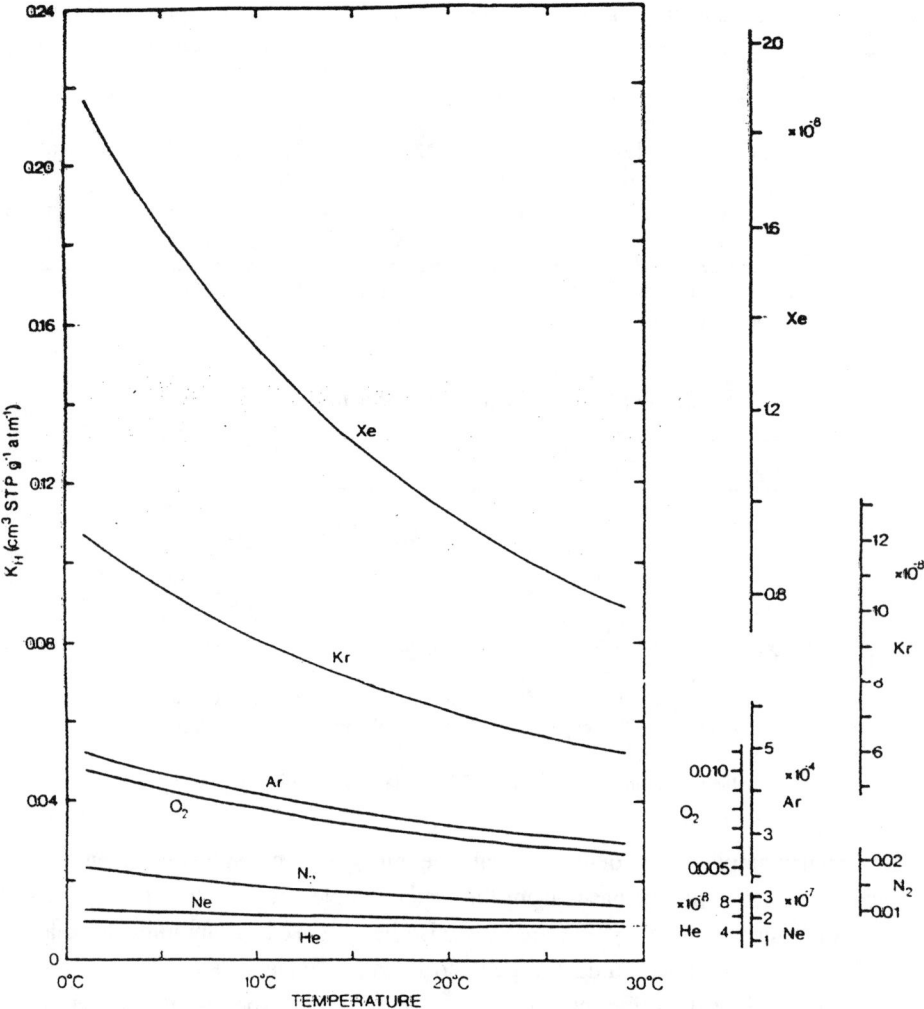

Figure 4.1 Solubility of gases in pure water, from Table 4.1 (N_2, Ar) and Table 4.3 (He, Ne, O_2, Kr, Xe). Ordinate of main figure is Henry's law solubility. Ordinate scales at right show concentrations (in cm³ STP/g) for equlibrium 1 atm air with composition given in Table 1.2.

Table 4.5. *Illustrative apparent wet-saturation anomalies (%) in sea water for various effects*

Effect		ΔHe	ΔNe	ΔAr	ΔKr	ΔXe
True partial pressures 1% higher than nominal; any temperature		1.0	1.0	1.0	1.0	1.0
Temperature increases 1°C after equilibration	at 0°C	0.5	0.9	2.6	3.5	4.2
	at 25°C	0.2	0.7	1.7	2.1	2.6
Air injection, 1 cm^3 (STP, dry)/l	at 0°C	12.9	10.0	2.4	1.2	0.6
	at 25°C	13.9	12.1	4.0	2.3	1.3
Mix equal amounts of water at 0°C and 10°C		0.2	0.3	1.3	2.0	2.8
Mix equal amounts of water at 20°C and 30°C		0.1	0.1	0.7	1.1	1.5

where c is observed concentration and c^* is the expected or equilibrium concentration. Generally, c^* is calculated for water at a specified temperature in equilibrium with vapor-saturated air at 1 atm total pressure, in which case Δ is the wet-saturation anomaly. For deep ocean water, the relevant temperature is the potential rather than the in situ temperature. Representative values of c^* are included in Table 4.6.

Failure of a given water sample to reflect equilibrium with the air with which it was last in contact can be ascribed to a variety of causes. An obvious and important case is nonconservation (e.g., addition of nonatmospheric gases; see Sections 4.5 and 4.6). Equally obvious is the simple lack of equilibration for kinetic reasons (e.g., rapid isolation of glacial melt), which would be difficult to treat quantitatively. This evidently is not too severe a problem, however, since equilibration times are apparently fairly short (e.g., a day or so for the top few meters of sea water; cf. Broecker, 1974); in any case, we are unaware of any saturation anomalies ascribed to kinetic equilibration failure. However, a number of additional effects are likely to be significant and can be described quantitatively. These are discussed here and illustrated in Table 4.5.

1. *Pressure effects.* Variations spanning a few percent or so can be generated by total barometric pressures different from 1 atm and by relative humidities different from 100%. A given pressure variation will produce the same Δ for all gases, independent of temperature.

2. *Temperature effects.* Anomalies will arise if the wrong temperature is used to calculate c^*. This could be due to evaporative effects at the air-water interface or to a change in water temperature after isolation from air (other than that from adiabatic compression). The apparent anomaly will be greater for the heavier gases and greater at lower temperature, reflecting the temperature dependence of c^*.

3. *Bubbles.* Bubbles will produce supersaturation because of the hydrostatic pressure excess. At one extreme, if only a small fraction of the air in a bubble

Table 4.6. *Selected data for noble gas abundances (cm^3 STP/g) in water*

Sample	Notes	Temperature (°C)	He (10^{-8})	Ne (10^{-7})	Ar (10^{-4})	Kr (10^{-8})	Xe (10^{-8})
Fresh water							
Air saturation	a	0	4.90	2.25	4.98	12.57	1.95
Air saturation	a	25	4.41	1.78	2.84	6.22	0.83
Groundwater (Israel)	b	53	997	2.03	3.06	7.61	1.01
Sea water							
Air saturation	a	0	3.97	1.77	3.80	9.05	1.43
Air saturation	a	25	3.70	1.47	2.26	4.79	0.65
Pacific surface water	c	22	3.84	1.55	2.42	5.30	—
Pacific deep water	d	1	4.42	1.86	3.73	9.04	—
Atlantic deep water	e	4	4.22	1.85	3.51	8.43	—
Pacific deep water	f	1	—	1.77	3.06	7.31	1.08
Red Sea brine	g	56	1400	1.4	—	—	—

^{a} Values for equilibration with 760 torr total pressure of air saturated with water vapor. He, Ne, Ar data from Table 4.2; Kr, Xe data from Tables 4.3 and 4.4. Sea water salinity taken to be 35‰.
^{b} Hamat Gader spring water; Mazor (1972).
^{c} Average of top seven samples, Piquero station 3 (14°S 102°W), by Bieri & Koide (1972); see Table 4.7.
^{d} Average of deep water (>2000 m) samples, three Piquero stations, Bieri & Koide (1972) (their table 3).
^{e} Average of bottom 3 samples, AII-20 Station 944 (16°N 59°W), by Bieri, Koide & Goldberg (1968); see Table 4.7.
^{f} Average of samples M-39 and M-26, Expedition Monsoon, by Mazor, Wasserburg & Craig (1964). NB These data indicate saturation anomalies ≈20%, which would probably be considered questionable at present.
^{g} Atlantic II sample 714-G7, Lupton, Weiss & Craig (1977a); data tabulated are cm^3 STP g^{-1} of H$_2$O.

disolves before it escapes to the surface, this will produce the same effect as an increase in pressure; see item 1. The other extreme, when the bubble completely dissolves, is denoted *air injection*. The resultant Δ will be inversely proportional to c^* and so be greater for the lighter gases and smaller at lower temperatures, although Δ for He and Ne will not be very sensitive to temperature.

4. *Mixing*. Because the solubility-temperature curves are concave from above (Fig. 4.1), mixing of waters equilibrated at different temperatures will produce apparent supersaturation relative to the temperature of the mix. The effect will be greatest for the heavier gases but will be minimal for He and Ne, and in the normally encountered range it is greater at lower temperatures.

Because of the variety of effects that can and apparently do contribute to saturation anomalies, it is not possible to determine the origin of a given anomaly from a single datum. Resolution becomes feasible if there are data for as many gases (in a single sample) as there are effects, however. If all the effects are relatively small, the net anomaly Δ is approximately the sum of the Δs from the individual effects. For

He, Ne, and Ar in particular, for which the most extensive data are available, plausibly likely mixing anomalies are small (cf. Table 4.5). If it is assumed that only pressure and temperature variations and air injection, items 1–3 as listed earlier, contribute to the apparent saturation anomalies of a given water sample, then for each gas the observed Δ is a homogeneous linear combination of ΔP, ΔT, and Δa, representing pressure, temperature, and air injection. The coefficients depend only on the solubility (and temperature) and are different for each gas (Table 4.5). Thus, knowledge of ΔHe, ΔNe, and ΔAr permits inversion and determination of ΔP, ΔT, and Δa. Formal application of this approach is illustrated by Craig and Weiss (1971) (cf. Figure 4.3).

Observed saturation anomalies can thus be resolved into contributions from the effects described earlier as long as the data are sufficiently precise. In practice, the effects cannot be resolved as clearly as might be hoped, and, as noted by Craig and Weiss (1971), the resolution is quite sensitive to experimental uncertainties.

4.2 Solubility Data

The early geochemical literature concerned with noble gases in water relied heavily on the solubility data of Morrison and Johnstone (1954) for pure water and König (1963) for sea water. Subsequent and more accurate experimental data differ from these early standards by a few to several percent.

Weiss (1970a) reviewed and evaluated the literature data then available for N_2, O_2, and Ar and fitted them to a smooth functional dependence. His results are reported in terms of the Bunsen coefficient β, defined as the quantity of gas (in cm³ STP) dissolved in unit volume (1 cm³) under unit partial pressure (1 atm). Pure water data were fitted to an integrated van't Hoff equation:

$$\ln \beta = a_1 + a_2/T + a_3 \ln T \tag{4.2}$$

Because of the "salting out" effect, gases are some 25% less soluble in sea water than in fresh water (Table 4.4). The effect is often rendered in terms of the Setchenow relation

$$\ln \beta = b_1 + b_2 S \tag{4.3}$$

where b_1 gives $\ln \beta$ in pure water and b_1 and b_2 are independent of salinity S.

Natural fluids commonly contain appreciable dissolved solids (typically NaCl), ranging from a few percent by weight (e.g., sea water) to roughly 30 wt % (e.g., brines). Smith and Kennedy (1983) measured argon solubility in NaCl solutions at temperatures ranging from 0°C to 65°C (Figure 4.2). For this temperature range, argon solubility decreases considerably with increasing salinity; the argon solubility in the most concentrated brines (25 wt %) is four to seven times less than in fresh water at 65°C and 0°C. Smith and Kennedy (1983) also calculated thermodynamic parameters for the solubility in the brines for other noble gases.

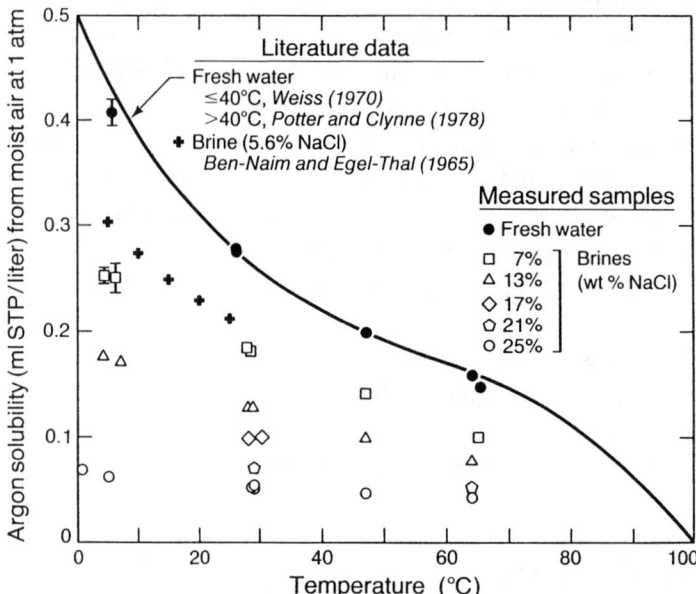

Figure 4.2 Argon solubility (cm^3 STP l^{-1} of solution) from moist air at 1 atm total pressure for fresh water and NaCl brines as a function of temperature. Note that the argon solubility decreases systematically with increasing temperature and NaCl content. After Smith and Kennedy (1983).

Combining relations (4.2) and (4.3) and scaling the temperature, Weiss (1970a) fitted the data to

$$\ln \beta = A_1 + A_2(100/T) + A_3 \ln(T/100) \\ + S\left[B_1 + B_2(T/100) + B_3(T/100)^2\right] \quad (4.4)$$

Parameters for this equation for N_2, O_2, and Ar solubility are shown in Table 4.1, scaled for salinity S in per mil (‰).

Subsequently, Weiss (1971a, 1971b) presented experimental data for He, Ne, and Ar solubilities, concluding that modification of the Ar solubility parameters was unnecessary and fitting his He and Ne data to Equation (4.4) with results also shown in Table 4.1.

Of more immediate practical utility for geochemical application are moist-air solubilities, the c^* in eq. (4.1). Weiss (1970a, 1971a) parameterized c^* by

$$\ln c^* = A_1 + A_2(100/T) + A_3 \ln(T/100) + A_4(T/100) \\ + S\left[B_1 + B_2(T/100) + B_3(T/100)^2\right] \quad (4.5)$$

with results given in Table 4.2, scaled for S in per mil (‰) and c^* in cm^3 STP/kg under 1 atm total pressure of water-saturated air. The data in Table 4.2 are the usual basis for evaluation of wet-saturation anomalies. Detailed tables of β and c^* at various

T and S are given by Weiss (1970a, 1971a), and Kester (1975) also gives tables of c^* calculated from Table 4.2 data.

Benson and Krause (1976) have also published high-precision solubility data for O_2, He, Ne, Kr, and Xe in pure water. Their data were fitted in terms of the Henry's law constant k in

$$f = kx \qquad (4.6)$$

where f is the gas fugacity above a solution in which the mole fraction of the gas is x. They fitted their data to the functional form

$$\ln k = A_0 + A_1/T + A_2/T^2 \qquad (4.7)$$

with the coefficients given in Table 4.3. The Benson and Krause O_2 data agree within about 0.2% with the experimental data which were smoothed by Weiss (1970a) to give the Table 4.1 parameters, but their He and Ne solubilities are both 1–2% higher than those of Weiss (1971a).

The basic solubility parameters β (Table 4.1) and k (Table 4.3) are both readily usable for calculation of equilibrium concentrations at arbitrary partial pressures. In principle, correction for volume change by solution and nonideal gas behavior must be made, and it should be noted that the solubility experiments are performed at partial pressure near 1 atm, but for geochemical purposes the corrections are negligible. It is unfortunate that different experimental determinations of the same physical quantity still frequently differ by more than their stated uncertainties. Still, the available solubility data appear sufficiently good, no worse than a percent or so uncertainty, and often considerably better, for all the noble gases (and N_2 and O_2) in pure water and all but Kr and Xe in salt water.

Kester (1975) tabulated salt water moist-air saturation values for Kr and Xe from the data of Wood and Caputi (1966), assuming the salinity dependence expressed in Equation (4.4). For Kr and Xe normalizations in this chapter, we will use the pure water solubilities of Benson and Krause (1976) modified by the tabulated salinity dependence in Kester (cf. Table 4.6). Since the Wood and Caputi pure water solubilities agree well with those of Benson and Krause at about 25°C but are higher by 3–4% at 0°C, and the salting-out coefficients do not smoothly continue the mass-dependent trend of the lighter gases (Table 4.4), the sea water solubilities of Kr and Xe must evidently be considered uncertain by at least a few percent.

The only report of experimental determination of noble gas isotopic fractionation upon solution in water is that of Weiss (1970b), who found that ^{3}He is *less* soluble than ^{4}He by $(1.2 \pm 0.2)\%$ at 0°C, the fractionation increasing in magnitude by about 0.1% per 10°C. The effect is in the same sense but much larger than the $\leqslant 0.1\%$ effects found for O_2 and N_2 by Klots and Benson (1963). On the basis of the reasonably inferred mass dependence, we can guess that isotopic fractionation will be quite small, around 0.1%, for Ne and negligible for the heavier gases.

4.3 Seawater

Materials, such as dissolved gases, without known sources or sinks in the ocean are said to be conservative. To first order, the conservative gases dissolved in sea water, specifically including the noble gases, are in solubility equilibrium with air, surface waters at ambient conditions, and deep waters at the conditions of their formation (when they were isolated from air). Dissolved O_2, because of its role in biological (and thanatological) processes, is nonconservative, and its variations in sea water are well known and widely studied. Dissolved N_2 is a marginal case: sources and sinks are evidently possible in special cases, but in the open ocean N_2 can apparently be considered nearly conservative. In any case, observations of dissolved N_2 are generally quite close to saturation values; saturation anomalies up to a few percent have been reported, both positive and negative and without clear correlation with other oceanographic variables. Reviews for O_2 and N_2, as well as other nonreactive gases, are readily available (e.g., Broecker, 1974; Kester, 1975).

Representative data for noble gases in sea water are given in Table 4.6 and 4.7. Much of the early data is apparently characterized by analytical difficulties, with interpretations further hampered by lack of adequate solubility data. As the experimental observations, and particularly the solubility data, became reliable at the 1% level, the general features of dissolved noble gas distribution became evident (cf. Craig & Weiss, 1971; Bieri, 1971). As noted, dissolved noble gases (and N_2) are present at approximately the moist-air saturation values, but there are saturation anomalies at the level of a few to several percent (Table 4.7), well beyond stated experimental uncertainties. The light gases He and Ne, in particular, show consistently positive anomalies. This is suggestive of air injection (Table 4.5), and indeed the consensus is that in surface waters in general, and in Atlantic deep waters, the anomalies are attributable to air injection by bubbles, with allowance for modest pressure variations and postequilibration temperature changes. Supersaturation due to air injection of the order of 0.5–$1.0\,\text{cm}^3/\text{l}$ is common in the studied waters. Air injection, or the other plausible mechanisms for producing anomalies described earlier (Section 4.1 and Table 4.5), can account for only about two thirds of the average ΔHe in Pacific deep waters, however (Craig & Weiss, 1971; Bieri, 1971). The balance, about 3% of the saturation concentration, is now generally attributed to nonconservative behavior of He, specifically to the addition of juvenile (radiogenic) He from the suboceanic mantle or crust (Section 4.5). This conclusion is probably the most important result to emerge from general noble gas elemental abundance studies of sea water.

Both Craig and Weiss (1971) and Bieri (1971) stress that analysis of the He anomalies must be based on comparison with those of Ne and Ar. Craig and Weiss (1971) made a formal analysis expressing each of ΔHe, ΔNe, and ΔAr as a linear sum of terms in ΔP, ΔT, and Δa (Section 4.1). A convenient graphical presentation (Figure 4.3) is the relation between ΔHe–ΔNe and ΔNe–ΔAr. Since pressure variations

Table 4.7. *Apparent wet-saturation anomalies[a] in marine waters*

Water (Expedition-Station)	Location	Surface Water				Deep Water				References
		ΔHe	ΔNe	ΔAr	ΔKr	ΔHe	ΔNe	ΔAr	ΔKr	
Atlantic (AII-20-944)	16°N 59°W	4.8	5.0	2.7	3	5.2	5.5	1.9	7	Bieri et al. (1968)
Drake Passage (Zapiola-C-14)	58°S 60°W	4.0	5.0	3.2	10	—	—	—	—	Bieri et al. (1968)
Pacific (NOVA-1)	34°N 146°W	3.3	2.0	−0.4	2	6.5	1.6	−5.6	−3	Bieri et al. (1968)
Pacific (NOVA-4)	9°N 178°E	2.3	1.6	0.0	4	7.6	3.8	−3.1	0	Bieri et al. (1968)
Pacific (Piquero-3)	14°S 102°W	4.3	5.0	3.1	5	11.9	6.3	1.7	5	Bieri & Koide (1972)
Pacific (Carrousel-14)	33°S 73°W	—	4.7	−2.5	—	8.6	3.8	−0.2	—	Craig et al. (1967)

[a] Anomalies in percent, normalized to Table 4.2 (He, Ne, Ar) or to Tables 4.3 and 4.4 (Kr).

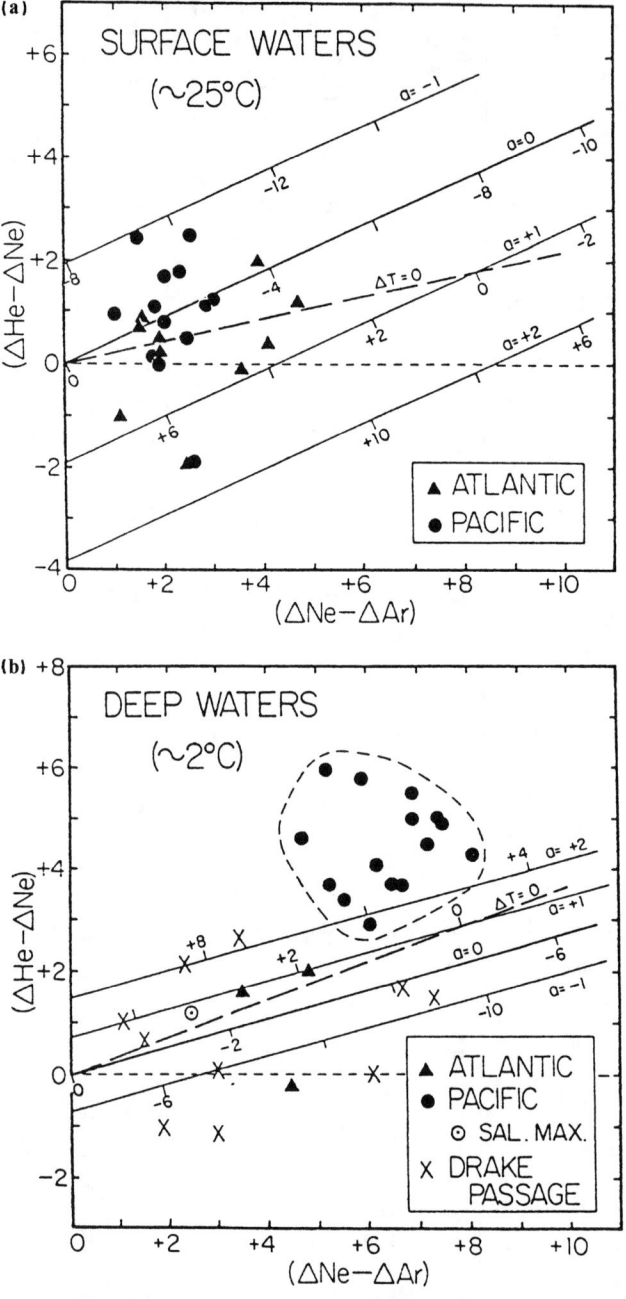

Figure 4.3 (a) Correlation of apparent saturation anomalies in sea water; Δ values given in percent. Solid lines are loci for constant amounts of air injection, labeled by a in cm³ STP/kg. A second family for constant postequilibrium temperature changes ΔT can be constructed; only the line for $\Delta T = 0$ is shown, but ΔT intersection values on the a = constant lines are indicated. Data are from Bieri et al. (1968). Reproduced from Craig and Weiss (1971). (b) Correlation of apparent saturation anomalies in sea water [cf. Figure 4.3(a)]. Drake Passage waters are shallow but cold. Data are from Bieri et al. (1968). Reproduced from Craig and Weiss (1971).

are the same for all the gases, ΔP drops out of these differences, and the graph can be analyzed in terms of a grid of intersecting loci of constant ΔT and constant Δa. [The two families are not orthogonal, and their positions and angle of intersection depend on temperature; they cross at around 10°C, so the relative slopes are different in Figures 4.3(a) and 4.3(b).] From the parameters in Figure 4.3 it is evident that inferred ΔTs (and sympathetically varying ΔPs) are quite sensitive to uncertainties of a percent or so. It is clear, however, that saturation anomalies in surface waters and Atlantic deep waters can be understood in terms of plausibly modest values for air injection and pressure and temperature variations, but that those for Pacific deep water cannot. The high ΔHe values cannot be generated by any plausible combination of these mechanisms and so must be interpreted in terms of juvenile (nonatmospheric) excess He.

As noted, the excess He is needed to account for only about one third of ΔHe in Pacific deep waters. It is disquieting when an effect of interest must be observed as a relatively small superposition on another, unrelated, effect, coincidentally of the same order of magnitude, particularly when the separation of the effects is not so very much larger than experimental uncertainties. The existence of excess He is not in doubt, as can be inferred from the observations described in Sections 4.4 and 4.5, but evaluation of its magnitude is quite sensitive to uncertainties in the data, both the observations and the solubilities. In both respects, the history of noble gas investigations in water has included many cases where systematic errors, as indicated by discrepancies with other experiments, have exceeded stated confidence limits. Assessment of excess He would change, for example, in response to revision of the solubility data. (The *fresh* water He and Ne solubilities of Benson & Krause, 1976, are 1–2% higher than those of Weiss, 1971a; if it is speculated that the salt-water solubilities should be increased by, say, 1%, this would reduce the inferred air injection term, *but* not the excess He, since ΔHe and ΔNe would both be reduced: cf. Figure 4.3.) As has been noted in the literature, it must also be presumed that Henry's law solubility can be accurately (to about 1%) extrapolated across more than five orders of magnitude from laboratory measurements at around 1 atm to the partial pressures in air. Revision of the He abundance in air (Table 1.2) by 1% or so would also change the overall picture.

Aside from identifying air injection as a common feature in most waters and excess He in Pacific deep waters, noble gas elemental abundances have also been used as tracers for water masses and to study mixing in intermediate waters. Discussions are given by Bieri et al. (1966, 1968), Craig, Weiss and Clarke (1967), Craig and Weiss (1968), and Bieri and Koide (1972). On the whole, the arguments cut finer and are therefore closer to the limit of experimental uncertainties than the air injection and excess He features discussed earlier, and so are less definitive; in any case, they seem not to have made any large impact on marine science.

To this point in this section, "noble gases" have meant He, Ne, and Ar, which are the principal data base. Among the more reliable recent data, Kr data are scarce (Bieri et al., 1968; Bieri & Koide, 1972), and Xe data are essentially nonexistent,

Table 4.8. *Sea water contribution to terrestrial noble gas inventory*

Gas	Approximate Average Concentration (cm^3 STP/g)	Total[a] amount (cm^3 STP)	Fraction[b] of air inventory (%)
He	4.0×10^{-8}	5.6×10^{16}	0.3
Ne	1.7×10^{-7}	2.4×10^{17}	0.3
Ar	3.5×10^{-4}	4.9×10^{20}	1.3
Kr	8.5×10^{-8}	1.2×10^{17}	2.7
Xe	1.1×10^{-8}	1.6×10^{16}	4.5

[a] For 1.41×10^{24} g of sea water. The fresh-water contribution will be negligible in comparison.
[b] Quantities in air from Table 1.2.

presumably because of analytical difficulties. Kr anomalies, normalized to the saturation concentrations described in Section 4.2, are included in Table 4.7. Good solubility data would be helpful in integrating ΔKr into the ΔHe–ΔNe–ΔAr analysis, where it might be useful since it is more sensitive to ΔT and less sensitive to Δa than the lighter gases (Table 4.5).

Xe data would be even more useful in this sense, if there were any. The best data are probably those of Mazor et al. (1964) (Table 4.6). Normalized as for Kr previously, their data yield ΔXe = +7% in surface water and ΔXe = −21% in deep water. The surface water result is similar to more recent ΔKr (Table 4.7). Although the Mazor et al. data also include approximately 20% saturation anomalies in Ne and Ar that would be considered spurious today, it is tempting to speculate that their deep water Xe undersaturation is real, possibly also the scattered but very low values of Bieri, Koide, and Goldberg (1964), since of all the noble gases Xe is the best candidate for an adsorptive sink in ocean water (cf. Chapter 2).

As a final note, the noble gases in sea water are derived from air and, like the water itself, are properly considered part of the terrestrial atmosphere. Particularly for the more soluble heavy gases, their contribution to the total inventory is a small but nontrivial, if generally ignored, addition to the air reservoir (Table 4.8).

4.4 Meteoric Water

Noble gases in rainwater are in solubility equilibrium with air and, although we know of no investigations addressed to the question, there is no reason to believe any differently for river and lake water. Most work on meteoric water has thus been on groundwater, which is removed from contact with air, especially geothermal water (see review by Mazor, 1975). Matters of interest are the temperature at which the water was last equilibrated with air, whether it has behaved as a closed system since air equilibration, and indeed whether or not it is actually meteoric.

Concerning the latter question, the alternative to meteoric water is juvenile water. Distinction between these two alternatives for the origin of geothermal waters has been attacked extensively through a variety of geochemical methods, generally with the result that the water in question is meteoric rather than juvenile. In the case of noble gases, there is in fact no clear idea of what the noble gas signature of juvenile water ought to be. If noble gases are observed in quantities appropriate for meteoric water, or in quantities sensibly related, this may be taken as presumptive evidence that the origin is indeed meteoric, although certainly moderate dilution by noble-gas-free juvenile water could not be excluded. In any case, no water has ever been identified or suggested to be anything other than meteoric on the basis of its noble gases.

The general rule is that, except for juvenile radiogenic gases (Section 4.5) frequently leached from the host rocks, noble gases are conservative in groundwater. An interesting set of examples (cf. Table 4.6) is the study of thermal waters in the Jordan Rift Valley, Israel, by Mazor (1972). Such waters have observed temperatures up to 63°C. Relative to their present temperatures, they are supersaturated in noble gases; the noble gas contents are, however, quite appropriate for air saturation at surface temperatures, 15–25°C (Mazor, 1972; also see Benson, 1973). This indicates that noble gas contents have been conserved during their history in groundwater, evidently in excess of 10^4 yr, and furthermore that Pleistocene surface temperatures in this region were much the same as they are today, rather than significantly closer to 0°C.

In other cases, however, nonconservation is evident. Mazor and Wasserburg (1965), for example, found that gaseous emanations from hot springs in Yellowstone National Park (United States) contained noble gases in relative proportions more or less correct for surface temperature air saturation, indicating, along with other lines of evidence, that the water was meteoric. Mazor and Fournier (1973), however, found that even though the same was generally true in most cases for the actual waters, the absolute concentrations in water were typically lower than the air-saturation values, down to 3% in one case. Reviewing this and other cases, Mazor (1975) points out that the absolute abundances are undersaturated not only for surface temperatures but also for observed temperatures and so could not be attributed to reequilibration at the elevated temperature. Instead, he suggests, this indicates loss of the noble gases through partitioning into a vapor phase, presumably steam; he also cites at least one case where relative abundances are correct for surface temperature equilibration, but the absolute abundances are too high, apparently reflecting recondensation of the vapor after isolation from its source. Mazor (1975) goes on to note that such noble gas saturation anomaly patterns might be useful for diagnosing steam formation in geothermal water systems, a feature of obvious utility in geothermal energy prospecting.

No data for noble gas solubility in ice are available, but solubilities can be expected to be much lower than for liquid water. The only noble gas observations are these of Matsuo and Miyake (1966), who analyzed Ar along with N_2, O_2, and CO_2 in natural ices. They found the major gases and Ar present in roughly atmospheric proportion, evidently contained principally in occluded gas bubbles. The bubbles were present in variable amount, but often of the same order as that expected for exsolution on freezing (about 3% by volume). By the same technique described in Sections 4.1 and 4.3, it is possible to partition observed gases into a solution component and an "air injection" component, with most of leverage coming from highly soluble CO_2. Matsuo and Miyake found the solution component to be generally quite small (<2%), so the dominant effect is evidently mechanical occulusion of air bubbles.

Sorption of the heavier noble gases on ice might be interesting (cf. Section 7.5), but there are no relevant data.

In many examples of hot springs, fumaroles, and the like, associated with tectonic activity, the water involved is meteoric but the noble gases are, in part, juvenile. Observations are described in Chapter 6.

4.5 Juvenile ^{4}He

The sea water data described in Section 4.3 lead to the conclusion that even though marine waters in general show "explainable" positive saturation anomalies up to several percent for He and Ne, the He abundances of deep water in the Pacific (but not the Atlantic) were higher still and could not be explained in terms of conservative behavior after isolation from air. A nonconservative effect, addition of nonatmospheric (i.e., juvenile) He, was invoked. On the basis of the best analytical observations and solubility data now available, the excess (juvenile) He contributes typically about 3% to the saturation anomaly ΔHe in Pacific deep water, about one third the total anomaly (the other two thirds are due mainly to air injection). Since the element He is mostly the isotope ^{4}He, and mass spectrometers measure ^{4}He anyway, the juvenile He is actually ^{4}He, and was quickly identified as radiogenic ^{4}He. This was not surprising: The effect was predicted before it was observed (Revelle & Suess, 1962), declared in the first experimental observations (Bieri et al., 1964; 1966) even before the more reliable experimental data, solubility data, and resolution techniques could be applied (Craig & Weiss, 1971; Bieri, 1971), and shown to be quite plausible quantitatively (Bieri, Koide & Goldberg, 1967). The average flux of juvenile ^{4}He into the atmosphere estimated by Bieri et al. (1967) was 2–3 × 10^6 atom cm^{-2} s^{-1}. A more recent value, 3 × 10^5 atoms cm^{-2} s^{-1} (Craig, Clarke & Beg, 1975), is described in Sections 4.6 and 6.12.

Although the 3% excess He is fairly small and, by itself, arguable, it should be pointed out that there are other cases where supersaturation effects due to addition of juvenile radiogenic ^{4}He are not at all subtle. Hundredfold or more enrichments of He, relative to air saturation, are frequently observed in geothermal groundwaters (Table 4.6), reflecting extraction of radiogenic ^{4}He from host rocks. More recently, Lupton et al. (1977a) have observed a striking and, for sea water, unprecedented 380-fold enrichment of He (Table 4.6) in Red Sea brines (discussed further later). In both these cases, note that the nonradiogenic gases are present at normal saturation levels.

Addition of juvenile radiogenic ^{4}He should be and (presumably) generally is accompanied by juvenile radiogenic ^{40}Ar, although the latter effect is observable only in the more extreme cases. For a nominal ^{4}He/^{40}Ar ≈ 10 in radiogenic gas, a hundredfold enrichment of He in, say, saturated 25°C sea water requires 3.7×10^6 cm^3 STP/g (Table 4.6), while the corresponding 3.7×10^{-7} cm^3 STP/g of ^{40}Ar produces only an 0.16% addition to the dissolved Ar. (The equivalent figure for ^{136}Xe is $\Delta \approx 0.002\%$.) Part of the imbalance is due to the composition of the source gas and part to the low solubility of He, but mostly this results from the scarcity of He in air in comparison with its sources (Section 6.6).

In practice, small enrichments of radiogenic ^{40}Ar can be determined with much greater precision and less ambiguity by examining the ^{40}Ar/^{36}Ar ratio for deviations from the air ratio than by trying to assess small deviations from solubility equilibrium. The same remark applies to the ^{4}He/^{3}He ratio, although, as described later, results proved otherwise.

4.6 Juvenile ^{3}He

Because of its low natural abundance and, in normal practice, severe interferences in mass spectrometric analysis, in the early noble gas analyses of sea water ^{3}He was either ignored or used only as the spike in isotopic dilution. In the past decade, however, ^{3}He data have assumed a position among the most important in terrestrial noble gas geochemistry.

The custom for reporting He isotopic composition is to use a ratio anomaly δ^3He, usually in percent:

$$\delta^3 \text{He} = \left[\frac{(^3\text{He}/^4\text{He})_s}{(^3\text{He}/^4\text{He})_{air}} - 1 \right] \times 100\% \tag{4.8}$$

Here (^{3}He/^{4}He)$_s$ is the observed sample ratio. Note that the normalization is the air ratio, not an expected ratio. The distinction is significant for He dissolved in water, for which there is a perceptible isotopic fractionation (Weiss, 1970b), so that the null value, for solubility equilibrium with air, is δ^3He ≈ -1.4%.

Uncertainties in sea water analyses and solubilities, and in resolving effects that can lead to apparent saturation anomalies (Table 4.5) make it difficult to identify small contributions of juvenile radiogenic ^{4}He by way of saturation anomalies. None of these impediments affects the ^{3}He/^{4}He ratio, however, but the addition of radiogenic ^{4}He nearly devoid of ^{3}He will create a compensating negative ratio anomaly, δ^3He ≈ $-\Delta_E{}^4$He (for small variations), where Δ_E designates the excess (= nonatmospheric = juvenile) saturation anomaly.

In the expectation of using this effect to verify and more accurately quantify the presence of juvenile radiogenic ^{4}He, Clarke et al. (1969) measured ^{3}He/^{4}He ratios in South Pacific (Kermadec Trench) deep waters. Instead of negative anomalies, they

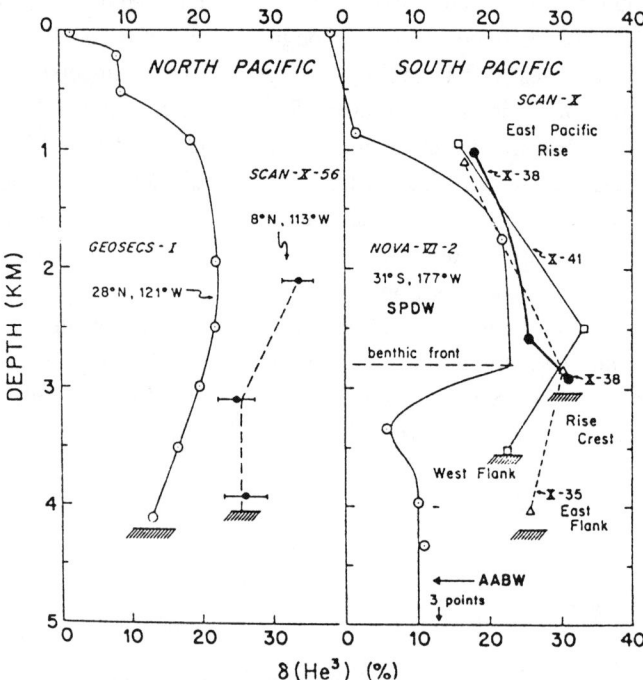

Figure 4.4 Excess ³He profiles at various locations in the Pacific Ocean. NOVA (Kermadec trench) data from Clarke et al. (1969), GEOSECS data from Clarke et al. (1970), SCAN data from Craig et al. (1975). SCAN station 38 (6°30′S, 107°24′W) is at the crest of the East Pacific Rise; nearby stations 35 and 41 are on opposite flanks. Reproduced from Craig et al. (1975).

found positive anomalies,* δ^3He averaging around +10% and ranging up to +22% (Figure 4.4). Thus the ³He saturation anomaly, Δ^3He, was found to be even larger than Δ^4He, and juvenile He to contain ³He as well as ⁴He, with ³He/⁴He greater than the air ratio. In the absence of alternative sources of ³He, Clarke et al. (1969) concluded that the ³He is primordial, and that the solid Earth is still degassing into the atmosphere volatiles it incorporated at its formation. Subsequent investigations have amply confirmed the existence of the effect, extended the range of observation, and supported this interpretation.

Since radiogenic He has a ³He/⁴He ratio less than the air value, the only serious candidate for an alternative to primordial ³He is production by decay of tritium (³H). Cosmic ray interactions in the atmosphere are a well-known source of ³He (Section 5.5), and some of this is channeled through tritium, which will enter surface water

---

* The report by Clarke et al. (1969) is an interesting example of the utility of the δ-value normalization. They identified *ratio anomalies* at the 2% level of precision, even though their corrected (for instrumental discrimination) *absolute* values were too low by 25%. Their uncorrected absolute ratios were much closer to presently accepted values.

and to some extent be entrained in deep water before its decay. Craig and Clarke (1970), however, have shown that this is only a minor perturbation, accounting for no more than about 2% of the primordial ^{3}He effect in deep waters. In surface waters (≤ 1 km), on the other hand, anthropogenic (nuclear weapons tests) tritium enrichments are large, and decay to ^{3}He can and does produce Δ^3He ratio anomalies up to a few percent. This situation can even be turned to advantage and be used as a chronometer for isolation (from air) on a timescale comparable to the 12-yr half-life of tritium (cf. Jenkins & Clarke, 1976).

The distribution of excess primordial ^{3}He in deep waters has been found to exhibit considerable structure in both vertical and horizontal directions, indicating that the sources of ^{3}He are not diffusely distributed about the ocean bottom but are concentrated in localized areas. The logical expectation is that the sources should be associated with accretional plate margins, where mantle material is brought to the surface. In the discovery report, Clarke et al. (1969) observed a mid-depth maximum in δ^3He and suggested its origin to be injection at the East Pacific Rise. Although this particular mid-depth maximum was later found to reflect a transition from one water mass to another and not association with the East Pacific Rise, the general idea was nevertheless found to be correct. Further work in the Pacific (see Figure 4.4) showed that δ^3He is indeed characteristically at a maximum at mid-depth rather than in bottom waters (Clarke, Beg & Craig, 1970), and can clearly be associated with the crest of the East Pacific Rise (Craig et al., 1975). The maximum anomaly found by Craig et al. (1975) in the vicinity of the Pacific-Cocos-Nazca triple junction is δ^3He = 33%, and since the same water exhibits a total saturation anomaly ΔHe (= Δ^4He) = 15%, the corresponding saturation anomaly for ^{3}He is Δ^3He = 55%. In a similar study, Lupton (1979) found higher (up to δ^3He = 68%, Δ^3He = 92%) and more localized anomalies in the Gulf of California.

In the Atlantic, deep water anomalies in ^{3}He were found to be considerably less than in the Pacific, in accord with the conclusions based on total saturation anomalies (Section 4.3), but nevertheless quite definitely present in a characteristic level δ^3He $\approx$ 5% (Jenkins et al., 1972). Albeit at a lower level than in the Pacific, the deep Atlantic ^{3}He excesses also show considerable structure; in a detailed study of the western Atlantic, Jenkins, and Clarke (1976) observed a maximum δ^3He of 13% and identified a localized source in the Gibbs Fracture Zone southwest of Iceland. To the south (at about 30°N), a section across the Mid-Atlantic Ridge shows no perceptible influence of the ridge itself on δ^3He (Lupton, 1976), a result in marked contrast to the comparable data for the East Pacific Rise (Figure 4.4).

Primordial ^{3}He concentrations are systematically lower in the Atlantic than in the Pacific (there are no data for the Indian Ocean). Part of this difference can be attributed to the more rapid flushing of Atlantic waters, and part, to the circumstance that deep waters entering the Pacific already have positive anomalies (from the Atlantic). Part of the difference is also evidently due simply to weaker sources in the Atlantic than in the Pacific, as would be expected on the basis of relative rates of crustal formation.

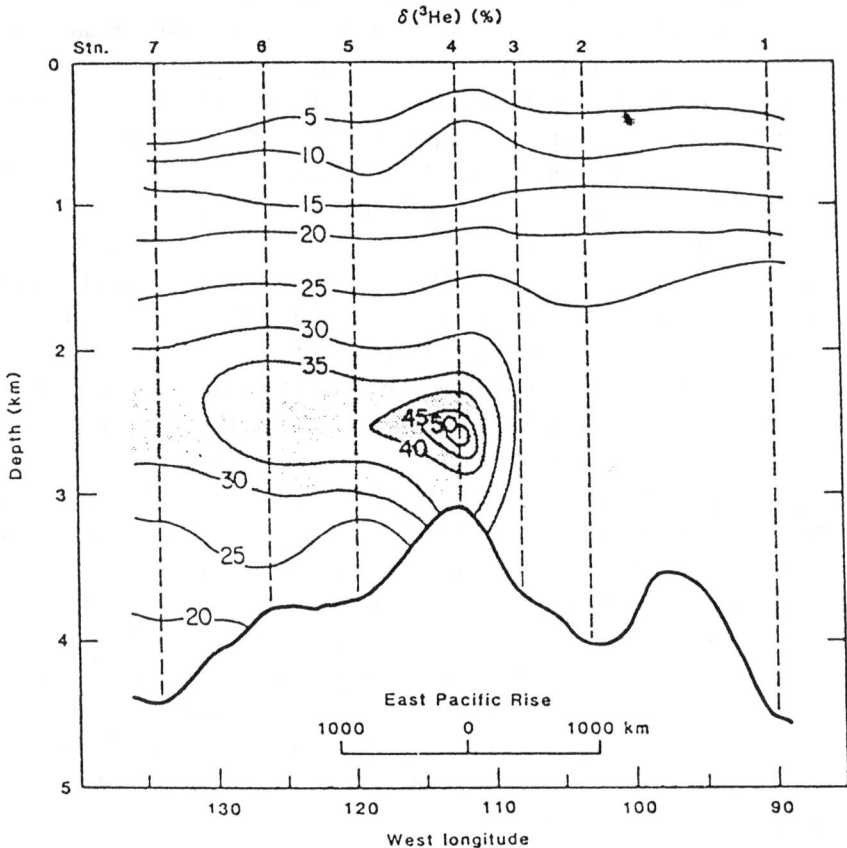

Figure 4.5 A section of ^{3}He concentrations in sea water over the East Pacific Rise at 15°S. The contour labels are ^{3}He ratio anomalies [Equation (4.7)] in percent. Reproduced from Lupton and Craig (1981).

Away from sources, excess ^{3}He will be conservative and so can be used for tracing movement and mixing of different water masses. Although experimental uncertainties are relatively high (about 1% in δ^3He), so too are the absolute magnitudes of the variations. Excess ^{3}He can thus be a useful adjunct to techniques employing temperature, salinity, or other dissolved species (cf. Jenkins & Clarke, 1976). A dramatic illustration of both the locational nature of juvenile ^{3}He sources and its application to tracing of water mass movements is shown in Figure 4.5 (Lupton & Craig, 1981).

An extremely fine localization of primordial ^{3}He injection, on a 10-m scale, has also been observed. It has been suggested that lower-than-expected conductive heat flow at oceanic ridges could be due to significant heat transport by hydrothermal circulation (e.g., Talwani, Windisch & Langseth, 1971), in which recently emplaced hot rock drives convection of local sea water. On the basis of temperature-salinity relationships, Weiss et al. (1977) made the first identification of hydrothermal circulation in the open ocean, observing several "plumes" (temperature differential ≤0.2°C)

4.6 Juvenile ^{3}He

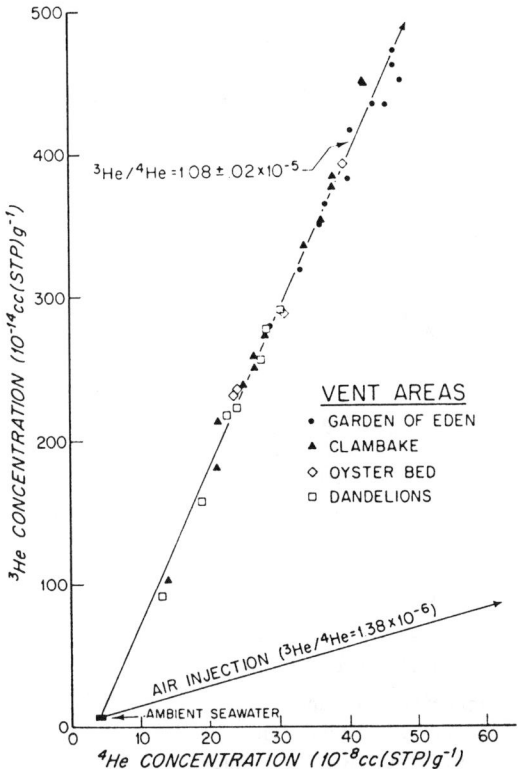

Figure 4.6 Large excesses of juvenile He in hydrothermal waters sampled at specific vent fields in the Galapagos Rift. Note ambient sea water concentration near origin (cf. Table 4.6). Reproduced from Jenkins et al. (1978).

above axial fissures in the Galapagos Rift (but not at a nearby East Pacific Rise site). Lupton, Weiss & Craig (1977b) analyzed water sampled in the plumes and found ^{3}He (and Rn) excesses (up to δ^3He = 99%) relative to background water already rich in ^{3}He (δ^3He = 30%). The primordial identification of ^{3}He provides strong support for the thesis that these plumes are indeed associated with recent emplacement of materials from the upper mantle.

More detailed examination and sampling allows association of hydrothermal circulation with specific vent fields. In such waters samples in the Galapagos Rift by the Alvin deep submersible, Jenkins, Edmond, and Corliss (1978) report juvenile He enrichments which dwarf the normal saturation concentrations by factors up to 11 for ^{4}He and 60 for ^{3}He (Figure 4.6). A particularly significant feature of this report is that added He occurs roughly in proportion to added heat (ΔT up to 12°C in sampled water), corresponding to about 7.6×10^{-8} cal/atom of ^{3}He (Figure 4.7). Jenkins et al. note that if this value is representative, hydrothermal circulation may indeed account for the depression of conductive heat flow relative to models for total heat flux. As

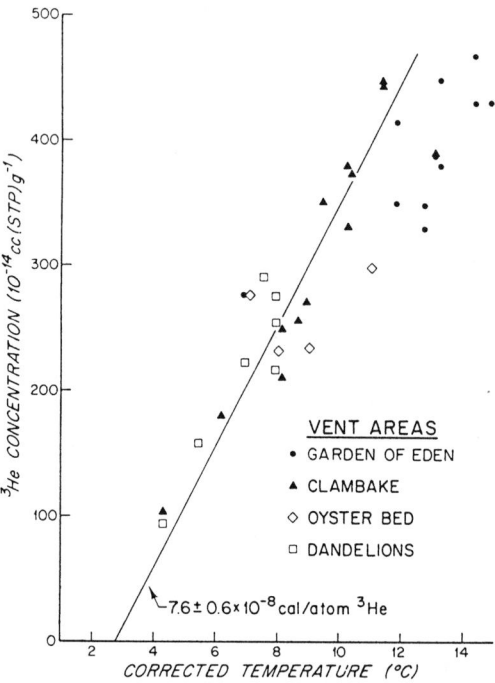

Figure 4.7 Relationship between ^{3}He and water temperature in hydrothermal waters sampled in specific vent fields in the Galapagos Rift (cf. Figure 4.5). The correlation suggests that both heat and juvenile He are brought to the surface by the same volcanic process, in the ratio 7.6×10^{-8} cal/atom (of ^{3}He). Reproduced from Jenkins et al. (1978).

is expected, however, more recent studies revealed that the relationship between heat and He-flux shows considerable temporal and spacial variations, corresponding to the volcano-tectonic stage of evolution of vents (e.g., Hammond et al., 1991).

An extreme example of hydrothermal circulation at an accretional plate margin (which was known first and served as an analog for the apparently more general case) occurs in the Red Sea (Craig, 1966, 1969). There the circulating water passes through evaporites, becoming brine ($S \approx 250$‰) and is thus stabilized against convective mixing and dilution in spite of being hot (around 50°C). The brines show striking enrichments of a number of elements, including He (Table 4.6): about 380 times saturation for ^{4}He, and a spectacular enrichment of ^{3}He, up to 3400 times saturation (Lupton et al., 1977a). While the other enriched elements evidently come from the evaporites, the He must originate in the ridge basalts, the only plausible source for such a high ^{3}He/^{4}He ratio.

The composition of juvenile He is obviously a parameter of considerable interest. In normal waters, the limitation on the calculation is identification of the excess ^{4}He, since the fractional ^{4}He excess is smaller than that of ^{3}He, and must be identified by an absolute concentration excess which must be resolved from larger effects due to air injection and temperature/pressure variations. Combining their ^{3}He data with the He, Ne, and Ar data of Bieri and Koide (1972), Craig et al. (1975) adopted, as characterizing the waters in the Pacific-Cocos-Nazca triple junction region, ΔHe = 12%, Δ_EHe = (3.5 ± 0.5)%, and δ^3He = (32.0 ± 0.5)%, which leads to ^{3}He/^{4}He = (16 ± 3) × 10^{-6} for the juvenile He, 11 times the air ratio. The equivalent calculations can be made only with large uncertainties for the other Pacific or the Atlantic locations, where the anomalies are smaller.

Identification of juvenile He composition in the more localized and more enriched waters is readily made. Compositions are apparently only narrowly variable at each location. Added ^{3}He/^{4}He is about 10 × 10^{-6} (cf. Figure 4.6) in both Galapagos Rift data sets and also for the Gulf of California (Lupton et al., 1977b; Jenkins et al., 1978; Lupton, 1979). The preceding Craig et al. (1975) value, which presumably provides a more broadly based average, is higher, but in view of its relatively large uncertainty can be imagined consistent with this composition. The Red Sea value (Lupton et al., 1977a) is 12 × 10^{-16}, significantly higher. It is interesting to compare these compositions to juvenile He in the rocks from which the water presumably extracts its excess He. The typical rock value is ^{3}He/^{4}He ≈ 13 × 10^{-6} (Section 6.7), quite in accord with the East Pacific Rise and Red Sea compositions. The "mantle component" He observed in oceanic rocks is perhaps not as uniform as was once thought (e.g. Farley & Nevoda, 1998), however, so pending further data it is probably possible to reconcile the lower Galapagos and Gulf of California water ratios with local rocks without invoking an alternative or additional source. (Lupton et al., 1977b, noted that the local rocks are too young for their in situ radiogenic He to significantly depress their inherited ^{3}He/^{4}He ratios.)

It is of considerable interest to determine the flux of juvenile He from the oceans into the atmosphere. This is a hazardous undertaking, since it requires making assumptions about the excess He distribution in the whole world ocean, mostly unsurveyed in this respect, and combining with not very well constrained estimates for the rate at which it is transported from deep water to the surface. Craig et al. (1975) calculate average fluxes of 3×10^5 atoms cm^{-2} s^{-1} of ^{4}He and 4 atoms cm^{-2} s^{-1} of ^{3}He for the oceanic input of juvenile He into the air (the cm^2 normalization is for the whole Earth, not just the oceans), estimating a likely 50% uncertainty. Uncertain though they are, these fluxes are important because they are the only globally significant volatile fluxes from the solid earth that can be calculated on the basis of actual observations. Their role in budget and inventory calculations is considered in Chapter 6.

4.7 Paleotemperature: Noble Gas Record

As seen in Figure 4.1, the noble gas solubility in water shows considerable variation with temperature. Therefore, noble gas contents in groundwater, which was in solubility equilibrium with the ambient, can be used to estimate the atmospheric temperature. On the basis of this temperature dependency of noble gas solubility in water, much work has been made to estimate paleotemperature or to obtain paleoclimate record since the pioneering work by Mazor (1972).

Reconstruction of faunal records in deep-sea sediments and oxygen isotope measurements have been yielding useful information of the ocean's paleotemperature. However, the lack of these tools in the case of continents has hampered the estimation of paleotemperature in continents. The latter estimate has mainly been made on somehow indirect approach such as pollen data, periglacial feature, and soil carbonate, all of which suffer from considerable uncertainty stemming from the assumption that must be made to convert the observation to temperature. The advantage of the noble gas thermometer, as compared to the other paleotemperature methods, is that it is based on the relatively simple physical principle that directly relates noble gas concentration to the ambient temperature (see Stute & Schlosser, 1993, for a recent review of the noble gas paleotemperature method). However, in actual practice, we need some cautions.

Among various cautions that must be exercised in applying the noble gas paleotemperature method, the following two points are particularly critical. These are (1) correction for excess air in water and (2) assessment of the preservation of noble gas record in groundwater. Noble gas concentrations of groundwater usually exceed those calculated for thermodynamic solubility equilibrium with air. This is due to the "excess air." The excess air in groundwater is attributed either to air contamination at the time of sampling or, more commonly, to the trapping of small pockets of air in the fractures, fissures, or intergranular pores in surrounding rocks (Andrew, 1991). Wilson and McNeill (1997) showed that in some favorable case air contamination may be identified by the anomalously high excess Ne content compared to duplicate samples.

After groundwater, which had been equilibrated with atmosphere and hence impregnated with paleotemperature record, leaves the water table (recharge area), the groundwater is likely to be isolated from the atmosphere and migrates to the discharge area. During this migration process, mixing (macrodispersion) and leakage (microdispersion) between aquifers would smooth the impregnated paleotemperature record. Therefore, to derive useful paleotemperature information from aquifers sampled in discharged area, we must assess these dispersion effects. Stute and Schlosser (1993) showed that the smoothing of the paleotemperature signal is a function of the relative importance of dispersion and advection in the aquifer, namely D/v^2, where D is the dispersion coefficient and v is the flow velocity. The smoothing effect then can be discussed in terms of a characteristic time constant τ after which the amplitude of an original sinusoidal climate oscillation (frequency ω) is reduced to $1/e$, that is, $\tau = v^2/(D\omega^2)$. For a large-scale (of the order of 50 km), confined aquifer characterized by a typical flow velocity of $v = 1$ m/yr, D/v^2 will be 100 years. Hence, the amplitude of the shortest Milankovitch cycle, characterized by a period of 23,000 years, would be reduced to $1/e$ within 134,000 years, and the paleoclimatic record could be resolved in the aquifer system (Stute & Schlosser, 1993). Several

4.7 Paleotemperature: Noble Gas Record

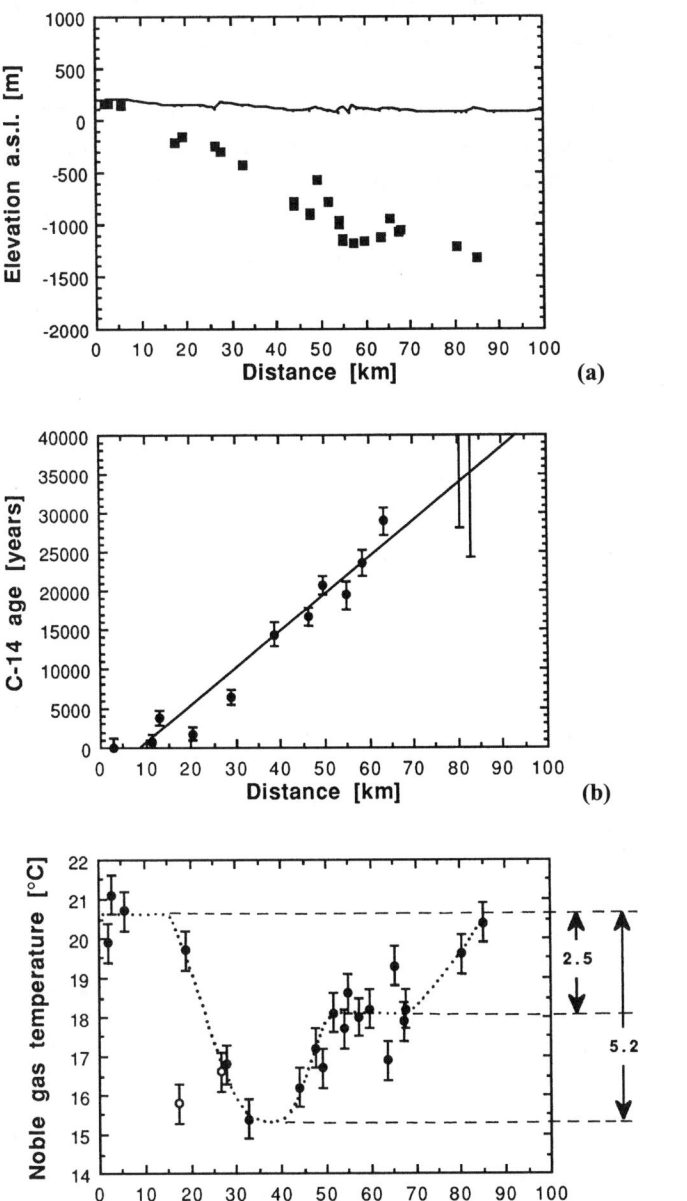

Figure 4.8 (a) Vertical section along the flow line (ragged horizontal line shows the surface topography). The horizontal scale (km) is the distance from the recharge area. The water sampling sites are shown by square. (b) ^{14}C ages of the waters as a function of distance from the recharge area. (c) Noble gas temperature as a function of the distance. Note that the large drop in the temperature around 30–40 km corresponds to the last maximum glacier period. After Stute et al. (1992).

successful applications of the noble gas paleotemperature method to aquifers are reported in the literature (e.g., Stute et al., 1995; Pinti, Marty & Andrew, 1997; Beyerle et al., 1998).

Figure 4.8 shows an example of the estimation of paleotemperature based on noble gas measurements in the groundwater of some aquifers in Texas (Stute et al., 1992). Paleotemperatures estimated from the noble gas concentration shows a large drop (about 5.2°C) at the distance 30 km away from the recharge area [Figure 4.8(c)], which corresponds to an age between 12,000 and 17,000 years, close to the last glacial maximum (about 18,000 years ago).

Chapter 5

Crust

5.1 Introduction

Except for radiogenic components, the amount of noble gases in the solid crust is safely assumed to be insignificant in terrestrial noble gas inventory. In general, except for in situ derived radiogenic and nucleogenic noble gases, trapped noble gases in crustal rocks are of an atmospheric origin. Because of the higher concentration of U, Th in the crust than in the mantle, however, nucleogenic Ne isotopes are often discernible in crustal materials. Near the Earth's surface, down to a few meters in depth, cosmic ray-induced cosmogenic Ne and He isotopes can also be observed. In deep-ocean bottom where sedimentation rate is extremely slow, extraterrestrial noble gases carried by cosmic dusts become conspicuous. In this chapter, we will discuss these unique features of noble gases in the crust.

5.2 Sediments

Noble gases in sediments once attracted much interest of noble gas geochemists for the following reasons. First, sediments could be a major reservoir for Xe and may account for the "missing Xe," a long-standing puzzle in noble gas geochemistry (cf. Section 7.5). Because of the fine particle size, sedimenting particles would very effectively adsorb noble gases, especially the heaviest noble gas Xe, during sedimentation. Adsorption of atmospheric noble gases would also be substantial after emplacing sediments on the surface. It is also conceivable that they may have trapped a considerable amount of noble gases dissolved in water in the rock fabrics. All these theories suggest that sediments are a potential reservoir of Xe. Second, noble gases trapped

Table 5.1. *Noble gases in sediments*

Sample	3He (10^{-12} cm^3 g^{-1})	20Ne (10^{-9} cm^3 g^{-1})	36Ar (10^{-9} cm^3 g^{-1})	84Kr (10^{-9} cm^3 g^{-1})	130Xe (10^{-9} cm^3 g^{-1})
Subaerial sediment	n.a.	20	60	8	3
Deep-sea sediment	10	3	20	0.4	0.02
Air-saturated sea water (25°C)	0.052	133	759	27.3	0.26

Noble gas amounts are median values calculated from the data in Figure 5.1(a)–(e). Reference sources for the data are given therein.

in the rock fabrics may be related to the contemporary atmospheric noble gases dissolved in the water. Hence, one may hope that the study of the trapped noble gases in sedimentary rocks would yield useful information about the noble gas state in the paleoatmosphere (e.g., Turner, 1988).

However, noble gas contents in sediments so far measured all indicated that the amounts were too small to account for the missing xenon (cf. Section 7.5). Attempts to seek for paleoatmospheric noble gas in sediments has so far been focused on Ar isotopic composition, since ^{40}Ar/^{36}Ar would show the most significant variation between the modern atmosphere and paleoatmosphere. However, these attempts have not been very successful because of difficulty in correcting for in situ radiogenic ^{40}Ar to derive indigenous trapped Ar, that is, the paleoatmospheric Ar. Hence, noble gas geochemists' interest in sediments has somehow waned and relatively little work has been done in the last decade. Lately, interest in noble gases in sediments has revived partly due to the finding of extraterrestrial noble gases which were implanted in cosmic dusts and deposited in deep-ocean bottom. This will be discussed in the next section.

We summarize noble gas amounts in deep-sea and subaerial sediments in Figure 5.1. From the data displayed here, we calculated median values which are shown in Table 5.1. Both Figure 5.1 and Table 5.1 show that even though there is little difference in the lighter noble gas concentration between subaerial and deep-sea sediments (He, Ne, and Ar), heavier noble gases are much more abundant in subaerial sediments than in deep-sea sediments. As in volcanic rocks (cf. Section 6.6), most sediments, either deep-sea or subaerial, show fractionation toward the heavier ones relative to air noble gas, although the mechanism for the fractionation may be different. Figure 5.2 shows noble a gas elemental abundance pattern relative to the air abundance; subaerial sediments show much more severe fractionation.

Although it seems clear that in general the trapped noble gases in sedimentary rocks are atmospheric, it is by no means clear by what mechanisms the gases were trapped. An obvious candidate mechanism is solution. Noble gas may achieve solubility equilibrium during growth of authegenic minerals and become trapped by further growth. The most relevant solubility data are for serpentine (Zaikowski & Schaffer, 1979), which are also shown in Figure 5.2. Serpentine solubility is indeed

5.2 Sediments

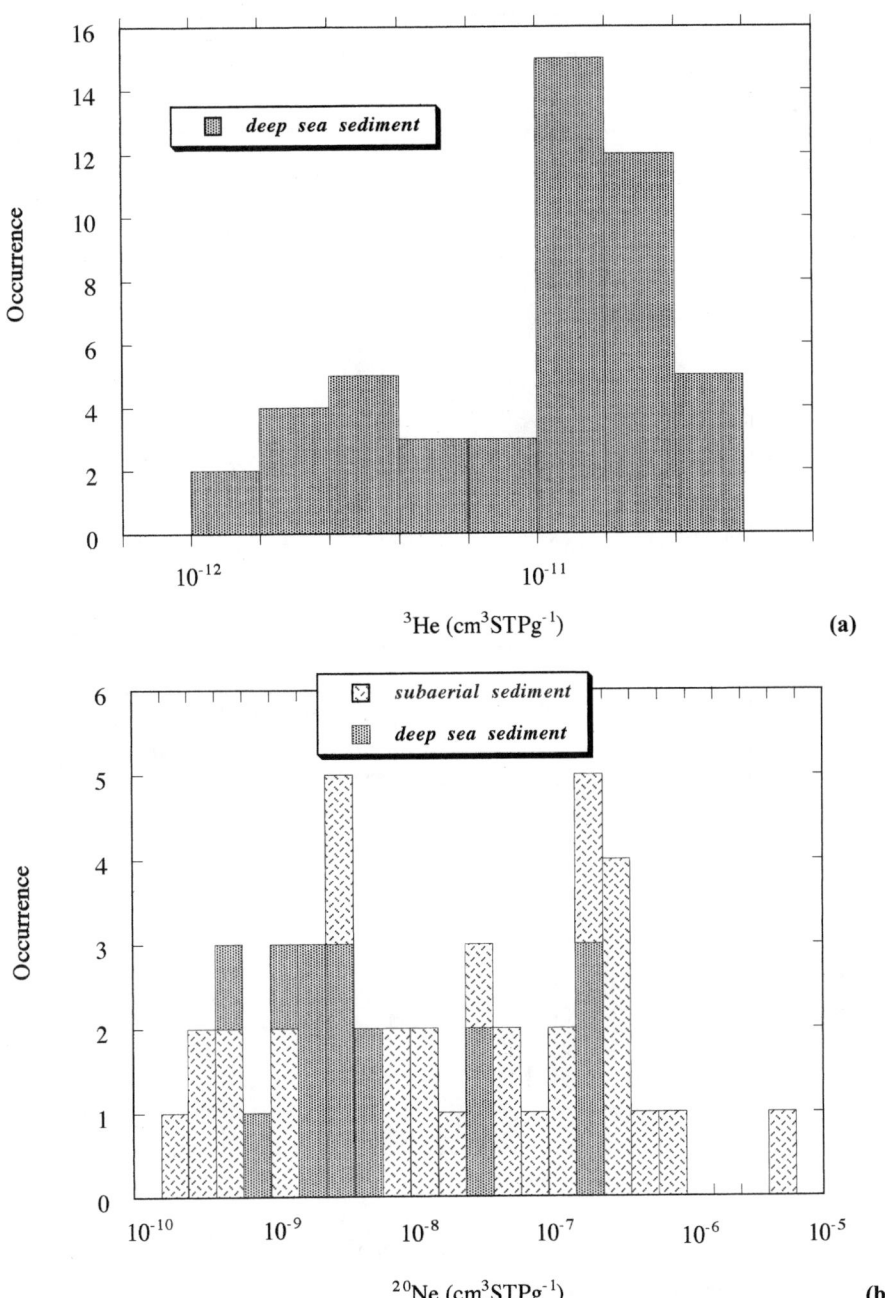

Figure 5.1 Noble gas elemental abundance in deep-sea sediments (dark shade) and in subaerial sediments (light shade). The vertical axis indicates the number of samples analyzed. Subaerial sediments: Bogard et al. (1965), Canalas, Alexander, and Manuel (1968), Phinney (1972), Frick and Chang (1977), Kuroda and Sherill (1977), Rison (1980a), Podosek, Honda, and Ozima (1980), and deep-sea sediments: Takayanagi and Ozima (1987). (a) ^{3}He concentration in sediments. (b) ^{20}Ne concentration in sediments. (c) ^{36}He concentration in sediments. (d) ^{84}He concentration in sediments. (e) ^{130}He concentration in sediments.

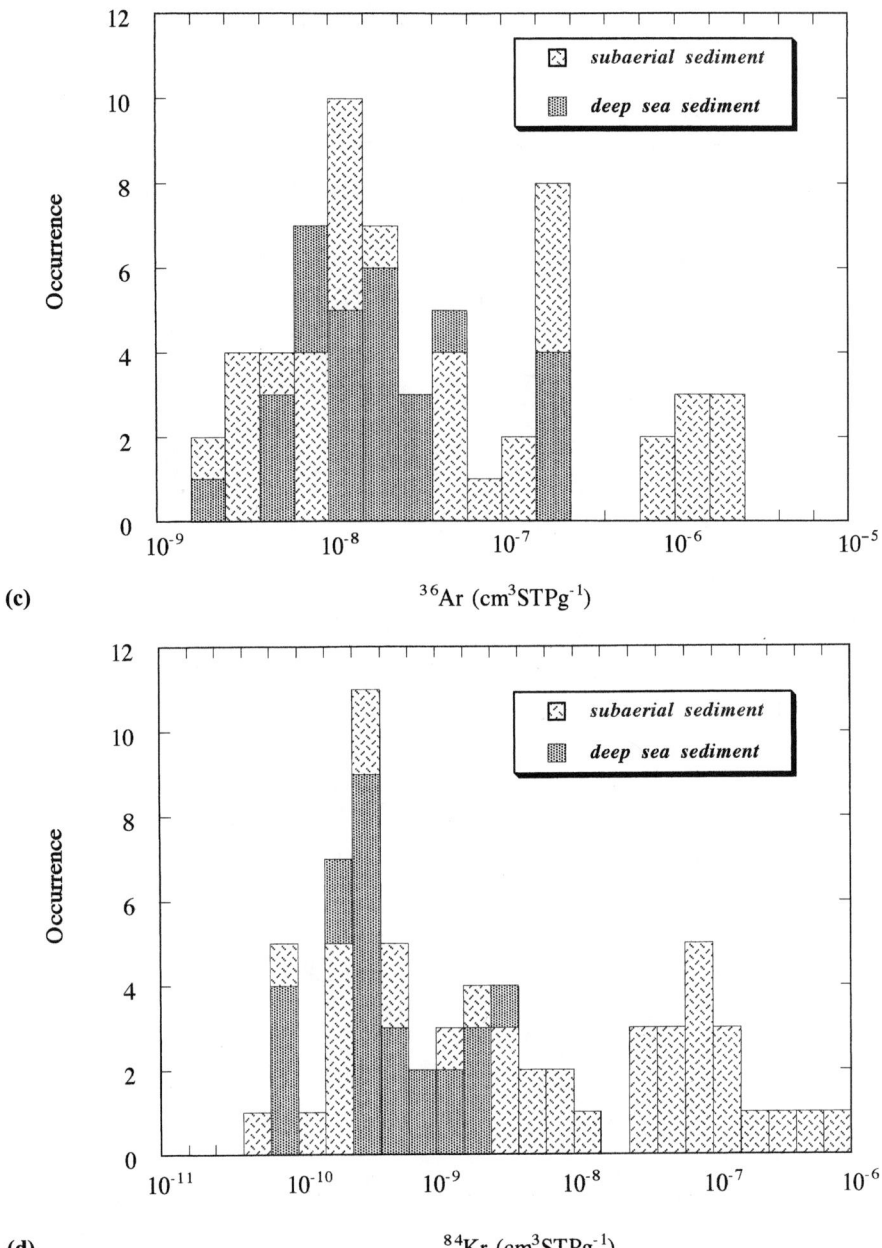

Figure 5.1 (*continued*)

5.2 Sediments

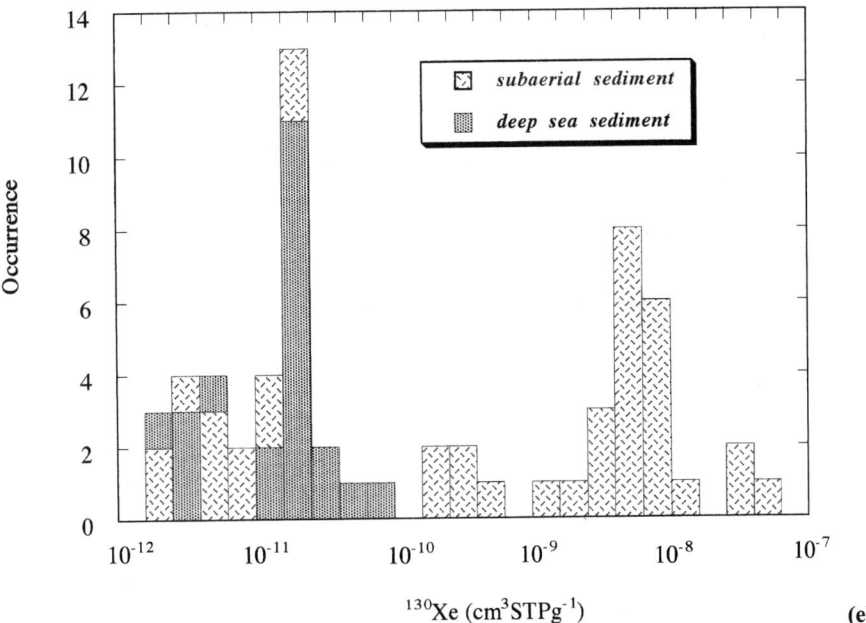

Figure 5.1 (*continued*)

relatively high (compared to igneous minerals) and fractionation in favor of the heavy gases. If this relation is representative of sedimentary minerals, then solubility may be important in cases of relatively low abundances and shallow fractionation patterns. By the same token, solubility appears inadequate to account for cases of much higher abundances and much more steeply fractionated compositions. Note, however, that the serpentine data are for 340°C, and that if the heats of solution are negative, the solubilities would be higher and probably more steeply fractionated at surface temperatures. Solubility seems a potentially viable mechanism for producing some of the trapped gases in sedimentary rocks, but there is very little basis for quantitative evaluation, and the role of solubility will remain essentially hypothetical until additional relevant solubility data are available.

For fractionation patterns progressively favoring the heavier gases, Ozima and Alexander (1976) suggested an association with water, in which dissolved gases also follow such a pattern. It is plausible that most authegenic sedimentary minerals form in water, in which case the immediate source of the gases is water, and it may also be that hydrous minerals are more efficient than other minerals in noble gas trapping. The solubility is important, if water and dissolved gases were assimilated into the rocks rather than exchanged with some larger reservoir. Assimilation may account for the cases of little fractionation such as those observed in deep-sea sediment (Figure 5.2), but the gases dissolved in water are considerably less fractionated than most of the trapped gases in subaerial sediments (Figure 5.2). The latter fact seems to suggest a different mechanism of noble gas trapping between deep-sea sediment and subaerial sediment.

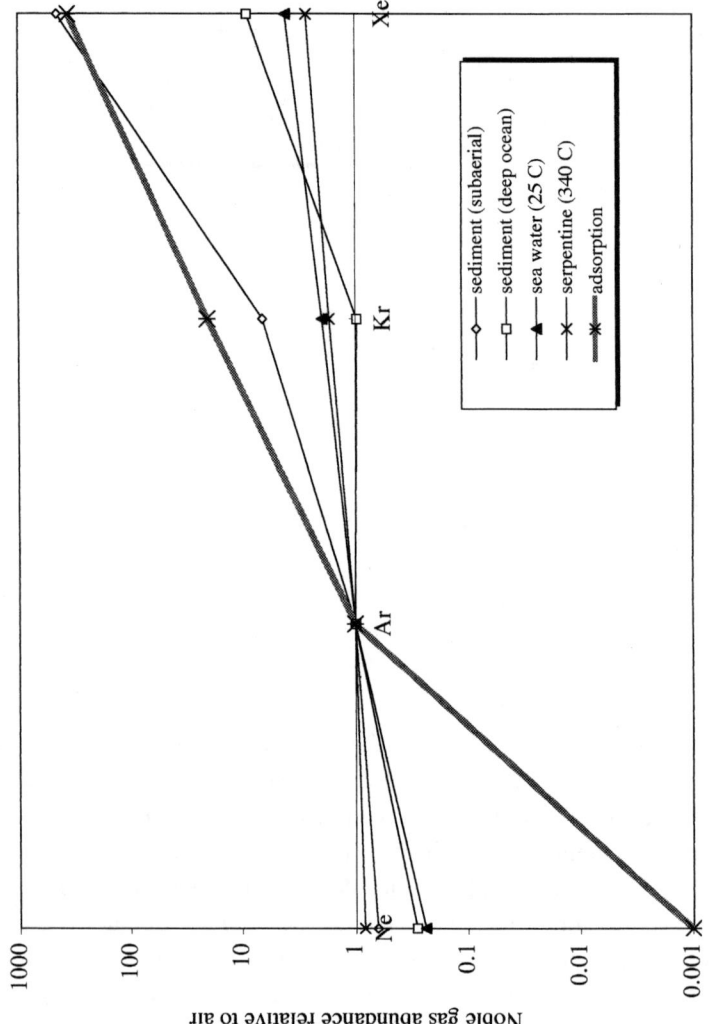

Figure 5.2 Noble gas abundance (normalized to ^{36}Ar) relative to air abundance. Data are from Table 5.1. The noble gas adsorption curve on pulverized Allende meteorite (Fanale & Cannon, 1972) is also shown for comparison.

The other proposed mechanism is adsorption (e.g., Fanale & Cannon, 1971; Podosek et al., 1981b), which has the immediate virtues of producing very high concentrations and very strong fractionation at typical surface temperatures. The strong fractionation in adsorption closely approximates the pattern observed in subaerial sediment (Fanale & Cannon, 1971). However, a hypothesis for adsorption as a trapping mechanism is incomplete in that adsorbed gas is not really trapped, and that an auxiliary fixation mechanism is also needed. This is not so much a problem. One possibility is that adsorbed gas could be trapped when covered over by grain growth, as also suggested for solution. As an example of such, Matsumoto et al. (1996) reported the case of noble gas trapping in magnetite. They synthesized magnetite by oxidizing Fe with water vapor in a noble gas atmosphere. They observed a negative correlation between the concentration of argon trapped in the magnetite and its synthetic temperature and suggested that argon was first trapped by adsorption and further fixed into the magnetite crystalline structure during crystal growth.

Questions of mechanism aside, Honda et al. (1979) noted that so simple a process as static compression can act as fixation agent, as demonstrated by an upward shift in characteristic degassing temperature. Podosek et al. (1981b) suggested that at least in some cases no specific trapping mechanism at all is needed: Gases in interior spaces of a sample must execute a random walk, making many collisions on surfaces, before escaping, and adsorption on these surfaces will slow down the process. Podosek et al. noted that even though the sticking time in an individual adsorption event is short, the characteristic time for removal of adsorbed gas could be long on a laboratory timescale if the total path length for escape is long (a few to several millimeters) and the individual step length in the random walk is short, 10–20 Å (a characteristic distance for interlaminar spacing in clays).

5.3 Cosmic Dusts

Merrihue (1964) first noted that magnetic separates from deep-sea red clays had enormously high ^{3}He/^{4}He ratio of more than 10^{-4}. He concluded that the high ^{3}He/^{4}He ratio is due to extraterrestrial debris that were implanted with solar wind helium (^{3}He/^{4}He $\approx 4 \times 10^{-4}$). He further suggested that the debris in the sediments may amount to 1%. Krylov et al. (1973) compiled ^{3}He/^{4}He isotopic data on ocean sediments measured in the former Soviet Union and found that the ^{3}He/^{4}He ratio is generally higher in pelagic sediments where the sedimentation rate is lower. They suggested that the high ^{3}He/^{4}He ratio is attributable to extraterrestrial materials that were more concentrated in the slowly accumulating ocean floor. Ozima et al. (1984) confirmed the inverse correlation between the ^{3}He concentration and the sedimentation rate for a large number of deep-sea sediment samples obtained from the Pacific and Atlantic ocean floors. They also observed a correlation between the ^{3}He content and the ^{3}He/^{4}He ratio for the samples (Figure 5.3) and concluded that the correlation

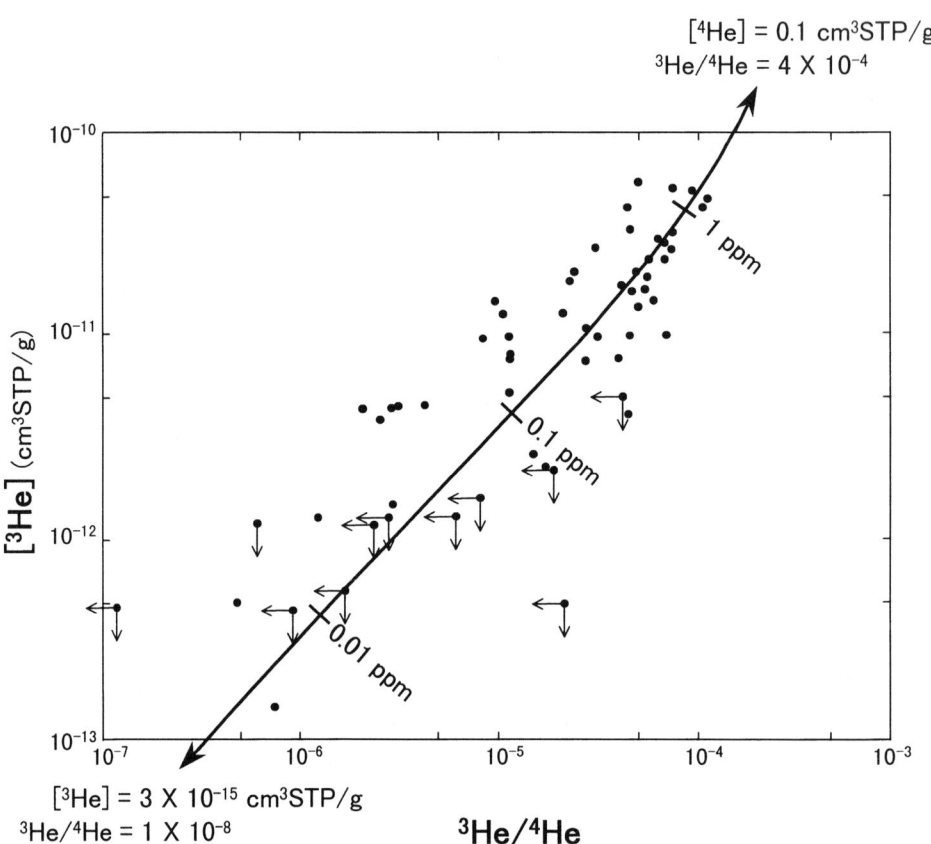

Figure 5.3 ³He contents in sediments are plotted against ³He/⁴He ratios. The curve can be interpreted to be a mixing curve between helium in cosmic dusts (⁴He = 0.1 cm³ STP/g, ³He/⁴He = 4 × 10⁻⁴) and that in sediment (³He = 3 × 10⁻¹⁵ cm³ STP/g, ³He/⁴He = 10⁻⁸). Numbers on the curve indicate the fraction (ppm) of the cosmic dust in sediments. Redrawm from Takayanagi & Ozima (1984).

can be reasonably explained as mixing between radiogenic helium produced in the sediments (³He/⁴He ≡ 1 × 10⁻⁸, ⁴He ≡ 3 × 10⁻⁷ cm³ g⁻¹) and solar He implanted in cosmic dusts (³He/⁴He ≡ 4 × 10⁻⁴, ⁴He ≡ 0.1 cm³ g⁻¹). In constructing the mixing curve, they chose the isotopic ratios representative of the solar helium and the typical radiogenic helium in the crust as the end members. The concentration of the extraterrestrial He component was then determined to make the best fit of the mixing curve to the observed data. Thus chosen, helium concentrations agree with those observed for cosmic dusts (e.g., Rajan et al., 1977) and deep-sea sediments (e.g., Takayanagi & Ozima, 1987), thereby giving a support for the mixing hypothesis.

If we assume that a fallout rate of cosmic dusts has been constant and ³He in sediments is entirely due to cosmic dusts, a ³He concentration in sediment can be expressed as a function of sedimentation rate (r) and of a cosmic dust fall-out flux (F),

$$(^{3}He)_{sed} = \alpha F/r\rho$$

5.3 Cosmic Dusts

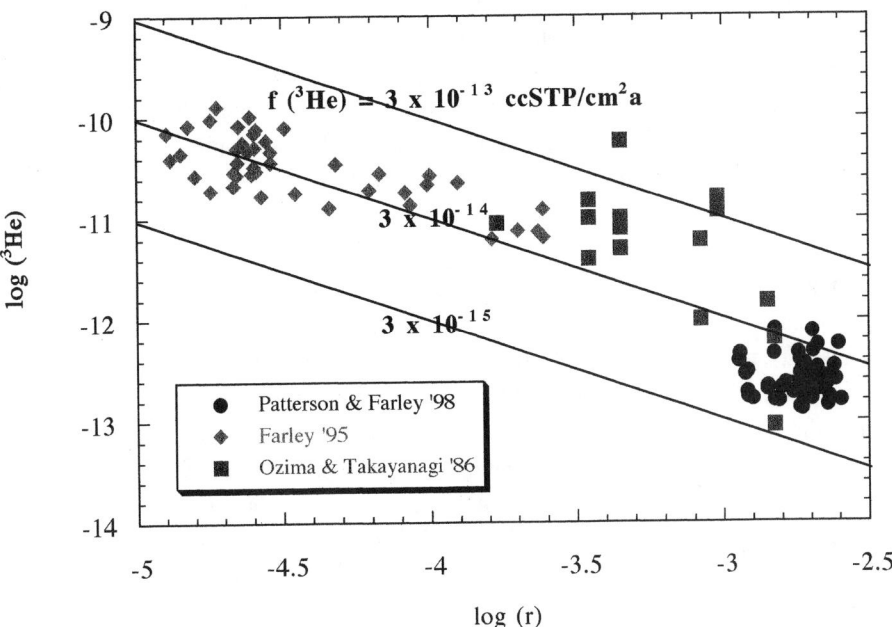

Figure 5.4 ^{3}He contents in sediments are plotted against sedimentation rates (r) of sediment samples in a log-log scale. Reference lines correspond to a constant fall-out rate of extraterrestrial ^{3}He [see Equation (5.1)]. Sedimentation rate for Farley '95 and Farley & Patterson '98 were calculated from data given in the reference by Farley (1995) and Patterson & Farley 1998.

where α and ρ indicate a concentration of ^{3}He in cosmic dust and density of sediment. Denoting a ^{3}He fall-out rate by $f(^3\text{He})$ which equals to αF, we have

$$(^3\text{He})_{sed} = f(^3\text{He})/r\rho \tag{5.1}$$

Therefore, if the ^{3}He fall-out rate has remained constant, ^{3}He concentration in sediment is inversely proportional to the sedimentation rate. In Figure 5.4 we plotted all the available published data in a log-log diagram in which the data cover a time span of about 60 Ma. The data roughly form a linear trend, suggesting that the assumption of a constant rate of ^{3}He fall out approximately holds. Reference lines in Figure 5.4 correspond to various values of $f(^3\text{He})/\rho$. From the plot, we infer that the ^{3}He fall-out rate roughly falls in a range of $3 \times 10^{-14} \sim 3 \times 10^{-15}$ cm^3 STP/cm$^2 \cdot$ a, or a global rate of $1.5 \times 10^4 \sim 1.5 \times 10^5$ cm^3 STP/a with an assumed value of $\rho = 3$ g/cm^3. This estimated ^{3}He flux corresponds to a mean cosmic dust flux of about 400–4000 tons/a, if we assume 0.1×10^{-1} cm^3 STP g^{-1} for the ^{4}He content (^{3}He/^{4}He $= 4 \times 10^{-4}$) in cosmic dust (Rajan et al., 1977). On the basis of an equation similar to Equation (5.1), Farley (1995) estimated a mean ^{3}He flux of 0.6×10^{-12} cm^3 STP cm^{-2} ka^{-1} from his data on Cenozoic pelagic sediment core from the Pacific Ocean. These estimated fall-out rates may favorably be compared with the values estimated from the satellite, radio and visual observations of about a few hundred tons per year for particle size below 10 μm in diameter (Hughes, 1978).

Here, it would be worth emphasizing that the preceding estimate of ^{3}He fall-out rate was made only from measurements of ^{3}He content in sediment samples with known sedimentation rate, but did not require knowledge of a cosmic dust fall-out rate or ^{3}He content in cosmic dusts. Moreover, the plot (Figure 5.4) enables us to make an objective appraisal for an averaging value of fall-out rate as well as for the underlying assumption on a constant fall-out rate.

Esser and Turekian (1988) estimated an accretion rate of extraterrestrial particles in ocean bottom and in varved glacial lake deposit on the basis of osmium isotope systematics and concluded a maximum accretion rate of between 4.9×10^4 and 5.6×10^4 tons/a. The discrepancy between this estimate and those derived from helium can easily be attributed to the difference in the size of the cosmic dust particles under consideration. Cosmic dusts of greater than a few ten micrometers may not be important in the helium inventory of sediments because the larger grains are likely to lose helium due to atmospheric impact heating (e.g., Brownlee, 1985). Stuart et al. (1999) concluded from studies on Antarctic micrometeorites that 50- to 100-μm micrometeorites may contribute about 5% of the total flux of extraterrestrial ^{3}He to terrestrial sediments. Therefore, the helium-based estimate deals only with these smaller particles.

It would also be interesting to note that sedimentation rate is one of the most important quantities in sedimentology. A conventional method to estimate a sedimentation rate is to measure ages at both ends of a sediment core. If we assume a uniform sedimentation rate, we can estimate a sedimentation rate by dividing the length of the core by the time duration of the core. However, it is usually difficult to ensure a uniform sedimentation rate; hence, the estimated sedimentation rate often involves considerable uncertainty. In contrast to this conventional method, the preceding "^{3}He method" (cf. Figure 5.4) would offer a straightforward way to estimate an absolute sedimentation rate, which can be obtained only from a laboratory measurement of ^{3}He content in a single sediment specimen [see Equation (5.1)] provided that the assumption of a constant fall-out rate holds.

An integration of an assumed ^{3}He fall-out rate of 10^5 cm^3 STP/a for 4.5 Ga gives a total ^{3}He input of 4.5×10^{14} cm^3 STP into the Earth. The integrated ^{3}He amount is about one ten thousandth of a minimum estimate of the primordial ^{3}He in the Earth, which is estimated from a total amount of the atmospheric ^{22}Ne with an assumption of the chondritic ^{3}He/^{22}Ne ratio (= 0.4). Hence, if the present fall-out rate has remained nearly the same throughout the history of the Earth, the extraterrestrial ^{3}He would not bear any significance in the Earth's ^{3}He inventory.

Contrary to the preceding simple inventory argument, Anderson (1993) and Allègre, Sarda, and Staudacher (1993) argued that the fall-out rate of cosmic dusts may have been much higher in the past, and that the cosmic dust-implanted ^{3}He (Anderson) and Ne (Allègre et al.) that were subducted into the mantle could be significant in the noble gas inventory in the Earth's interior. In support of this argument, Allègre et al. called attention to the much higher fall-out rate of cosmic material in the ancient lunar surface as suggested from the lunar crater systematics. They also

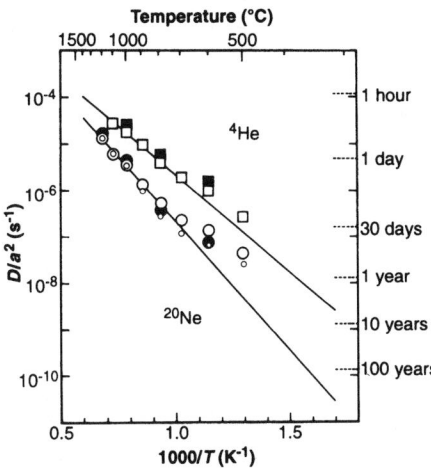

Figure 5.5 Arrhenius plots for diffusion of solar ^{4}He and ^{20}Ne in a magnetic separate from a Pacific Ocean sediment. The results of two duplicate samples (solid and open symbols) are shown. For ^{20}Ne, both the data corrected (large open circles) for the atmospheric component, i.e., excess ^{20}Ne and the data uncorrected (small open circles) are shown. The times required for 99% gas release are also shown for corresponding D/a^2 values. After Hiyagon (1993).

pointed out that the outgassing rate of ^{20}Ne from the Earth appears to be balanced by the estimated ^{20}Ne input rate by cosmic dusts, in that the outgassing rate of ^{20}Ne was estimated from the degassing rate of ^{3}He (cf., Section 6.12) and the assumed mantle ^{3}He/^{20}Ne ratio of 0.77 (Staudacher et al., 1989). Both sets of authors assumed much higher concentrations – by a few orders of magnitude – for the implanted ^{3}He (Anderson) and Ne (Allègre et al.) in the cosmic dusts than those generally observed in magnetic separates from ocean sediments (see Table 5.2).

Before concluding the significance of the noble gas input by cosmic dusts in Earth's interior, however, it is important to examine whether the noble gases in cosmic dusts were retained in the grains during subduction. Diffusion experiments on He and Ne from the magnetic separates in ocean sediments (Fukumoto, Nagao & Matsuda, 1986; Amari & Ozima, 1988; Hiyagon, 1993) seem to indicate that the diffusion coefficients are too large to allow the grains to retain any meaningful amount of He and Ne during subduction. Figure 5.5 shows an Arrhenius plot for diffusion of solar ^{4}He and ^{20}Ne in magnetic separates from a Pacific Ocean sediment (Hiyagon, 1993).

The slopes of the linear correlation then gives activation energies $E = 80 \pm 4$ and 106 ± 4 kJ/mol for ^{4}He and ^{20}Ne, for the diffusion coefficient $D = D_0 \exp(-E/RT)$. It is worth noting that the activation energies are much less than those estimated for olivine, a common rock-forming mineral (see Table 2.8). The time required for a gas

Table 5.2. Extraterrestrial He and Ne in some terrestrial samples

Sample/References	^{3}He (10^{-9} cm^3 g^{-1})	^{3}He/^{4}He (10^{-4})	^{20}Ne (10^{-9} cm^3 g^{-1})	^{20}Ne/^{22}Ne	^{21}Ne/^{22}Ne
Magnetic separates in ocean sediment[a]					
Amari & Ozima[b]	3.2 (0.8–4.0)	1.4 (1.1–2.2)	69.1 (52–183)	11.6 (11–12.2)	0.036 (0.031–0.043)
Matsuda et al.[c]	1.5 (0.05–2.9)	2.3 (0.9–2.7)	17.9 (0.1–39)	12.5 (10.5–13.5)	0.0345 (0.029–0.004)
Nier & Schlutter[d]	22.2 ± 2.5	3.1 ± 0.5	340 ± 50	11.7 ± 0.3	0.034 ± 0.002
Stratospheric dusts[e]					
Rajan et al.[f]	(^{4}He ~ 0.11 cm^3 g^{-1})				
Hudson et al.[g]			$(4.4 ± 1.6) \times 10^5$	11 ± 3	0.03 ± 0.01
Nier et al.[h]	13,000 ± 4,800	2.4 ± 0.3	1.6×10^6	12.0 ± 0.5	0.035 ± 0.006
Micrometeorites[i]					
Maurette et al.[j] (Antarctica blue ice)			>10,000	11.5 ~ 13.5	0.035 ~ 0.04
Olinger et al.[k] (Greenland ice sheet)			1,500–50,000	11 ~ 12	
Pelagic sediment (bulk)					
Farley[l]	0.02 ~ 0.1	0.05 ~ 2			
Patterson & Farley[m]	0.0001–0.001	0.02–0.7			
SW (solar)[n]		4.26–4.57		13.7–13.8	0.0328–0.0333

[a] The magnetic fraction generally comprises 1–0.1% (wt) of bulk sediment.
[b] Amari & Ozima (1988). Numerical values indicate a median value and a range (in a bracket).
[c] Matsuda, Murota & Nagao (1990). Numerical values indicate a median value and a range (in a bracket).
[d] Nier et al. (1990). Numerical values indicate average values for the sum of 850°C and 1600°C fractions.
[e] Dust particles were collected by a NASA high-flying (20km) U-2 plane.
[f] Rajan et al. (1977).   [g] Hudson et al. (1981).   [h] Nier & Schlutter (1990).
[i] Micrometeorites are generally larger than cosmic dust particles, some up to a few hundred micrometers. Some of them show partially melted features but still contain measurable amount of noble gases.
[j] Maurette et al. (1991).
[k] Olinger et al. (1990). ^{20}Ne/^{22}Ne shows an approximate range.
[l] Farley (1995).   [m] Patterson & Farley (1998).
[n] Ozima et al. (1998) and also see Table 3.3(a).

to diffuse out of a spherical grain (radius = a) can be calculated as $(D/a^2)^{-1}$. In the right axis of Figure 5.5, the times needed for 99% gas release are indicated for corresponding D/a^2 values. If the linear trends are extrapolated to lower temperatures, one may expect that 99% of ^{4}He and 99% of ^{20}Ne will be released from cosmic dusts within 10^5 years, even at 100°C and 200°C, corresponding to a depth of 12 and 24 km, respectively. Because an oceanic plate moves at a speed of about 10 km in 10^5 years, these calculation suggests that solar He and Ne implanted in the cosmic dusts would be released at the very beginning of subduction. At such shallow depths in subduction zones, the solar He and Ne released from cosmic dusts would be easily carried away with fluids and would not subduct into the mantle. On these grounds, Hiyagon (1993) concluded that the solarlike He and Ne identified in mantle-derived materials is unlikely to be the result of subduction of cosmic dusts but would be more likely to reflect a primordial solarlike noble gas component in the mantle. The similar conclusion was also given by Trull (1994) and Stuart (1994). However, we could also argue that diffuse-out of noble gases from grains does not necessarily mean that they are returning to the surface; they may be carried down to the Earth's interior with subducting materials and remain trapped in pore spaces and/or accommodated in authegenic mineral fabrics.

Extraterrestrial noble gas components mostly reside in a magnetic fraction in deep-sea sediments, which constitutes less than 0.2 wt % of the sediment. Cosmic dusts or interplanetary-dust particles (IDP) generally contain a substantial amount of magnetite. However, the major phase of IDP is amorphous silicates (Brownlee, 1979), and they must also have been implanted by solar wind noble gases. Hence, it is puzzling to see that the extraterrestrial noble gases are characteristically found in the magnetic fraction. Amari and Ozima (1988) postulated that amorphous silicates in cosmic dusts were transformed to magnetite due to shock heating during their atmospheric entry, although the shock heating must have been moderate enough to retain most of the noble gases. The transformation of amorphous silicate in IDPs to olivine and magnetite due to shock heating was first suggested by Fraundorf, Brownlee & Walker (1982) on the basis of their observation that there is a positive correlation between the amount of magnetite in IDPs and the degree of the shock heating. The observed high He retentivity in lunar ilmenite suggests that the extraterrestrial He would be trapped tightly in magnetites of which crystalline structure is very close to ilmenite.

Extraterrestrial components have so far been identified only in the case of the lighter noble gases – helium and neon – because of their relatively larger abundance in the magnetic separates. Since IDPs larger than a few ten micrometers could not retain noble gases due to the atmospheric heating (e.g., Brownlee, 1985), extraterrestrial noble gases in deep-sea sediments would likely be carried mostly by the smaller particles. This small size of noble gas-carrying particles suggests that noble gases in IDPs in sediments would be dominated by surface-implanted solar wind noble gases rather than by body-correlated trapped or spallogenic components. As is expected, both helium and neon isotopic compositions, ^{3}He/^{4}He and ^{20}Ne/^{22}Ne, are closer to the solar wind noble gases (cf. Section 7.3) than to planetary (i.e., chondritic) or spallogenic components. However, the ^{3}He/^{4}He and ^{20}Ne/^{22}Ne ratios are still considerably smaller than the solar wind components (Table 5.2).

The difference may be due to a preferential retention of noble gas components of which energy falls in a specific range. Amari and Ozima (1988) suggested that the size of IDP particles acted as a sieve to retain preferentially SEP (solar energetic particle; 10 KeV/amu < E < 0.1 MeV/amu) components whose energy is high enough to penetrate into the particle to be trapped firmly but low enough not to pass through it. The solar wind component, which has a much lower energy ($\approx$ 1–10 KeV/amu) and hence was implanted in the shallower surface region, would be relatively easily removed from the particle due to surface erosion in sea water. Magnetic separates from deep-sea sediments generally show fairly uniform ^{20}Ne/^{22}Ne isotopic ratios of about 11.6 (Amari & Ozima, 1988), which are close to that of SEP of about 11.3 (Wieler et al., 1986) but significantly lower than that of SW of 13.7 (Geiss et al., 1972). However, Olinger et al. (1990) cast a doubt on this interpretation, since they found that particles larger than 100 μm still retain a significant amount of implanted solar Ne even after atmospheric entry.

The lower ^{3}He/^{4}He ratios in the magnetic separates may also be due to the predominance of SEPs, which have a lower ^{3}He/^{4}He ratio < 2.5×10^{-4} (Benkart et al., 1988). Recent studies of lunar ilmenite grains by Nichols, Hohenberg, and Olinger (1994) seem to favor this interpretation; they found that the majority of the ilmenite grains were enriched in SEP.

In some deep-sea sediment columns, records of IDP fall-out can be traced back to several ten million years with the help of extraterrestrial He as a tracer. Such a remarkable example was reported by Farley (1995). Farley measured ^{3}He/^{4}He ratios in a pelagic sediment core drilled from the central North Pacific. The core is thought to carry a nearly complete sedimentation record to 72 Ma. Figure 5.6 shows records of the He concentration and the isotopic ratio in the drilled core.

A sharp increase in ^{3}He and ^{3}He/^{4}He ratio around 35 million years is nearly synchronous with the well-known tektite horizons observed around the world. However, the Cretaceous/Tertiary boundary (K/T) which is easily recognized by a large Ir peak in the core is associated with essentially no ^{3}He peak. Farley argued that decoupling of ^{3}He and Ir in the core illustrates the fact that the two tracers of extraterrestrial fall-out were delivered to the sea floor from different sources and in different ways, perhaps by size bias of the carrier dusts.

The preservation of He in the deep-sea sediments for more than a few ten million years appears to be contrary to what we would assume from the He diffusion experiments of deep-sea sediments (e.g., Hiyagon, 1993). The laboratory experiments suggest almost complete loss of He from magnetic separates from deep-sea sediments in less than 10^7 years. Farley suggested the existence of an extraterrestrial silicate phase, which has much higher He retentivity than magnetite, as a possible explanation for the enormous longevity of He in the Mesozoic sediment core. However, the existence of such hypothetical silicate is open to future investigation.

5.3 Cosmic Dusts 137

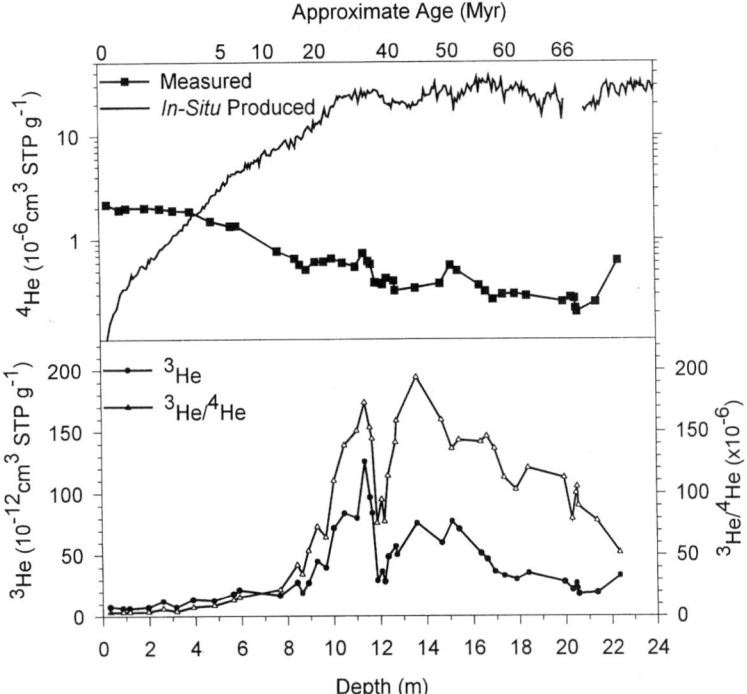

Figure 5.6 (top) The measured ^{4}He contents in a deep-sea sediment core compared with expected in situ production given the age and U and Th content of the sediments. (bottom) ^{3}He contents and ^{3}He/^{4}He ratios in the core. After Farley (1995).

An example of enormous He retentivity was recently reported in the case of fullerenes by Becker et al. (1996). Becker et al. identified about 1–10 ppm fullerenes (C_{60} and C_{70}) in a shock-produced breccia associated with the 1.85-Ga Sudbury impact structure. The impact shock could produce the fullerenes, but because of poor carbon content in the target rocks, Becker et al. (1994) suggested that the fullerenes were extraterrestrial in origin. The fullerenes contained about 10^{-8}–10^{-9} cm^3 STP/g of ^{3}He with ^{3}He/^{4}He ratio ranging from 3.5×10^{-4} to 8.5×10^{-4}. Since the helium isotopic ratios far exceed the solar wind ratio (~4×10^{-4}), but are similar to those in meteorites and IDP, Becker et al. (1996) suggested that the He was of extraterrestrial origin. If the He in the fullerenes were proved to be of extraterrestrial origin, this would give strong support for He retention over a billion years in the microcrystals under the Earth surface condition. Becker, Bunch, and Allamandolla (1999) also reported fullerenes in Allende meteorite. As discussed in Section 2.6, the high He retentivity in the fullerenes may be expected. It is then not entirely unreasonable to speculate that the high He retentivity suggested by the ancient IDP might be attributable to the fullerenes in the IDP, although this would require anomalously high content of the fullerenes in deep-sea sediments to account for the observed extraterrestrial He and Ne contents. Also, the possibility of nucleogenic origin of ^{3}He may not be ruled out because of the very long surface exposure duration, unless Li is shown to be totally absent in the fullerenes (cf. Section 5.6).

5.4 Noble Gases in Aquifer System

As a recent useful application of noble gases as a geochemical tracer, we will discuss in this section the dynamics of a large-scale aquifer system within the continental crust. We see a typical example of such a large-scale aquifer system in the northeastern part of Australia known as the Great Artesian Basin (GAB) aquifer system. The GAB is one of the world's largest confined groundwater basins and occupies $1.7 \times 10^6 \text{ km}^2$ of the Australian continent extending from northeast to south Australia. The basin consists of a multilayered confined aquifer system of continental quartzose sandstones with intervening silstones, mudstones, and shales in a relatively simple bowllike structure. Rainwater, which is supplied mainly in the eastern region, recharges the aquifer system (recharge zone). Thus supplied water then flows underground over a thousand kilometers toward the south and southwest, where natural springs emerge and provide water for extensive farming because the potentiometric surface is above ground level in most of the areas (discharge zone). Torgersen (1980) and Torgersen and Clarke (1985, 1992) applied noble gas geochemistry to study the aquifer system. They measured ^{4}He, Rn, and Ra contents in the aquifer system along the flow lines over several hundred kilometers. They found that variations in the ^{222}Rn and ^{226}Ra were rather small and, hence, concluded that an in situ source function for the excess ^{4}He in the GAB was fairly constant. The ^{4}He content in the fluid, however, showed systematic increase with increasing distance from the recharge area. With ^{14}C ages of the water, which is also proportional to the distance from the recharge area, Torgersen and Clarke (1992) estimated the accumulation rate of ^{4}He in the fluid. Thus, the estimated accumulation rate suggested that the in situ ^{4}He production cannot account for the accumulation rate in the fluid. Therefore, they concluded that the ^{4}He accumulating in the groundwaters of the GAB must come from a source outside the confines of the aquifer system, very likely involving the whole crustal section.

In some favorable cases, the age of fluid in an aquifer system may be estimated from measurements of a radiogenic ^{4}He concentration ([^{4}He]) and a flux (J_{He}) of radiogenic ^{4}He into a fluid system with the use of a relation $t = [^4\text{He}]/J_{He}$. The observed ^{4}He in fluid generally consists of radiogenic ^{4}He accumulated in the fluid and an initially air-saturated component. However, the latter component is generally much smaller than the former and can be safely neglected. Hence,

$$t = [^4\text{He}]_{rad}/J_{He} \approx [^4\text{He}]_{obs}/J_{He}$$

To estimate J_{He}, Torgersen and Clarke (1992) assumed that a fraction Γ_{He} of radiogenic ^{4}He produced by U, Th series decay in rocks entered the pore fluid phase. J_{He} is then related to the ^{4}He production rate in rocks, which is a function of U, Th

5.4 Noble Gases in Aquifer System

contents in rocks. Assuming that U, Th decay series were in radioactive equilibrium, U, Th contents can be related to the activity of Rn. Assuming also that a fraction Γ_{Rn} of ^{222}Rn was released by rocks into the fluid system, they derived the following expression for the age of a fluid:

$$t(\text{year}) = (^4\text{He}/\Gamma_{He})/(^{222}A_{Rn}/\Gamma_{Rn}) \times 3.12 \times 10^{12}/\{1 + 0.123(\text{Th}/\text{U} - 4)\} \quad (5.2)$$

where $^{222}A_{Rn}$ and ^{4}He are measured activity (dpm) of ^{222}Rn and total ^{4}He concentration (cm^3 STP/cm^3 of liquid) in the fluid phase, and (Th/U) is a ratio in rocks. In deriving Equation (5.2), a parameterized expression for J_{He} [see Equation (5.9)] was used.

When the age of the crustal water is known, it is possible to estimate a fluid transport velocity. With the measured values of He concentration at different levels in the Cajon Pass deep scientific drill hole, Torgersen and Clarke (1992) estimated the age of the waters at each level and then deduced the vertical velocity from a relation, $v = h/\Delta t$ where h and Δt indicate the vertical distance and the age difference between the two levels. Thus estimated vertical velocity was 0.04–6 cm a^{-1}, which was much larger than the instantaneous flow velocities of about 10^{-5} cm a^{-1} that was calculated from the pressure and permeability measurements with the use of Darcy's law; $v = -k/\mu(\nabla p - \rho g)$, where v is velocity, k is permeability, μ is viscosity, p is pressure, ρ is density, and g is gravity. The discrepancy may be simply a reflection of the difference between the instantaneous and the long time averaged ones as argued by Torgersen and Clarke (1992). However, it also seems important to examine whether the free parameters used in these above calculations (e.g., Γ_{Rn} and Γ_{He}) are truly representative values.

Stute et al. (1992) made numerical calculation to study flow dynamics of the Great Hungarian Plain (GHP) aquifer system with the use of He concentration data. They solved the diffusion-advection equation for He transport for a two-dimensional case.

$$D\left(\frac{d^2C}{dx^2} + \frac{d^2C}{dz^2}\right) - v_x \frac{dC}{dx} - v_z \frac{dC}{dZ} + A = 0 \quad (5.3)$$

where C is the excess ^{4}He concentration ($\equiv$ total ^{4}He − atmospheric ^{4}He), D is the dispersion coefficient (~1 m^2 a^{-1}), and A denotes the in situ production rate of radiogenic ^{4}He. The numerical calculation showed that in the discharge area the excess He ($\equiv$ mantle He + non-in-situ crust He) concentration is nearly constant throughout the whole depth, whereas it decreases considerably from the deeper region to the surface in the recharge area. The results were well supported by the observed He concentration data. From the comparison with the observed He concentration data with the numerical calculation, they also estimated vertical flow velocities in the discharge area of 1.5 mm a^{-1}, which suggests a turnover time of $(0.5 - 2.3) \times 10^6$ years for the groundwater flow system. The model calculation is crude in the sense that it is limited to a two-dimensional case. Nonetheless, it is important to note that the approach yields insights into fluid flow in the crust that may not be obtained by other methods.

5.5 Cosmogenic Noble Gases

When cosmic rays impinge into the Earth, the interactions of the cosmic ray particles with Earth materials produce many "cosmogenic" nuclides (including noble gases). The primary cosmic ray flux at the Earth's orbit has two components: galactic cosmic rays (GCR), which come from the galaxy, and solar cosmic rays (SCR), which are emitted from the sun. The GCR particles are a mixture of energetic protons (~87%), alpha particles (~12%), and some heavier particles (~1%), whereas about 98% of the SCR particles is proton. Because of the relatively low energy (typically 1–100 MeV), the SCR particles are essentially shielded by the atmosphere, and its nuclear effects or the production of cosmogenic nuclides are limited only at the very top of the atmosphere. Hence, except for unusually high solar activity periods such as the October 1989 solar particle event, the cosmogenic nuclides by the SCR can scarcely be important in geochemical discussions. However, nuclear effects of cosmic rays (essentially GCR) are observable to great depths, up to $10^6 \, g \, cm^{-2}$ in the Earth's surface, due to the penetrating muons, which were derived from cosmogenic pi-mesons produced in the atmosphere.

Cosmic rays attenuate very rapidly during the passage through the atmosphere. The nuclear absorption mean free paths for the atmosphere are fairly well known (about $150 \, g/cm^2$). The mean free paths at high latitudes are smaller than at the low latitudes because the energy spectrum of the high-energy parent nucleons ($E > 400 \, MeV$) is harder at lower latitudes. In addition to the atmospheric absorption, the geomagnetic field deflects cosmic rays, both of which result in variation of the nuclear effects at the Earth's surface. In the polar region, cosmic rays will travel to the Earth's surface along the geomagnetic field that is perpendicular to the surface, but in the equatorial region cosmic rays will be deflected by the geomagnetic field that runs parallel to the surface. If geomagnetic field intensity (in terms of a geomagnetic dipole moment) increases by a factor of 2, cosmic ray-induced nuclear reaction rate will reduce by about a factor of 2 at low latitudes. On the other hand, the reaction rate will increase by about 70% if geomagnetic field decreases by a factor of 2. The shielding effect of cosmic rays due to the geomagnetic field is most effective at the equator (and, more precisely, at the geomagnetic equator). Because the geomagnetic dipole moment has been fluctuating by about a factor of 2 during the last ten thousand years (e.g., Merrill & McElhinny, 1996), a caution must be exercised when using cosmogenic nuclides in discussing earth science problems. As a notable example, we may mention that the variation of the geomagnetic dipole moment necessitates a significant correction in ^{14}C ages.

The interaction of cosmic ray particles within the Earth's atmosphere produces a cascade of secondary particles, many of which have enough energy to reach the

5.5 Cosmogenic Noble Gases

Table 5.3. *Cosmogenic noble gases*

Isotopes	Half-Life (yr)	Main Targets
^{3}H	12.3 (decays to ^{3}He)	O, Mg, Si, Fe, (N, O)
^{3}He, ^{4}He	Stable	O, Mg, Si, Fe, (N, O)
^{20}Ne, ^{21}Ne, ^{22}Ne	Stable	O, Mg, Si, Fe, (N, O)
^{36}Ar, ^{38}Ar	Stable	Fe, Ca, K
^{37}Ar	35 days	Fe, Ca, K(Ar)
^{39}Ar	269	Fe, Ca, K (Ar)
^{81}Kr	2.1×10^5	Rb, Sr, Zr (Kr)
^{78}Kr, ^{80}Kr, ^{82}Kr, ^{83}Kr	Stable	Rb, Sr, Zr
$^{124-132}$Xe	Stable	Te, Ba, La, Ce, I

Target elements in parentheses are for the Earth's atmosphere. Data are from Lal (1988).

Earth's surface and induce nuclear reactions to produce cosmogenic nuclides. Table 5.3 shows cosmogenic noble gases in the Earth and their principal target elements. Although the amount of the cosmogenic noble gases bear little significance in the total inventory of the terrestrial noble gases, they are important as a unique geochemical tracer of various geological processes.

There are three principal types of nuclear reactions due to the interactions of terrestrial materials with cosmic rays: (i) by high-energy spallation of nucleons ($E > 40$ MeV), principally by neutrons, (ii) by thermal neutron capture, and (iii) muon-induced nuclear disintegrations. Muon reactions become important only at depths below sea level. The estimation of the production ratio is difficult because of lack of knowledge of the probabilities of formation of nuclides in the different reactions.

The theoretical estimation of the production rates of ^{3}He, Ne, and Ar (Table 5.4) on various target elements has been made by several authors (e.g., Lal, 1991; Masarik & Reedy, 1995). The values shown in Table 5.4, however, should be regarded to be tentative and may be subject to uncertainties >25 ~ 30% (Lal, 1991).

With the use of the production rates, we can then estimate production rates at sea level and at latitudes higher than 60° for any material composed of the elements given in Table 5.4. After the production rate of nuclear interactions at sea level is known, the total rate of nuclear interactions in any target, for example in rocks, can be obtained by scaling the values by the $A^{2/3}$ factor, where A is the atomic mass number of the target (Lal, 1991).

The first identification of cosmogenic noble gas in crustal rocks was reported by Kurz (1986). Kurz measured a series of basaltic-drilled core samples from Mauna Loa volcanoes in the Hawaiian Islands. The lava flow (^{14}C age $\approx 20,000$ a) has retained its surface flow structure and, therefore, was concluded to have experienced little surface erosion. Because the ^{3}He concentration in the surface of the lava flow (10^{-12}–10^{-13} cm^3 STP g^{-1}) gave reasonable agreement with a theoretical production

Table 5.4. *Production rate of cosmogenic noble gases from selected targets*

Target	Cosmogenic Nuclides					
	^{3}He	^{20}Ne	^{21}Ne	^{22}Ne	^{36}Ar	^{38}Ar
O	83					
Mg	50	80 (3.16)	110 (4.36)	(3.97)		
Al	47	28 (1.37)	24 (1.21)	(1.92)		
Si	66	33 (1.21)	18 ($\equiv$1)	(1.14)		
			45*			
Ca					69	200
Fe	28	0.22	0.24		1.2	1.7

Production rate: atoms g^{-1} (target) a^{-1}. All data are at sea level and at higher latitudes (>60°). Data are from Lal (1991) and (*) are from Niedermann et al. (1994). Numerals in parentheses are neon production rate ratios normalized to ^{21}Ne in Si for the main target elements (Schäfer et al., 1999).

rate, Kurz concluded that the ^{3}He in the lava flow was essentially cosmogenic. Figure 5.7 shows the ^{3}He concentration against the depth in the core.

As is seen from Figure 5.7, the ^{3}He variation with depth shows a large departure from a simple exponential depth dependence, which is seen in the case of the atmosphere. Kurz attributed this departure to the change in production mechanism. At greater depths in the lava, the nucleon component is greatly decreased, and the muons become the dominant cosmic ray particles. Nuclei that capture muons have a high probability of emitting neutrons, which in turn can produce ^{3}He via ^{6}Li(n, α)^{3}H → ^{3}He. The cosmogenic (^{3}He)$_c$ production is then expressed empirically as the sum of the simple exponential decay term and the muon dependence term (Kurz, 1986),

$$^3\text{He}_c(z) = {}^3\text{He}_c(z_0)\exp(-z/l) + K\mu^-(z) \tag{5.4}$$

where l is the attenuation length ($\approx$ 160 g cm^{-2}), K is a constant, and $\mu^-(z)$ is the depth dependence of the muon stopping rate.

Such nonexponential decrease of the cosmogenic ^{3}He production is consistent with the neutron flux profile with depth calculated by Masarik and Reedy (1995) (Figure 5.8). The interaction of cosmic ray particles with the Earth's atmosphere produces a cascade of secondary high-energy particles, of which the most important are neutrons and protons in view of cosmogenic-nuclide production. The particle fluxes attenuate exponentially on passing through the atmosphere, in which the proton fluxes decrease much faster than neutron fluxes; hence, the neutrons become dominant for cosmogenic-nuclide production. Near the air-surface interface, the total neutron flux shows considerable variation due to the difference in the production and transport in these two materials.

5.5 Cosmogenic Noble Gases

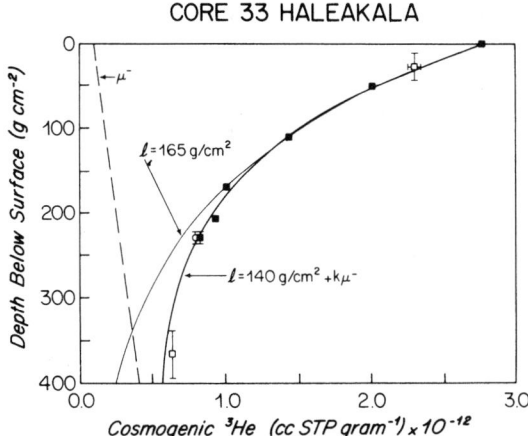

Figure 5.7 Depth dependence of cosmogenic ^{3}He in a drilled core in Haleakala volcano. The diagram shows a very large deviation from an exponential depth dependence (i.e., mean absorption free path $l = 165\,\mathrm{g\,cm^{-2}}$) below $170\,\mathrm{g\,cm^{-2}}$. The curve that yields a better fit to the experimental points is a combination of a simple exponential and the effect of muon-produced ^{3}He via ^{6}Li(n, α)^{3}H. The dashed line is the assumed depth dependence of the muon stopping rate. After Kurz (1986).

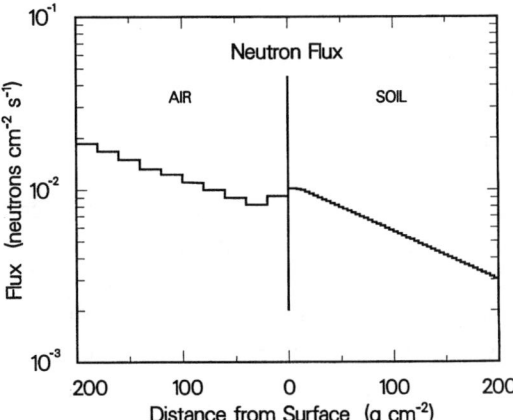

Figure 5.8 The calculated total neutron flux (cosmogenic) on either side of the air-surface interface (after Masarik & Reedy, 1995). Note that the average neutron flux in the crustal interior due to U, Th decays ($\sim 5 \times 10^{-4}$ neutrons cm^{-2} s^{-1}; cf. Section 5.6) is more than an order of magnitude less than the cosmogenic flux at the air-surface interface.

Table 5.5. *Cosmogenic ^{3}He and ^{21}Ne at the Earth's surface*[a]

Nuclide	Target	Calculated [atoms a^{-1} (g-target)$^{-1}$]	Measured [atoms a^{-1} (g-target)$^{-1}$]
^{3}H	Atmosphere	(0.26 atoms cm^{-2} s^{-1})[1)]	
^{3}He	SiO$_2$ (1)	124	
	Fo$_{81}$ olivine (1)	105	
	Mg$_2$SiO$_4$ (2)	68.4	
	Fe$_2$SiO$_4$ (2)	50.8	
	MgSiO$_3$ (2)	70.0	
	FeSiO$_3$ (2)	56.4	
	Volcanic rocks, Utah (3)		107 ± 4, 87 ± 6
	Olivine, Hawaii (4)		187 ± 45, 82 ± 22, 190 ± 28
	Volcanic rocks, Western U.S. & France (5)		115 ± 4
^{21}Ne	SiO$_2$ (1)	18.41	
	Fo$_{81}$ olivine (1)	41.1	
	Volcanic rock, Western U.S. (6)		45
	Quartz, Sierra Nevada (7)		21 ~ 45
	Quartz, Antarctica (8)		86.5 (alt. ~1000 m)
	Garnet lehrzolite, Central Asia (8)		245.6 (alt. ~1150 m)
	Spinel peridotite, Central Asia (8)		375 (alt. ~2000 m)

[a] Values are reduced at sea level with no geomagnetic cutoff except Reference 7.

Table 5.5 References: 1, Masarik & Reedy (1995); spallation reactions only; 2, Lal (1991); 3, Cerling (1990); 4, Kurz et al. (1990); 5, Cerling & Craig (1994); 6, Poreda & Cerling (1992); 7, Niedermann et al. (1994); 8, Staudacher & Allègre (1991).

In Table 5.5, we show the production rates of ^{3}He and ^{21}Ne, both calculated and measured, in minerals and volcanic rocks. The production rate of ^{3}H in the atmosphere is also shown. It is interesting to note that the two Hawaiian samples show much higher ^{3}He concentration than other samples. These anomalously high ^{3}He contents indicate that there may be some other source(s) of ^{3}He production not considered in the calculation (Lal, 1991). Only a part of the discrepancy could be attributed to uncertainties in cross-sectional values (Lal, 1991). In this regard, it may also be important to examine the possible effect of the nuclear tests in the central Pacific in the late 1950s. During that period, an enormous amount of ^{3}H produced in the atmosphere must have fallen with rain and penetrated into surface rocks through numerous veins or cracks. It is then conceivable that some fraction of the rainwaters was assimilated into rock fabrics; thus, trapped ^{3}H decayed to ^{3}He in the rocks. Kurz (1986) ruled out this possibility on the grounds that, in order to account for the ^{3}He observed in the Hawaiian volcanic rocks ($\approx 10^{-12}$ cm^3 g^{-1}), they must contain 40% of water that had a tritium concentration of 1000 TU (1 TU = 10^{-18} T per H), which Kurz

Table 5.6. *Production ratios of cosmogenic $^{3}He/^{21}Ne$, $^{22}Ne/^{21}Ne$, and $^{21}Ne/^{26}Al$ in some mineral separates*

Target	Samples	Locality
$(^{3}He/^{21}Ne)_{cosmogenic}$		
Olivine (1) ~2.1	Basalts	Maui, Hawaii; altitude = 2780 m
Olivine (2) 2.44 ± 0.29	Lava flow	Western U.S.; altitude ~ 1500 m
Quartz (3) 2.76	Orthoquartzite	Antarctica; altitude = 1300 m
$(^{22}Ne/^{21}Ne)_{cosmogenic}$		
Quartz (4) 1.243 ± 0.022	Sandstone	Antartica, altitude = 1882 m
$(^{21}Ne/^{26}Al)_{cosmogenic}$		
Quartz (5) 0.65 ± 0.11	Granites	Sierra Nevada; altitude = 3556 m, 2452 m

Table 5.6 References: 1, Marti & Craig (1987); 2, Poreda & Cerling (1992); 3, Staudacher & Allègre (1991); 4, Niedermann, Graf & Marti (1993); 5, Niedermann et al. (1994).

considered to be unrealistic. He also pointed out that a 1790 lava flow did not have any excess in ^{3}He. However, from the 1950s to the 1960s, the atmospheric tritium concentration in the northern hemisphere often amounted to more than a few thousand TU. For example, about 3000 TU in rainwater was observed in Boston during the aerial nuclear tests in the central Pacific in 1954. In Europe, the maximum tritium concentration in precipitation was 2000 in the 1960s (Schlosser et al., 1988). It is almost certain that tritium concentrations in rainwaters in Hawaii, which was much closer to the test sites, was higher by more than an order of magnitude. It would be possible that some fraction of the ^{3}He in the Hawaiian lava is due to the anthropogenic ^{3}He.

Marti and Craig (1987) first identified cosmogenic ^{21}Ne in olivines and clinopyroxenes in lavas from the summit of Haleakala volcano in the Hawaiian islands. To elucidate the cosmogenic component, they first removed almost all the mantle-derived noble gases from the lava samples by vacuum crushing, a procedure that releases the volatiles trapped in fluid inclusions within the phenocrysts (cf. Section 6.2). The cosmogenic ^{21}Ne was then identified as the excess relative to the atmospheric ^{21}Ne/^{22}Ne ratio. This gave a cosmogenic production ratio of $(^{3}He/^{21}Ne)_{c} = 2.1$, which is close to the ratios commonly observed in meteorites. Since then, several authors reported cosmogenic ^{21}Ne in surface rocks. Production ratios of cosmogenic ^{21}Ne relative to cosmogenic ^{3}He are given in Table 5.6.

After the production rates for the cosmogenic ^{3}He and ^{21}Ne are accurately known, they can be used to date the exposure age of surface rocks to cosmic ray radiation. With the use of these cosmogenic nuclides, a few attempts have been made to determine the exposure age of surface rocks (e.g., Kurz, 1986; Staudacher & Allègre, 1991; Cerling, Poreda & Rathburn, 1994; Wells et al., 1995). However, the ages thus determined have considerable uncertainty. This is because in addition to the uncertainty in the production rates, surface erosion disturbs the fundamental premise in the exposure age dating, that is, an assumption of a fixed geometry of the sample with respect to cosmic ray radiation. Trull et al. (1995) also called caution to possible ^{3}He diffusion loss from samples in applying ^{3}He exposure dating. They found that ^{3}He exposure ages in metamorphic quartzite rocks are dramatically lower than ^{10}Be ages of the same sample, which they attributed to diffusion loss of ^{3}He. Trull et al. concluded that "massive" quartzite rocks are often microcrystalline, which is responsible for a high degree of ^{3}He diffusion loss. They, therefore, recommended that "macroscopically" crystallized quartz should be used for ^{3}He exposure dating.

As we noted earlier (cf. Table 5.5), there is considerable discrepancy between the theoretically estimated ^{3}He production rate and the observed values, and the discrepancy may partly be attributed to erosion of the surface rock. Conversely, the discrepancy can be used to infer erosion rate (e.g., Lal, 1991), one of the most important but one of the hardest parameters to estimate in geomorphology. Assuming that production rate decreases exponentially with depth (x) in rock, the concentration of cosmogenic ^{3}He at time t and depth x is expressed as (Lal, 1991),

$$C(x,t) = \int_0^t P_0 \exp(-mx)\, dt \tag{5.5}$$

where P_0 is the production rate at the surface and $1/m$ is an absorption mean free path. For a constant erosion rate of ε, we can write $x_t(t) = x_0 - \varepsilon t$, where x_0 denotes the original depth and x_t is the present-day depth. Integrating Equation (5.5), we have

$$C(x_t, t) = (P_0/m\varepsilon)\exp(-mx_t)\{1 - \exp(-m\varepsilon t)\} \tag{5.6}$$

Hence, knowing the age of the rock and the concentration of cosmogenic ^{3}He or ^{21}Ne, we may estimate the erosion rate ε from Equation (5.6). The procedure appears straightforward, but in applying it to an actual case, we have difficulties such as the nonlinear erosion rate (Lal, 1991). Some interesting preliminary results were reported by Graf et al. (1991), who attempted to use the cosmogenic ^{21}Ne/^{26}Al ratio measured in basement rocks to infer variation in the Antarctic ice cover. From ^{3}He data, Kurz (1986) estimated the erosion rate of about 10 m/Ma for Haleakala lava.

Another geochemically interesting application of cosmogenic noble gas can be seen in the case of radiogenic ^{129}I, which decays to ^{129}Xe with a half-life of 17 Ma. ^{129}I is continuously being produced by cosmic rays in the upper atmosphere and from spontaneous fission of ^{238}U in the crust. Because of the long half-life, ^{129}I would then be in secular equilibrium through geochemical cycle in the crust, and the ratio of ^{129}I/I would become constant. On the basis of detailed study on the source, sink and transfer functions of iodine in the crust, Fabryka-Martin et al. (1985) concluded that the present-day equilibrium ratio of ^{129}I/I is about 6×10^{-13} in the hydrosphere. Although the abundance ratio of ^{129}I is extremely small, as is the radiogenic ^{129}Xe, the latter might become noticeable in the environment where iodine was highly enriched. Fabryka-Martin et al. reported the use of natural ^{129}I to estimate ages of some brines.

5.6 Crustal He and Ne – Nucleogenic Components

Besides cosmic ray-produced noble gases, which are discernible only at the very surface region of the Earth (down to a few meters from the surface), there are several geochemically important nucleogenic noble gases produced by nuclear reactions due to U and Th decays in the solid Earth. Because of high concentrations of U and Th in the crust, U, Th-induced noble gases are more important in the crust than in the mantle. Ne and He in the crust contain substantial fractions of these nucleogenic components. Because of their high diffusivity, these lighter noble gases tend to diffuse out of minerals and migrate into aquifer system or to the atmosphere. It has been shown that those noble gases in a large aquifer system were acquired from a large crustal section (e.g., Torgersen & Clarke, 1992; Kennedy, Hiyagon & Reynolds, 1990; also see Section 5.4) and hence, would represent an average noble gas composition in the large part of the crust. The fact that nucleogenic He and Ne isotopic compositions in aquifer systems from widely separated areas are quite alike (Kennedy et al., 1990) further suggests that these isotopic compositions would be a good representative of the average crustal He and Ne. We will show that such characteristic "crustal He" isotopic component is clearly observable on the histogram of a large amount of terrestrial He isotopic data (Figure 6.8). In an analogous way, we may define a "crustal Ne," which represents the average nucleogenic components in the crust. Ne observed in crustal samples are then understood to be a mixture of the "crustal Ne" with the atmospheric Ne.

^{4}He and ^{3}He: ^{4}He, not only in the crust but also in the Earth, is essentially radiogenic, and has been produced from radioactive decay of U, Th series elements. Only a significant source for nucleogenic ^{3}He in the crust is a reaction ^{6}Li(n, α)^{3}H($T_{1/2}$ = 12.3 a) → ^{3}He, where neutrons are derived from a spontaneous fission of ^{238}U and from reactions of light elements such as Na, Mg, Al, and Si with α particles emitted from U, Th decays. However, in a very shallow surface region (less than a few meters), the secondary cosmic ray neutrons would be more important.

A neutron production rate (not including the cosmogenic components) was calculated by Andrews et al. (1986) for a granitic rock as a function of U, Th contents,

$$P_n = \rho\{0.4764[U] + 1.49[U] + 0.60[Th]\} \text{ neutrons cm}^{-3}\text{a}^{-1} \tag{5.7}$$

where ρ denotes the density of a rock, and [U], [Th] are in parts per million. The first term represents a ^{238}U spontaneous fission neutron production rate, and the second and third terms represent those due to (α, n) reactions due to U, Th decays. Taking values of U = 3 ppm, Th/U = 3.5, and ρ = 2.5 g cm^{-3} for granite, Equation (5.7) gives

$P_n = 1.04 \times 10^{-6}$ neutrons cm^{-3} s^{-1}. Thus, produced neutrons are then thermalized and captured mostly by Li to produce ^{3}He.

^{3}He production rate (cm^3 STP s^{-1}) in rocks is expressed as (Morrison & Pine, 1955)

$$P(^3He) = P_n \times p_{th} \times f_{Li} \tag{5.8}$$

where p_{th} denotes a probability of thermalization of neutrons and f_{Li} represents a fraction of the thermal neutrons absorbed in ^{6}Li; $f_{Li} = n_{Li}\sigma_{Li} / \Sigma n_Z \sigma_Z$, where σ_{Li}, σ_Z, are cross sections of Li and element Z for thermal neutron capture and n_{Li} and n_Z are the number of Li and other target atoms per gram of rock. Putting this estimated value of $P_n = 1.04 \times 10^{-6}$ cm^{-3} s^{-1}, $p_{th} = 0.8$, and $f_{Li} = 0.05$, which corresponds to an average Li content in granite (20 ppm), we have $P(^3He) = 4.16 \times 10^{-8}$ atoms cm^{-3} s^{-1}.

^{4}He production per gram of a specimen can be expressed as a function of U content (Craig & Lupton, 1976),

$$P(^4He) = 0.2355 \times 10^{-12} U^* \quad (cm^3 STP g^{-1} a^{-1})$$
$$U^* = U\{1 + 0.123 \times (Th/U - 4)\} \tag{5.9}$$

where U is in ppm. Again, taking values of U = 3 ppm, Th/U = 3.5, and ρ = 2.5 g cm^{-3} for granite, we estimate a ^{4}He production rate $P(^4He)$ = 1.4 atoms cm^{-3} s^{-1} from Equation (5.9). Therefore, from Equations (5.8) and (5.9), we have nucleogenic He isotopic ratio ^{3}He/^{4}He = 2.97 $\times$ 10^{-8} for granite. This He isotopic ratio is close to the average He isotopic ratio for crustal materials (Figure 6.8).

Although the average chemical composition of the crust differs from granite, granite essentially accounts for U and Th inventories in the upper region of the crust. Therefore, we may assume that the average He isotopic composition in the crust (at least in the upper region in the crust) would be close to that produced in granite. In Chapter 6, we will show that the upper mantle is characterized by a well-defined ^{3}He/^{4}He isotopic ratio of about 10^{-5}, which is markedly different from the crustal ratio of about 10^{-8}. Because of this distinct difference in the isotopic compositions, the helium isotopic ratios have been widely used as a tracer to identify mantle-derived materials (cf. Chapter 6).

The production of ^{3}He in the crust is essentially due to the neutron capture reaction of Li [cf. Equation (5.8)], in that the neutrons were derived from U and Th [cf. Equation (5.7)]. Therefore, the production of ^{3}He is concomitant with the production of ^{4}He, giving a nearly constant production ratio of ^{3}He/^{4}He $\approx$ 10^{-8} in the average crust. However, within a small region whose dimension is comparable to the mean free path of neutrons (about 50 cm in common crustal rocks; Andrew et al., 1986), the production ratio ^{3}He/^{4}He would vary considerably depending on the Li/(U, Th) ratio of the region. This relation occurs because even though neutrons can easily penetrate into the region from the surrounding medium and produce nucleogenic ^{3}He by the ^{6}Li(n, α)^{3}H reaction, all ^{4}He in the region must be only in situ produced reactions because of the very short range of α particle (a few ten micrometers). Therefore, if the region contains little (U, Th) but mostly Li, there would be essentially nucleogenic ^{3}He, and we would observe extremely high ^{3}He/^{4}He ratio. As such, Tolstikhin (1978) reported very high ^{3}He/^{4}He ratio (up to 10^{-3} or even higher) in some Precambrian topaz, which had a very high Li/(U, Th) ratio. Tolstikhin, therefore, named topaz as a "natural neutron detector."

As given in Equation (5.7), the neutron production rate can be expressed as a function of U content. From the production rate, we can calculate the neutron flux (f_n) with the relation $f_n = v \times n$, where v denotes the mean velocity of neutrons and n is an equilibrium concentration of neutrons. The latter quantity is related to the neutron production rate (p_n) as $n = (p_n/\tau)$ where τ denotes the time constant for the neutron absorption in the medium ($\sim$2500 s^{-1}). Andrew et al. (1986) estimated the average neutron flux in the Stripa granite to be 5.5×10^{-4} neutrons cm^{-2} s^{-1}, which is in good agreement with the measured flux of 4.7×10^{-4} neutrons cm^{-2} s^{-1} in the borehole in the granite.

Since the mean free path of neutrons in rocks is considerably larger than average mineral grain size in crustal rocks, we expect that the neutron flux would reflect the average U content in the bulk rock body rather than U contents in individual mineral grains. This implies that the spatial variation of the neutron flux in the crust would generally be modest, mainly reflecting the variation in the average U contents among major crustal rock types. However, as we discussed in Section 5.5, the neutron flux in the atmosphere shows nearly exponential decrease with depth, and near the air surface the neutron flux shows anomalous variation. It is worth noting that the calculated cosmic ray-induced neutron flux (total) at the Earth's surface (Figure 5.8) is almost two orders of magnitude higher than the average neutron flux (4.7×10^{-4} n cm^{-2} s^{-1}) estimated in the Stripa granite (cf. Section 5.5). In fact, the cosmic ray secondary neutron flux observed at Stripa was about 32×10^{-4} cm^{-2} s^{-1} (Andrew et al., 1986). It would, therefore, be important to examine the cosmogenic (spallation as well as neutron-induced) ^{3}He as well as U, Th-induced ^{3}He, whenever one found an anomalously high ^{3}He/^{4}He ratio in near-surface materials.

Ne: Both ^{21}Ne and ^{22}Ne can be produced in a significant amount in crustal materials such as granite by nuclear reactions involving α particles and neutrons derived from U, Th radioactive decays. (The reactions and the target elements are listed in Table 5.8.) Except for ^{18}O(α, n)^{21}Ne and ^{19}F(α, n)^{22}Na($T_{1/2}$ = 2.605 a) $\rightarrow$ ^{22}Ne reactions, however, other reactions are unlikely to be important in Earth's noble gas inventory.

Figure 5.9 shows a neon three-isotope plot (Kennedy et al., 1990) where data were obtained from natural gases and brines in North America representing a broad geographical distribution. A fairly well-defined linear correlation on which air Ne is situated at the upper-left end suggests that the observed neon isotopic data resulted from a mixing between air Ne and another end member Ne characteristic to the crust. It is then reasonable to attribute the latter component to the nucleogenic component produced in the crust.

Since we can safely neglect nucleogenic ^{20}Ne in the Earth, an extrapolation of the linear trend to ^{20}Ne/^{22}Ne = 0 would give the production ratio of (^{21}Ne/^{22}Ne)$_{nucl}$ =

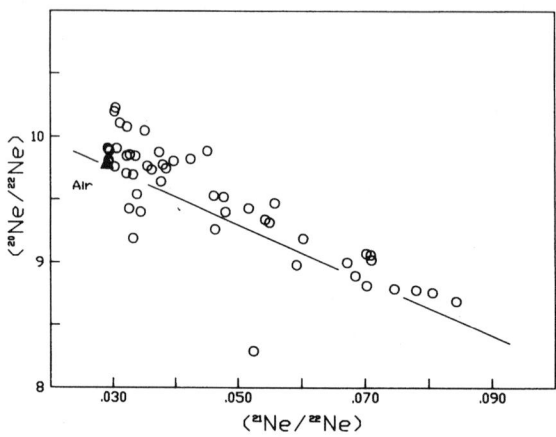

Figure 5.9 Neon three-isotope plot for data set from brines in North America. The solid line represents the least-squares fit to the data set, and the solid triangle is the neon isotopic composition in air. After Kennedy et al. (1990).

(0.469 ± 0.006), in that the ^{21}Ne and ^{22}Ne are derived from the nuclear reactions ^{18}O(α, n)^{21}Ne and ^{19}F(α, n)^{22}Ne, respectively.

Hünemohor (1989) obtained an empirical relationship between the nucleogenic isotopic ratio (^{21}Ne/^{22}Ne)$_{nucl}$ and an O/F ratio from measurements of U-rich minerals:

$$\log(^{21}\text{Ne}/^{22}\text{Ne})_{nucl} = (1.038 \pm 0.021)\log(^{18}\text{O}/^{19}\text{F}) + \log(2.2 \pm 0.3) \quad (5.10)$$

The relation holds over three orders of magnitude variation of this ratio observed in U, Th minerals. Applying this equation to the linear trend in Figure 5.9, we infer O/F (atomic ratio) = 110, which is 4 ~ 10 times smaller than that expected for the average crustal chemical composition. Kennedy et al. (1990) attributed this discrepancy to the limited range (~ 10–40 μm) of α particles. The pertinent O/F ratio is not that for average crustal composition but that within the reaction sphere defined by an α-particle range. Because the "crustal" Ne is observed on a widely spread geographic area, the reaction sphere is likely to be represented by common and widely distributed minerals. As likely candidate minerals, Kennedy et al. suggested micas or amphiboles, which are potential hosts for U and Th, and it is common for F to substitute for the hydroxyl anion.

Anorthosite, which is globally distributed in old cratons, may also be a possible source for the crustal Ne. Archean anorthosites in west Greenland, one of the oldest crustal rocks, have been studied for noble gases (e.g., Zadnik & Jeffery, 1985; Azuma & Ozima, 1993) and shown to contain a significant amount of nucleogenic Ne. An Ne three-isotope plot (Figure 5.10) shows a linear trend. The slope of the linear trend yields (^{21}Ne/^{22}Ne)$_{nucl}$ = 0.79. From Equation (5.9), we then estimate O/F (atomic ratio)

5.6 Crustal He and Ne – Nucleogenic Components

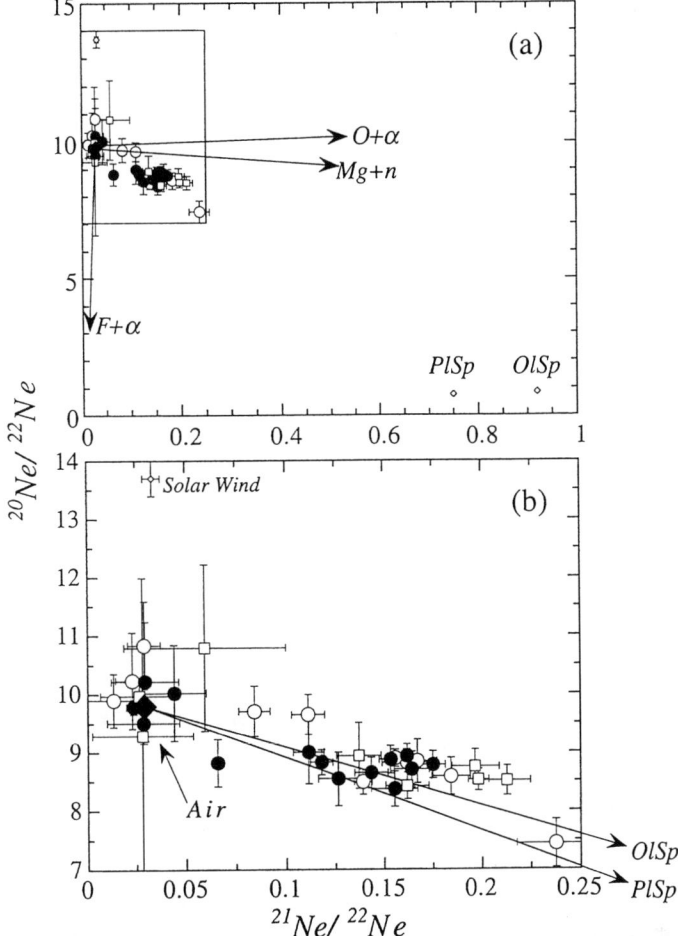

Figure 5.10 Neon three-isotope plot for anorthites (hollow circles), hornblendes (solid circles), and whole rock (hollow square). OlSp and PlSp indicate cosmic ray-induced spallogenic neons from olivine and plagioclase, respectively. Arrows (O + α, F + α, Mg + n) indicate the mixing lines between the air Ne and the nucleogenic Ne isotopes produced by $^{18}O(\alpha, n)^{21}Ne$, $^{19}F(\alpha, n)^{22}Ne$, and $^{25}Mg(n, \alpha)^{22}Ne$, respectively. Redrawn from Azuma and Ozima (1993). The enclosed part in panel (a) is enlarged in panel (b).

= 181. The ratio is within the range estimated for the "crustal Ne" from the data in Figure 5.9.

In Table 5.7, we compiled production ratios of $(^{21}Ne/^{4}He)_{nucl}$ and $(^{22}Ne/^{4}He)_{nucl}$ isotopes observed in various crustal samples. The water and gas samples cover a wide range of sampling sites and give fairly consistent ratios, being in harmony with the foregoing assumption of a well-defined "crustal He" and "crustal Ne."

Table 5.8 shows nucleogenic noble gases that would have been produced over a whole history of the Earth (4.5 Ga). The estimations are made for the assumed

Table 5.7. *Production ratio of $^{21-22}Ne/^4He$ and $^{21}Ne/^{22}Ne$ in selected crustal materials due to (α, n) reactions from U, Th decays*

Sample	$(^{21}Ne/^4He)_{nucl}$ ($\times 10^{-8}$)	$(^{22}Ne/^4He)_{nucl}$ ($\times 10^{-6}$) $[^{21}Ne/^{22}Ne]_{nucl}$	References
Natural water and gas			
Russia	3.7 ± 2.1		Tolstikhin (1978)
North America	4.6 ± 0.8	[<0.47]	Kennedy et al. (1990)
Crust		[0.47–0.76]	Ballentine (1997)
Minerals and rocks			
Anorthosite	7.4, 6.0	$[10^{-2} \times O/F]^a$	Zadnik & Jeffery (1985)
Uranium oxides	5.2^b	7.5^b	Eikenberg et al. (1993)
Xenotimes (YPO$_4$)	6.0^b		Eikenberg et al. (1993)

^{a} O/F: Atomic ratio of oxygen to fluorine.
^{b} Normalized to oxygen (^{21}Ne) and fluorine (^{22}Ne) (i.e., $^{21}Ne/[O] \cdot \alpha$ [g-O/g-sample]; $^{22}Ne/[F] \cdot \alpha$ [g-F/g-sample]); see text.

chemical compositions of the crust and the mantle (shown in Table 5.8) with empirical production rates (Eikenberg, Signer & Wieler, 1993). In Table 5.8, theoretically estimated production rates (Yatsevitch & Honda, 1997; Leya & Wieler, 1999) are also shown. Significant difference in $^{22}Ne^*$ production rate in the mantle among different estimates is due to the fact that Eikenberg et al. and Leya and Wiler considered only the α reaction, but did not take into account a neutron-induced reaction $^{25}Mg(n, \alpha)^{22}Ne$. However, even taking the larger estimate of ^{22}Ne production in the mantle by Yatsevitch & Honda (1997), the nucleogenic ^{22}Ne is still likely to be negligible in the mantle; if the average concentration of Ne in MORB were assumed to represent the mantle Ne concentration, ^{22}Ne in the mantle would be about three orders of magnitude as large as the previously estimated nucleogenic ^{22}Ne (cf. Table 6.1).

Heavier Nucleogenic Noble Gases: Besides He and Ne, U-enriched minerals generally contain a noticeable amount of nucleogenic isotopes of heavier noble gases that were produced by U fissions and various nuclear reactions involving α particles derived from U, Th decays. Typical nucleogenic noble gases are also shown in Table 5.8. Eikenberg et al. (1993) made extensive studies on nucleogenic noble gases in U-rich minerals (pitchblendes) that are geologically undisturbed as suggested from their concordant U-Pb ages. From the observed nucleogenic noble gases, they obtained the following production rates for nucleogenic ^{21}Ne, ^{22}Ne, and ^{38}Ar relative to the α-dose and the target element concentrations in the minerals.

$$^{21}Ne/[O] \cdot \alpha = (5.2 \pm 0.5) \times 10^{-8} \quad [g\text{-}O/g\text{-}sample]$$

$$^{22}Ne/[O] \cdot \alpha = (7.5 \pm 1.2) \times 10^{-8} \quad [g\text{-}O/g\text{-}sample]$$

$$^{38}Ar/[Cl] \cdot \alpha = (8.7 \pm 0.9) \times 10^{-8} \quad [g\text{-}O/g\text{-}sample]$$

5.6 Crustal He and Ne – Nucleogenic Components

Table 5.8. *Noble gases produced in the Earth interior for 4.5 Ga*

	Crust	Mantle	Remark
Chemical composition assumed for a calculation of isotope production			
O (wt %)	47	44	
F (ppm)	625	16	
Mg (wt %)	2	21	
Cl (ppm)	320	30	
K (wt %)	1.7 (*)		
U (ppm)	2.0	0.02	
Th (ppm)	6.0	0.06	
weight (g)	2.4×10^{25}	4.0×10^{27}	
Nucleogenic noble gases (cm^3 STP)			
^{4}He	5.6×10^{22}	9.2×10^{22}	U-decay
^{4}He	2.0×10^{22}	4.5×10^{22}	Th-decay
^{21}Ne	1.9×10^{15}	3.2×10^{15}	^{18}O (α, n)
	2.9×10^{15} ($)	4.6×10^{15} ($)	^{18}O (α, n)
	3.6×10^{15} (#)	5.8×10^{15} (#)	^{18}O (α, n), ^{24}Mg (n, α)
^{22}Ne	3.6×10^{14}	1.6×10^{13}	^{19}F (α, n)
	9.72×10^{14} ($)	4.5×10^{15} ($)	^{19}F (α, n)
	9.84×10^{14} (#)	1.2×10^{14} (#)	^{19}F (α, n), ^{25}Mg (n, α)
			^{19}F (α, p)
^{38}Ar	2.1×10^{13}	3.5×10^{12}	^{35}Cl (α, p)
^{40}Ar	5.04×10^{21} (*)		^{40}K decay
^{86}Kr$_{SF}$	2.9×10^{13}	5.5×10^{13}	^{238}U fission
^{136}Xe$_{SF}$	1.8×10^{14}	3.3×10^{14}	^{238}U fission

Table is modified from Eikenberg et al. (1993); estimations are made with empirical production rates on uranium-rich minerals (see text). ($) Leya & Wieler (1999) and (#) Yatsevich & Honda (1997) are theoretical estimates. (*) Allègre et al. (1996).

Although the primary processes for the production of ^{21}Ne, ^{22}Ne, and ^{38}Ar are α reactions of ^{18}O, ^{19}F, and ^{35}Cl, neutron capture is important in the production of ^{36}Ar. Eikenberg et al. also observed in many U-enriched samples nucleogenic ^{36}Ar, which they attributed to neutron capture on ^{35}Cl.

Since granites are generally enriched in U and Th, they are likely to contain a noticeable amount of the nucleogenic noble gases. Irwin and Reynolds (1995) studied noble gases in fluid inclusions in Stripa granite. They observed high abundances of nucleogenic Ar, Kr, and Xe as well as fissiogenic Xe. However, they found that measured amounts of these nucleogenic Ar, Kr, and Xe are a few orders of magnitude greater than the values expected from the chemical compositions of the host granite. Therefore, they concluded that most of these nucleogenic components were derived from outside of the Stripa granite, probably from older sedimentary rocks. The non-in-situ characteristic of nucleogenic noble gases seems to be rather common in crustal

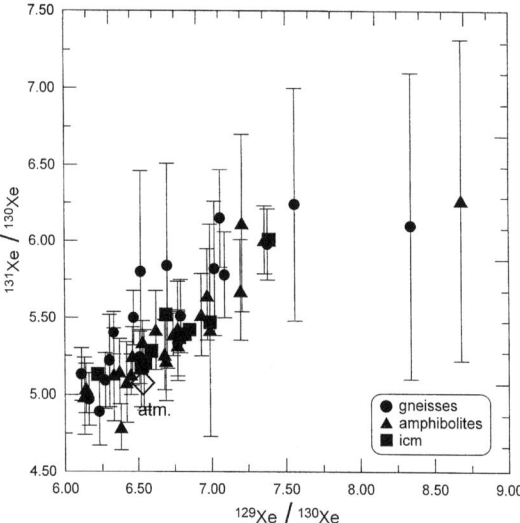

Figure 5.11 Xe three-isotope plot for samples from KTB deep-drilling core in Germany: icm = intrusive + cataclastic + mobilized (after Drescher et al., 1998). Note that $(^{129}Xe/^{131}Xe)exc \approx 1.1$ deduced from the linear trend in the diagram is close to the production ratio of nucleogenic Xe produced by neutron capture on Te.

environments and may bear important implications on geochemical problems. As another remarkable example, we discuss the case of KTB deep bore hole later.

Drescher, Kirsten, and Schäffer (1998) studied whole rock samples and mineral separates from the German Deep Drilling Project (KTB) drill core for noble gases. The drill site is located in northeast Bavaria, Germany, and the cores were obtained from a few hundred to 9000 meters below the surface. The drilled section is composed of two major rock types – amphibolites and paragneisses – and is supposed to represent the typical central European continental basement. Noble gas concentrations in the core samples are relatively low with modest excesses in ^{21}Ne due to (α, n) reactions with O and F. However, distinct nucleogenic Xe components are apparent. This is shown in Figure 5.11; excesses in ^{131}Xe and ^{129}Xe relative to the atmospheric Xe are apparent. The excess components are well correlated to each other, which gives $(^{129}Xe/^{131}Xe)_{exc} \approx 1.1$. The ratio is well within the production ratio of nucleogenic Xe from Te by neutron capture reactions: 0.64 for thermal neutron and 3.7 for epithermal neutron. Although there seems little doubt that the excess Xe was derived from Te, the amount of Te required to produce the observed amount of the nucleogenic Xe is unrealistically high; with a generally assumed neutron flux of 3000 neutrons $cm^{-2} a^{-1}$ in the crust and 370 Ma for the basement rock age, a few percent of Te would be required. Although there was no measurement of Te in the core samples, the content could hardly exceed a few parts per million. Therefore, Drescher et al. concluded that a mechanism other than in situ production must be at work in producing the excess nucleogenic Xe.

Such non-in-situ or parentless nucleogenic noble gases as demonstrated in the preceding examples may not be exceptional in a crustal environment. We note that there are a few reports on the similar "non-in-situ-nucleogenic" excess in ^{129}Xe and ^{131}Xe as those observed by Drescher et al. (e.g., Pinti et al., 1997) Although these anomalous results have often been dismissed as an experimental artifact, the preceding examples may suggest that nucleogenic noble gases, even the heaviest one, can migrate over a fairly wide range in a crustal environment. This might have important implications on nuclear waste disposal problems.

A remarkable example for migration of nucleogenic noble gases was demonstrated in the Oklo natural reactor zones by Shukolyukov and his colleagues (e.g., Shukolyukov et al., 1994; Meshik et al., 1995). Shukolyukov, Ashkinadze, and Verkhovskii (1976) observed isotopically anomalous Xe enriched mainly in ^{132}Xe and ^{131}Xe in zone 2 of the Oklo mine in Gabon. The occurrence of a natural nuclear chain reaction about 2 Ga in Oklo was first proved by French scientists, who thus confirmed the earlier suggestion for a natural nuclear reactor by Kuroda (1956). From detailed step-heating experiments, Shukolyukov et al. (1994) concluded that the isotopically anomalous Xe resulted from chemical fractionation of fission fragments that were produced by the natural nuclear chain reactions. Fast diffusion of fission Xe and its radioactive Ba precursors occur in fine-grained minerals at elevated temperatures and pressure. Anomalous Xe and Kr are therefore generated when fission fragments (Xe, Sn, Sb, Te, I, Kr, Se, and Br) migrate from the parent U-bearing mineral into adjacent phases where they subsequently decay into stable Xe and Kr isotopes. This natural process results in a pseudo-component whose isotopic composition is determined by relative residence times of the fission fragments in the environmental medium. In support of this process, Shukolyukov et al. (1994) showed that there was a clear correlation between the amount of isotopic anomaly and the sum of lifetimes of the respective β^- radioactive precursors. The anomalous Xe was named CFF (*c*hemically *f*ractionated *f*ission)-Xe by Shukolyukov et al. Recently, Meshik et al. (2000) suggested that the average isotopic composition of ^{136}Xe/^{134}Xe/^{132}Xe/^{131}Xe/^{130}Xe/^{129}Xe/^{128}Xe = 1/1.25/1.73/0.89/0.0045/0.274/0 for CFF-Xe.

Recently another precursor-lifetime-controlled isotopic anomaly is reported for the case of ^{4}He/^{3}He by Tolstikhin et al. (1999). They found anomalously low ^{4}He/^{3}He in some sedimentary rocks, down to 1/100 lower than those expected from radiogenic production (about 10^{-7}; cf. Section 5.6). They attributed the anomalously low ^{4}He/^{3}He ratio to a shorter residence time of ^{4}He than that of a precursor nucleus ^{3}H of ^{3}He in rocks.

5.7 The ^{4}He/^{40}Ar Ratio

Essentially all the ^{4}He and ^{40}Ar now observable on the Earth are radiogenic. Their ratio in any reservoir is governed by the relative abundance of U, Th, and K and the epoch and duration of their accumulation. The time dependence of the ^{4}He/^{40}Ar ratio is shown in Figure 5.12. For representative values such as Th/U = 3.3 and K/U = 10^4

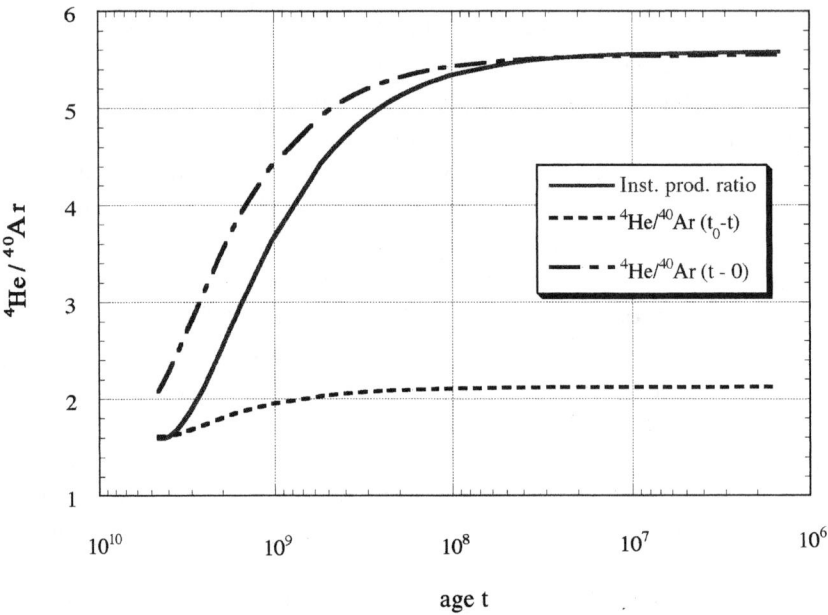

Figure 5.12 Variation in ^{4}He/^{40}Ar production as a function of age t. Full curve: the instantaneous production ratio. Dotted curve: the integrated production ratio for accumulation from t_0 ($t = 4.55$ Ga) until t. Semidotted curve: the integrated production ratio for accumulation from t until the present ($t = 0$). Input parameters are K/U = 10^4 (mass ratio) and Th/U = 3.3, with decay parameters in Table 1.4; different choices for K/U correspond to scale changes on the ^{4}He/^{40}Ar axis. The format is that of Zartman, Wasserburg, and Reynolds (1961).

(in mass ratios but not in atomic ratios!), the current production ratio is ^{4}He/^{40}Ar = 5.5, and integrated production from 4.5 Ga ago to the present is ^{4}He/^{40}Ar = 2.1. As seen in Figure 5.12, the instantaneous production ratio has not changed much over the last 10^8 years, and the integrated production ratio over such times varies even less from the current instantaneous rate. The ratio for radiogenic gas accumulated since 4.5 Ga ago has changed relatively little over the entire history of the Earth and hardly at all in the last 10^9 years or so.

Moreover, because of geochemical coherence among U, Th and K, of which validity has been well documented in various geochemical settings, and of the insensitivity of the ratio on time, the ^{4}He/^{40}Ar ratio in any closed system reservoir is confined in a narrow range from 2 to 6. Therefore, the ^{4}He/^{40}Ar ratio may be used as a parameter to assess a closed system condition of a given system with respect to noble gases. As such, studies by Elliot et al. (1993) and Ballentine and O'Nions (1991) may be instructive. Elliot et al. studied noble gas systematics in samples from the Dosso degli Angeli hydrocarbon gas field, between 3000 and 3750 m in the Po Basin, northern Italy. The formation of the Po Basin is due to sediment loading. Estimated radiogenic (^{4}He/^{40}Ar)$_{rad}$ ratio after correcting for air Ar ranges from 3 to 124, with an average value of 33 ± 4. The value is much in excess of the closed system ratio. There-

Table 5.9. *Trapped noble gases in cyclo-silicate minerals and in other common crustal minerals*

Mineral	^{4}He	^{20}Ne	^{40}Ar (^{36}Ar)	^{84}Kr	^{130}Xe
Cyclo-silicate					
Tourmaline	10^{-4}	10^{-10}	10^{-5} (10^{-9})	10^{-11}	10^{-12}
Cordierite	10^{-2}	10^{-7}	10^{-3} (10^{-6})	10^{-10}	10^{-12}
Beryl	10^{-2}		10^{-3}	10^{-10}	10^{-12}
Other minerals					
Apatite		10^{-10}		10^{-11}	10^{-12}
Quartz	10^{-7}	10^{-9}	10^{-9}	10^{-11}	10^{-12}
Feldspar		10^{-8}	(10^{-9})	10^{-11}	10^{-12}
Muscovite		10^{-8}	(10^{-8})	10^{-10}	10^{-11}
Magnetite (*)	10^{-5}	10^{-9}	(10^{-9})		10^{-11}

Data are from Saito et al. (1984) unless otherwise noted. Noble gas amounts are in $cm^3 STP\,g^{-1}$ and are representative values. (*) Markov et al. (1990): magnetites were separated from Precambrian banded iron formation (BIF) in Russia.

fore, Elliot et al. (1993) suggested the local differential release of ^{4}He over ^{40}Ar from the source minerals. In contrast to the Po Basin, the (^{4}He/^{40}Ar)$_{rad}$ ratios in the extensional Vienna and Pannonian Basins showed nearly the average crustal production ratio. Elliot et al. attributed this to the difference in tectonic setting between the Po Basin and the Vienna and Pannonian Basins in that the former has a regional thermal gradient less than half that of the latter; the tectonic activity reflected in the high heat flow may facilitate a more thorough extraction of both the radiogenic ^{4}He and ^{40}Ar from the reservoir, thereby giving nearly a closed system (^{4}He/^{40}Ar)$_{rad}$ ratio. In contrast, in the loading and stable crustal regions such as the Po Basin, incomplete extraction of noble gases gave rise to the fractionated ratio.

5.8 Cyclo-Silicate Minerals

The mineral beryls, cordierite, and tourmaline, collectively designated as cyclo-silicates, have large open channels in their crystal structures. These channels are large enough to accommodate atoms and molecules extraneous to the structure, and these minerals are well known as hosts of such extraneous elements as noble gases. Beryl and tourmaline are characteristic minerals of pegmatites, and cordierite is a common metamorphic mineral.

Cyclo-silicates, especially beryl and cordierite, are known to have an abnormally large amount of ^{4}He and ^{40}Ar, which cannot be accounted for the in situ production from the radioactive decay of U, Th, and ^{40}Ar. Lord Rayleigh (Strutt, 1908) first discovered that beryls contained excess ^{4}He, which cannot be accounted for by in situ radiogenic ^{4}He from U and Th. Subsequently, excesses of both ^{4}He and ^{40}Ar in all

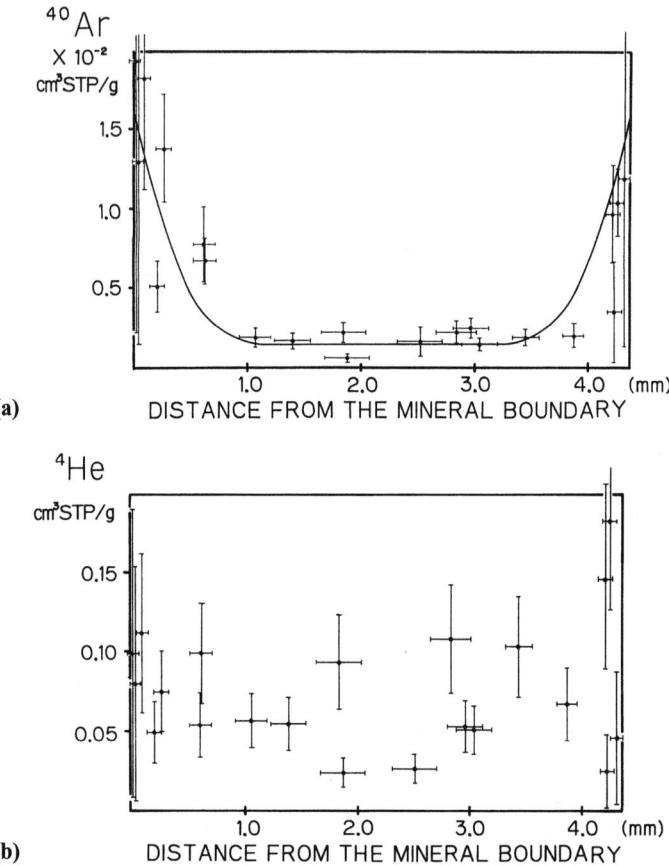

Figure 5.13 Concentration of ^{40}Ar (a) and ^{4}He (b) in a beryl crystal. The horizontal axis denotes a distance from the left-side boundary of a beryl crystal. The right-side boundary is situated at about 4.2 mm. After Toyoda and Ozima (1988).

these minerals have been amply documented (e.g., Saito et al., 1984, and references therein). Table 5.9 shows noble gas concentration in cyclo-silicates. Noble gas concentrations in other minerals common in crustal rocks are also shown for comparison. It is evident that the former has a few orders of magnitude larger ^{4}He and ^{40}Ar than the latter. Judging from a comparatively low concentration of U, Th, and K in the cyclo-silicates, it is very likely that ^{4}He and ^{40}Ar are essentially trapped components, but not in situ radiogenic components.

For both ^{4}He and ^{40}Ar, an age effect, first noted by Rayleigh (1933), appears. Damon and Kulp (1958), for example, found that beryls older than 2.5 Ga contain almost a factor of 10^2 more excess ^{40}Ar than those of ages younger than 1.0 Ga. In general, later investigations have supported the age effect, but it may still be possible that it is a sampling bias rather than a reflection of some geological process. The major discussion today concerns how the excess ^{4}He and ^{40}Ar were trapped in cyclo-silicates.

5.8 Cyclo-Silicate Minerals

One hypothesis is that the excess ^{4}He and ^{40}Ar diffuse into the cyclo-silicates throughout the geologic interval between the formation of the cyclo-silicates and the present. The age effect is a natural consequence of this model. The second hypothesis is that the cyclo-silicates trapped noble gases as they crystallized. This model would follow from the model that gases were trapped by water molecules or alkali ions such as Cs^+, incorporated during crystallization, which plugged the structural channels. Saito et al. (1984) observed that $^{40}Ar/^{36}Ar$ and $^{136}Xe/^{130}Xe$ ratios in the beryls are much higher than the ratios trapped in the cogenetic minerals, which they found difficult to explain if these excess noble gases diffused into the beryls after the crystallization of the pegmatite.

With the use of a laser gas-extraction system, Toyoda and Ozima (1988) measured a helium and argon concentration profile in a beryl crystal. The sample is from a pegmatite in South Dakota, United States, and the K-Ar isochron age of the pegmatite is 1.86 Ga (Saito et al., 1984). The concentration profiles are shown in Figure 5.13(a). Even though ^{4}He shows nearly a uniform concentration across the whole grain, ^{40}Ar is highly concentrated near the grain boundaries, resembling a diffusion profile. They showed that the observed ^{40}Ar concentration profile was fairly well simulated, if ^{40}Ar diffused into the beryl from the environment across the grain boundaries. The best-fitting diffusion curve [shown in Figure 5.13(a)] to the observed profile is obtained for $\sqrt{Dt} = 0.025$ cm with the initial ^{40}Ar concentration of about 10% of the presently observed amount. Since the choices of some realistic values for D and t, such as $D = 2 \times 10^{-18}$ cm^2 s^{-1} (experimentally determined diffusion coefficient of Ar in beryl at about 400°C, Toyoda & Ozima, 1988) and $t = 10^7$ years, satisfies the required value of the diffusion length, they favored postformational noble gas acquisition. The nearly uniform He concentration is then attributed to the much faster diffusion of He in the beryl. However, their explanation does not give a straightforward explanation for the age effect. The problem is still open to future investigation.

Chapter 6

The Mantle

6.1 Introduction

Their chemical inertness and low abundance make noble gases a unique geochemical tracer, and their application to the study of mantle geochemistry, differentiation, and volatile separation is one of the major successful developments in recent earth science research. Some noble gas isotopic compositions are quite different between the crust and the mantle, and some of these, such as $^{40}Ar/^{36}Ar$ and $^{129}Xe/^{130}Xe$, involve radiogenic isotopes and thus undergo predictable variations in time. Noble gases thus not only provide information on mantle processes and their spatial variations but also provide chronological constraints. In applying noble gases to the study of mantle geochemistry and dynamics, the subject of this chapter, we must first appraise the noble gas state of the mantle and how it differs from the atmosphere. Despite much recent progress, this issue is not yet settled, and some difficulties should be kept in mind during the following discussion. One difficulty is evaluating how representative of the mantle as a whole the various mantle-derived samples actually are. A second difficulty is that noble gases in mantle samples frequently occur in such low concentrations that there can be technical problems in their analysis. A problem corollary to low abundances is that it is not always easy to distinguish true mantle gases from potential atmospheric contamination.

To obtain information on the noble gas state in the mantle, it is necessary to analyze mantle-derived materials that have trapped mantle noble gases. Accessible samples include volcanic rocks, volcanic gases, mantle xenoliths, and diamonds. Among various mantle-derived materials, submarine volcanic rocks are particularly useful because of their wide occurrence and their relatively large (for mantle samples) amounts of trapped noble gases. So far, information has been obtained mainly from noble gas analyses of MORBs (Mid Oceanic Ridge Basalts) and OIBs (Ocean Island Basalts). Although both kinds of rocks are derived from the mantle, their mantle sources are quite different from each other. There is a widely held view that MORBs are derived from depleted (in incompatible elements) and degassed mantles, whereas OIBs are hot-spot (mantle plume) volcanic rocks, which represent less depleted and less degassed mantles (see Sections 6.6 and 6.7 for further discussion). Hence, studies on MORB and hot-spot volcanic rocks constrain the noble gas state in different zones in the mantle. It is important to note that because of their relatively young ages, these oceanic volcanic rocks provide information about only the present or recent mantle.

Recently, precise noble gas isotopic analyses of diamonds have also become available. Diamonds are derived from the mantle. However, diamonds are unique in that most known diamonds are of Precambrian age; hence, we can hope for noble gas records of the very ancient mantle, complementary to the constraints on the present mantle provided by recent volcanics.

Subaerial (i.e., erupted on the surface, rather than under water) volcanic rocks, even though mantle derived, evidently contain mostly atmospheric noble gases, trapped from the atmosphere during eruption. They are thus unlikely to offer useful information on the noble gas state of the mantle. Apparently there are exceptions, however. Dodson, Kennedy, and De Paolo (1997) report that olivine phenocrysts in Columbia River basalts contained typical mantle-type He and Ne, although heavier noble gases are indistinguishable from atmospheric components (e.g., $^{40}Ar/^{36}Ar < 350$). This suggests that the lighter noble gases might somehow have behaved differently from the heavier gases in the magma source.

Sections 6.2–6.5 describe in more detail the major kinds of materials from which information on the state of the mantle may be derived, with emphasis on the character of the materials and their mantle sources. Noble gas data from these materials are summarized in Sections 6.6 and 6.7.

6.2 Mantle-Derived Volcanic Rocks

Volcanic rocks erupted under water, particularly under deep water, characteristically have glassy margins or rinds that represent quenching by the water. The glassy margins contain comparatively large amounts of trapped noble gases that have isotopic and elemental compositions different from those in air and that are commonly held to best represent noble gases in their mantle sources (Dymond & Hogan, 1973).

Most submarine volcanic rocks contain CO_2-filled vesicles (bubbles) in glassy margins. Because noble gases in silicate melts partition very effectively into a gas phase (i.e., their solubilities are low), it would be expected that noble gases in submarine volcanics would be found in the bubbles; as will be discussed later, this seems indeed to be the case (e.g., Kurz & Jenkins, 1981; Marty & Ozima, 1986; Sarda & Graham, 1990; Graham & Sarda, 1991). The popping rocks, so-called because of

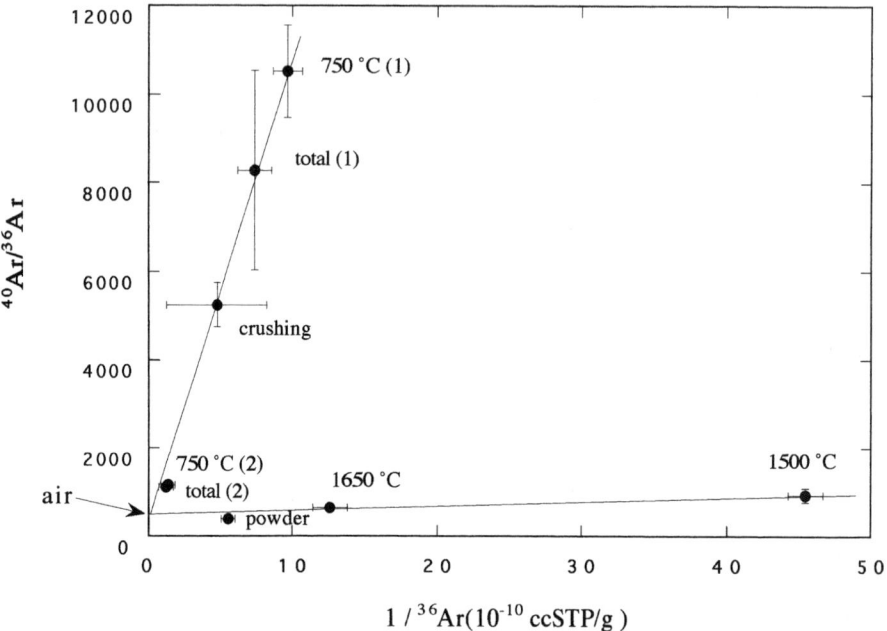

Figure 6.1 An ordinate-intercept diagram with lines indicating mixing of atmospheric Ar with nonatmospheric (mantle) Ar located off the figure to the upper right. Slope is proportional to the amount of mantle Ar (see text). The sample is a Mid-Atlantic Ridge tholeiite. Redrawn from Marty and Ozima (1986).

audible vesicle breakage when the rocks are brought from great depth to the surface, are well known for their unusually high concentration of bubbles (and noble gases) and are especially useful for precise noble gas isotopic measurements (e.g., Burnard, Graham & Turner, 1997). Noble gases in vesicles can be extracted either by crushing the rocks in vacuum or by heating a sample to a temperature high enough to destroy vesicles but not so high as to totally melt the sample.

Marty, Zashu, and Ozima (1983) observed that there are two noble gas components in submarine volcanic rocks: one residing in vesicles, mostly in glassy chilled margins, and another in the matrix. From argon isotopic analyses on submarine volcanic samples, Marty and Ozima (1986) concluded that the component in the vesicles represents essentially a pristine mantle noble gas, whereas the component in the matrix is more contaminated by air. This is demonstrated in Figure 6.1, where argon isotopic data for an oceanic tholeiite from the Mid-Atlantic Ridge are plotted on an ordinate-intercept ($^{40}Ar/^{36}Ar$ versus $1/^{36}Ar$) diagram. The total fusion (melting) gas, the gas released during crushing, and the gas extracted by heating to 750°C lie on a line, whereas the higher temperature fractions and the fusion of crushed sample (crushed to less than the minimum size of vesicles, i.e., to about 200 μm before fusion, and hence essentially free of vesicle-trapped gases) form a different line. The upper line in Figure 6.1, which corresponds to the Ar in vesicles, is interpreted to reflect mixing of air Ar and Ar with a very high $^{40}Ar/^{36}Ar$ ratio (a mantle component). The lower line, which corresponds to the matrix component, is then understood to represent mixing of air argon with a very small amount of mantle Ar.

To show this quantitatively, ordinate-intercept analysis (e.g., Takaoka & Nagao, 1980) of mixing between air and nonatmospheric (mantle) Ar gives

$$(^{40}\text{Ar}/^{36}\text{Ar}) = (^{40}\text{Ar}/^{36}\text{Ar})_a + S/(^{36}\text{Ar}) \tag{6.1}$$

where $S = {}^{36}\text{Ar}_m \times \{(^{40}\text{Ar}/^{36}\text{Ar})_m - (^{40}\text{Ar}/^{36}\text{Ar})_a\}$
where subscripts m and a denote the mantle component and air components, respectively, and unsubscripted variables denote the (measured) mix of these two components. It is safe to assume that $(^{40}\text{Ar}/^{36}\text{Ar})_m \gg (^{40}\text{Ar}/^{36}\text{Ar})_a$, in which case Equation (6.1) can be approximated

$$(^{40}\text{Ar}/^{36}\text{Ar}) = (^{40}\text{Ar}/^{36}\text{Ar})_a + \{^{36}\text{Ar}_m(^{40}\text{Ar}/^{36}\text{Ar})_m\}/{}^{36}\text{Ar} \tag{6.2}$$

Equation (6.2) shows that if different samples contain different amounts of air Ar but the same amount of mantle Ar, which can be assumed to have fixed isotopic ratio $(^{40}\text{Ar}/^{36}\text{Ar})_m$, they produce a linear array with a positive slope on an ordinate-intercept diagram such as Figure 6.1. The slope of the mixing line in the diagram is then proportional to the amount of $^{36}\text{Ar}_m$ (equivalently, $^{40}\text{Ar}_m$) in each sample. The much steeper slope for the vesicle component (Figure 6.1), compared to the high-temperature and crushed samples, indicates the much larger contribution of mantle Ar in the vesicles.

In an analogous way, Equation (6.1) can also be rearranged interchanging the roles of $^{36}\text{Ar}_m$ and $^{36}\text{Ar}_a$, with the same approximation yielding

$$^{40}\text{Ar}/^{36}\text{Ar} = (^{40}\text{Ar}/^{36}\text{Ar})_m - \{^{36}\text{Ar}_a(^{40}\text{Ar}/^{36}\text{Ar})_m\}/{}^{36}\text{Ar} \tag{6.3}$$

If a suite of samples contains the same amount of air Ar, we see from Equation (6.3) that the data will form a linear array with a negative slope on an ordinate-intercept diagram; the slope will be proportional to the amount of air Ar in the samples.

Note that even if the condition for a uniform amount of either mantle or air ^{36}Ar does not hold exactly, a positive or negative correlation on an ordinate-intercept diagram will still result if either $^{36}\text{Ar}_a \gg {}^{36}\text{Ar}_m$ or $^{36}\text{Ar}_a \ll {}^{36}\text{Ar}_m$. In the case of $^{36}\text{Ar}_a \gg {}^{36}\text{Ar}_m$, for example, variation in total ^{36}Ar corresponds to variation in $^{36}\text{Ar}_a$, but it is largely independent of $^{36}\text{Ar}_m$ (i.e., $^{36}\text{Ar}_m$ is effectively constant and the data should show a positive correlation slope). Conversely, $^{36}\text{Ar}_a \ll {}^{36}\text{Ar}_m$, in which the trapped noble gases are overwhelmingly from the mantle component, the data should show a negative correlation.

Most of the Ar data so far available for mantle-derived samples exhibit an approximate positive linear trend on an ordinate-intercept plot, indicating that in most samples air Ar is the dominant contribution (to ^{36}Ar). A few cases show a negative trend, however. Figure 6.2 shows an example obtained for mantle xenoliths from Salt Lake Crater (Hawaii) lava by Rocholl et al. (1997). Following the preceding discussion, we may argue that the Ar observed in these samples is dominantly of mantle origin.

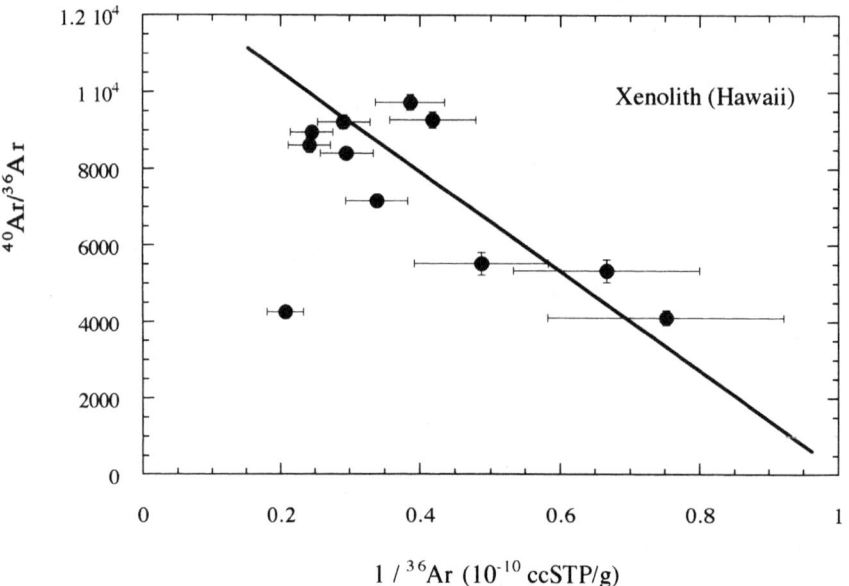

Figure 6.2 An ordinate-intercept diagram for Salt Lake Crater (Hawaii) xenoliths (cf. Figure 6.1). The negative slope of the correlation suggests that mantle Ar dominates over atmospheric Ar in these samples (see text). Reproduced from Rocholl et al. (1997).

It is generally believed that MORBs represent depleted (upper) mantle, and that volcanic rocks from major hot-spot regions such as Hawaii were derived from less-depleted (lower?) mantle. Noble gas isotopic data support this generalization in that the MORB source is more degassed than the OIB source(s) (see Section 6.7), as reflected in MORBs having a lower ratio of primordial ^{3}He and ^{36}Ar to radiogenic ^{4}He and ^{40}Ar, respectively.

It is important to note, however, that even though MORBs have relatively uniform ^{3}He/^{4}He ratios, hot-spot volcanics show wide variation in ^{3}He/^{4}He, ranging from sub-MORB values up to values a few times higher than in MORBs (e.g., Zindler & Hart, 1986; Kurz, 1993). Zindler and Hart (1986) suggested that the diversity in the ^{3}He/^{4}He ratios in OIB volcanics could be attributed to radiogenic ^{4}He produced in a pre-eruptive magma source. The diversity might also be due to the variable admixture of MORB-type He in hot-spot magmas or simply to heterogeneity in the OIB-source, less-degassed mantle. From He-Ne-Ar systematics, Hiyagon et al. (1992) concluded that contamination of MORB-type He is unlikely to be important in Loihi hot-spot volcanic rocks and suggested that mantle source regions for hot-spot volcanics are more heterogeneous than those for the degassed mantle sources of MORBs.

In addition to these two major types of oceanic volcanic rocks (MORBs and OIBs), back-arc volcanic rocks represent a third type of source mantle, one that has been subjected to admixture of subducting crustal materials. Honda et al. (1993b) analyzed pillow basalt glasses from the Lau Basin, a typical active back-arc basin with a half-spreading rate of approximately 38 mm/yr. They found that the average ^{3}He/^{4}He ratio is similar to the MORB value, but with large variation, and that ^{20}Ne/^{22}Ne ratios show a range similar to MORBs, concluding that He and Ne in the Lau Basin samples were similar to that observed in MORBs but with an additional slab-derived component contributing radiogenic He without any associated neon. Hilton et al. (1993), however, concluded that He-Ar systematics in Lau Basin volcanics were most satisfactorily explained by assimilation of the Lau crust by previously degassed magma prior to eruption, and that there was no evidence for slab-derived He in the samples.

6.3 Diamonds

Diamonds are also mantle samples and can thus also provide information about the noble gas state of the mantle complementary to that obtained from oceanic volcanic rocks. For this purpose, they also have the virtue of antiquity. The oldest oceanic crust is Mesozoic, so submarine volcanics sample the mantle with a history covering only a relatively small fraction of the age of the Earth. Diamonds are considerably older and, thus, offer the prospect of sampling ancient mantle over a significant fraction of Earth's history. Crustal emplacement ages of diamonds range from Cretaceous to early Precambrian, but their crystallization ages may be much older. Radiometric dates so far determined from supposedly syngenetic inclusions in several diamonds all show Precambrian ages, some being older than 3 Ga (Richardson et al., 1984, 1994; Burgess, Turner & Harris, 1991). It seems reasonable to assume that most diamonds are of Precambrian age.

The provenance of diamonds is still not very well understood, however. The experimentally determined pressure-temperature stability field for diamond indicates crystallization at mantle depths greater than 100 km, but some diamonds may have crystallized at much deeper levels, or diamond source material, possibly in fluid form, may have migrated from the deeper mantle and crystallized at shallower levels (Boyd, Pineau & Javoy, 1994).

Some diamonds are characterized by a "core" and "coat" morphology, notably including the Zaire cubic (microcrystalline) diamonds (Navon et al., 1988), which provide a large part of available noble gas data for diamonds. The coat of a coated diamond typically contains significant amounts of noble gases, often more than the amounts found in the glassy margins of submarine basalts. Ozima (1989), for example, found that in the coat of a clearly zoned Zaire cubic diamond, the Ne and Xe concentrations were nearly an order of magnitude higher than MORB glass averages. The high concentrations are important in improving the precision of isotopic analysis. Navon et al. (1988) reported that the outer coat contains inclusions of "mantle fluids," which seem to be ubiquitous in the upper mantle. Noble gases in the cubic diamonds reside mainly in the fluid inclusions (Ozima et al., 1989) and may thus represent fluid extraction of noble gases from a fairly extensive region of the mantle.

Since diamonds crystallized in the mantle at temperatures likely to exceed 1000°C and may well have resided there for a substantial time (e.g., more than 1 Ga), it is important to consider whether diamonds have retained their original mantle noble gases (at crystallization) or whether they have undergone some exchange with the local mantle.

To judge this issue, we must know noble gas diffusion coefficients in diamond. There are a few reports of experiments on noble gas diffusion in diamonds, mostly of He because of the relative ease of experiments. Luther and Moore (1964) reported a surprisingly high He diffusion coefficient in a synthetic diamond that was doped with B to produce He by neutron irradiation, but this high diffusion coefficient is now generally attributed to damage due to the very high radiation dose applied to the sample. Lal et al. (1987) measured He diffusion coefficients in several industrial-grade diamonds by a gas-soaking method in the temperature range 1200°C to 1700°C; similar results were also suggested by Wiens et al. (1994). The Lal et al. and Wiens et al. are much higher than those estimated by the standard laboratory gas-release method for the temperature range up to 2000°C (Honda et al., 1987; Ozima, 1989). The discrepancy can evidently be attributed to the different temperature ranges over which the measurements were performed. Laboratory experience suggests that noble gases in diamond reside in two different characteristic sites, one (accounting for no more than a few percent of total gas content) releasing below 1500°C and the other releasing only just below the graphitization temperature (around 2000°C in vacuum). The former is characterized by a much larger diffusion coefficient than the latter (Zashu & Hiyagon, 1995). The high-temperature components, which generally comprise more than 90% of the noble gases in diamonds and hence are more representative of the trapped component, have extremely low diffusion coefficients (cf. Figure 2.10); simple extrapolation of the experimental data obtained at 1500–2000°C (Zashu & Hiyagon, 1995) to an ambient mantle temperature of 1300°C yields $D/a^2 = 10^{-16}\,\text{s}^{-1}$ for He and $10^{-20}\,\text{s}^{-1}$ for Ar (Ozima, 1989), suggesting that at least Ar and also a substantial fraction of He can be retained in diamond grain for a geologically long time under mantle conditions.

Another important (and contentious) issue regarding noble gases in diamonds, particularly He, is the extent to which their indigenous mantle gases may be contaminated by extraneous components introduced during crustal residence, because the He isotopic ratio (^{3}He/^{4}He) observed in some diamonds is higher than the primordial He (i.e., planetary He) in meteorites (see Section 7.3). Lal et al. (1987) and Lal (1989) (also see Begemann, 1994) called attention to the circumstance that some diamonds, emplaced near the Earth's surface or in an environment with high U content, may have been subjected to cosmic ray irradiation or to α particle (^{4}He) injection from surrounding U-rich minerals. Some diamonds may have acquired significant cosmic ray-induced spallation He, which would act to raise the ^{3}He/^{4}He ratio; others may have acquired significant radiogenic ^{4}He from their surroundings. In evaluating these issues, it is important to know the circumstances of the diamonds' crustal history. Unfortunately, many of the diamonds analyzed for noble gases were acquired through commercial channels so that their geological settings, considered proprietary information, are usually not well documented.

Lal (1989) derived an expression for the implantation rate I (atoms/g/s) of radiogenic ^{4}He: $I = 3Q\rho_k S/(4\rho_d r)$, where Q and ρ_κ are the rate of ^{4}He production in and the density of the surrounding medium, S is the mean range of ^{4}He (about 20 μm), and r and ρ_d are the radius and density of the diamond. In a study of 12 diamonds whose crustal histories are well known, Ozima (1990) found that the amounts of implanted ^{4}He predicted by this expression were one to two orders of magnitude lower than the He actually observed in these diamonds and thus concluded that the observed ^{4}He is primarily a trapped mantle component. In contrast, McConville and others (McConville & Reynolds, 1989; McConville et al., 1991) concluded that most of the ^{4}He in a group of Western Australian gem-quality diamonds must have been implanted, as suggested by Lal (1989), because their ^{4}He/^{40}Ar ratios are around 10^3, far in excess of values expected for common mantle materials.

The possibility of spallation ^{3}He is a particular concern because of the possibility that the high ^{3}He/^{4}He ratios observed in some diamonds might be an artifact and not representative of the mantle. The issue remains unresolved. Note, however, that for cosmic ray irradiation to produce the amount of ^{3}He observed in some diamonds (more than 10^{-11} cm^3 STP/g), the diamonds must have been very near the surface (within about a meter) for a long time (several million years). This precludes diamonds recovered from mines. Since alluvial diamonds are commonly buried in gravel deposits, the required geological setting would be expected to be rare, and the significance of cosmogenic ^{3}He thus would be likely to be exceptional.

Another reason for concern that high ^{3}He/^{4}He ratios in diamonds do not actually represent a mantle source is the possibility of ^{3}He production by neutron-induced nuclear reactions. A number of reactions are possible, but chief among them in this context is ^{6}Li(n, α)^{3}H $\rightarrow$ ^{3}He (Zadnik et al., 1987). Unambiguous quantitative evaluation of this possibility is difficult because neutron fluence depends on environment, often largely unknown, and Li concentration, for which scant data are available. Ozima (1989) has reviewed the issues relevant to neutron-induced production of ^{3}He in diamonds, concluding that possibly, excepting very rare circumstances, there are not enough neutrons and not enough Li for this suggested mechanism to be significant.

Much of this discussion has focused on the He in diamonds. For the heavier gases, spallation should be negligible (for want of appropriate targets), diffusion coefficients should be much lower than for He, and implantation from outside should be negligible (the only known possibility is fission Xe). One may thus be more confident that the heavier gases in diamonds are representative of the mantle at the time of formation.

It is implicit in all the preceding discussion that diamonds actually are formed in the mantle. We note that a type of diamond occurrence called carbonado is likely to be of crustal rather than mantle origin (Ozima & Tatsumoto, 1997); these are not considered in connection with characterization of the mantle.

6.4 Volcanic Gases, Fumaroles, and Hydrothermal Waters

Because magmas are directly or ultimately derived from the mantle, volcanic gases must be expected to contain mantle-derived noble gases. This relation is seldom evident, however, except for He, which is so strongly depleted in the atmosphere; radiogenic ^{40}Ar; and occasionally nucleogenic ^{21}Ne and ^{22}Ne. Otherwise, the noble gas isotopic compositions in volcanic gases are generally indistinguishable from atmospheric composition. This is because volcanic gases consist mainly of meteoric water (generally 70% to essentially 100% by volume; e.g., see Kusakabe & Matsubaya, 1986), which was saturated with atmospheric noble gases when it entered the local circulation system, and these atmospheric gases typically overwhelm the mantle component. Radiogenic Ar and He and nucleogenic Ne may be introduced into volcanic gases from the mantle source but also possibly from the country rocks around the magma chamber. Although volcanic gases thus do not offer useful information on the heavier noble gases of the mantle, their He isotopic compositions have been applied to volcanology as a mantle tracer, especially in arc environments where other suitable materials are unavailable (e.g., Craig, Lupton & Horibe, 1978; Sano & Wakita, 1987; Poreda & Craig, 1989; Hiyagon & Kennedy, 1992).

Noble gases in fumaroles and hydrothermal waters also have atmospheric isotopic compositions except for radiogenic (^{4}He, ^{40}Ar) and nucleogenic (^{21}Ne, ^{22}Ne) isotopes, which were produced in the crust (e.g., Craig et al., 1978; Kennedy et al., 1985). Relative noble gas elemental abundances in hydrothermal waters show a considerable degree of fractionation due to differences in equilibrium solubility in water at various temperatures (e.g., Mazor & Verhagen, 1983; Kennedy et al., 1985; Pinti et al., 1997; Ballentine, 1997). In favorable circumstances, the temperature dependence of noble gas solubility in water may be used to infer the paleotemperature of the recharge area for the aquifer (e.g., Stute et al., 1992) (see Section 4.7).

6.5 Mantle Xenoliths

Basaltic volcanic rocks occasionally contain ultramafic xenoliths of unambiguous mantle provenance. Noble gases in these xenoliths may thus be studied as another avenue to information about mantle noble gases. In this respect, xenoliths might have an advantage over basaltic volcanics also originating in the mantle, since they crystallized at depth and may be less susceptible to gas loss or air contamination on the surface. Moreover, some xenoliths with fluid inclusions are known to have higher noble gas concentrations than basaltic volcanic glasses (Poreda & Farley, 1992) thereby allowing more precise analyses.

To interpret xenolith data, however, it is important to consider the general question of how well they can be expected to preserve the record of their mantle environment independently of the record in their host magmas (cf. Hofmann & Hart, 1978, for solid elements; Trull & Kurz, 1993, for noble gases; and Kellog, 1992, for a general discussion). Studies of Sr, Pb, and Nd isotopic compositions in various xenoliths indicate that these elements generally have not equilibrated with their host lavas. Experimental studies of diffusion coefficients of Sr, Mg, Al, and other cations indicate that this is reasonable, and that isotopic disequilibrium on a centimeter scale could persist for periods at least of the order of 10^8 to 10^9 year in dry crystalline mantle. It is not clear to what extent this generalization extends to the noble gases, since they are expected to have higher diffusion coefficients, particularly the lighter gases.

Bernatowicz (1981), for example, found the same gas compositions (except for plausibly in situ radiogenic ^{40}Ar) in xenolith minerals and their host basalts; this finding might indicate that the sources of both had the same gas composition or that the xenoliths equilibrated with their hosts. There are, however, data indicating lack of isotopic equilibration. Rison (1980b), for example, found differences among the minerals in a South African kimberlite xenolith. Also, Staudacher, Sarda, and Allègre (1990) observed that dunite, harzburgite, and wehrlite nodules (5–10 cm in size) in 2300-year-old lava flows had the same He isotopic ratio as the host lava but that their Ne and Ar isotopic ratios were quite different; therefore, they concluded that the original He in the xenoliths had exchanged (by diffusion) with He from magma during contact on a timescale of the order of a year. In contrast, Farley and Poreda (1991) observed that He as well as heavier gas isotopic compositions in spinel lherzolite xenoliths in Samoan volcanics are very different from compositions in the host lava. Dunai and Baur (1995) studied He-Ne-Ar systematics in a large number of mantle xenoliths from various European Cenozoic volcanic provinces such as the Massif Central in France, the Eifel in Germany, and Spitsbergen and the Graz Basin in Austria. They concluded that He systematics is decoupled from the Sr-Nd systematics of the xenolith but closely related to the Sr-He, Nd-He, and U-He systematics of the European subcontinental magma sources. These results suggest that noble gases in these xenoliths are equilibrated with the host magma and do not retain the information of the xenolith source. An even more perplexing case was reported by Matsumoto et al. (1997) for subcontinental xenoliths in recent alkali basalts from southeastern Australia. They found that even though the fact that Ne in fluid inclusions in apatite separated from spinel lherzolites shows an isotopic signature similar to that of plume-related OIB (e.g., as in Hawaii), Ne released from the crystalline structure of the apatite showed atmospheric Ne mixed with nucleogenic ^{21}Ne and ^{22}Ne. Amphiboles and olivines in the same sample had typical MORB-type He and Ne. As is evident from these examples, noble gases in xenoliths may reflect multistage histories, and caution is required before conclusions about the noble gas state in the mantle source of the xenoliths can be drawn.

6.6 Mantle Noble Gas (Abundance)

In Sections 6.6 and 6.7, we consider how the data for mantle-derived sources described in Sections 6.1–6.5 can be used to constrain the noble gas state of the mantle.

Noble gas concentration data for various mantle-derived materials are listed in Table 6.1 and illustrated in Figure 6.3. Concentrations are diverse, for a given element ranging over some one to two orders of magnitude. The values in Table 6.1 and Figure 6.3 are means, however; for individual samples, the range is more characteristically three and in the extreme four orders of magnitude. Even so, considering the range of

Table 6.1. *Noble gas abundances[a] in mantle-derived materials*

	N[b]	^{3}He 10^{-9} cm^3 STP g^{-1}	^{20}Ne 10^{-9} cm^3 STP g^{-1}	^{36}Ar 10^{-9} cm^3 STP g^{-1}	^{84}Kr 10^{-12} cm^3 STP g^{-1}	^{132}Xe 10^{-12} cm^3 STP g^{-1}
MORB						
Ozima & Zashu (1983)	10	0.16 (0.20)	0.48 (0.74)	0.57 (0.28)	14 (9)	2.0 (5.7)
Sarda & Graham (1990)	32	0.18	0.16	0.053	2.4	0.28
Hiyagon et al. (1992)	9	0.15 (0.12)	0.66 (0.64)	1.88 (2.5)	55 (78)	5.6 (3.7)
Niedermann, Bach & Erzinger (1997)	11	0.145 (0.044)	0.198 (0.188)	0.113 (0.112)	4.9 (5.8)	1.17 (2)
OIB (Loihi)						
Sarda & Graham (1990)	14	0.024	0.77	1.3	33	1.9
Hiyagon et al. (1992)	5	0.062 (0.0057)	0.51 (0.19)	0.79 (0.46)	65 (8.8)	1.9 (1.7)
Honda et al. (1993a)	16	0.014 (0.015)	0.78 (0.83)	1.54 (2.2)	46 (58)	2.8 (2.8)
Valbracht et al. (1997)	12	573[c] (700)	0.03[d] (0.058)	0.562 (1.1)	13.9 (22)	0.84 (0.11)

	b					
Diamond						
Ozima & Zashu, 1983	16	0.013 (0.019)	0.19 (0.20)	0.30 (0.52)	10 (19)	1.3 (1.7)
Ozima & Zashu, 1991	9	0.050 (0.0027)	0.54 (0.52)	2.75 (1.35)	42 (2.3)	7.0 (3.7)
Xenolith						
Poreda & Farley, 1992	25	0.0030 (0.0042)	0.29 (0.02)	0.54 (0.47)	8.4 (1.3)	1.8 (1.1)
Atmospheric abundance[e]						
Table 1.3			10.9	20.83	431	2.4

[a] Concentration (cm^3 STP g^{-1}) is given in units shown in the column headers. The figures given are the mean of the number of samples analyzed; figures in parentheses indicate one-sigma dispersion about mean (not analytical error).
[b] Number of sample analyzed.
[c] ^{4}He.
[d] ^{22}Ne.
[e] Total air inventory divided by mass of the Earth (5.976×10^{27} g).

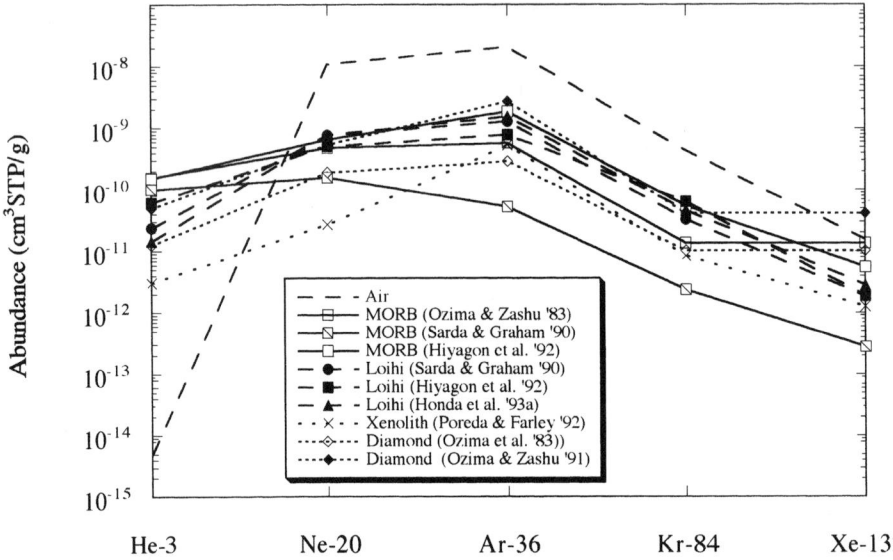

Figure 6.3 Absolute noble gas concentrations in selected mantle-derived samples. (Data are from Table 6.1. Air data are from Table 1.3.) The ordinate scale is in cm³ STP/g.

geographic locations, mantle provenance, and type of sample, it seems noteworthy that the range of concentrations is not larger than it actually is.

Another feature that we find noteworthy is that systematic differences between various types of samples is not especially striking. For example, given the widely based generalization that MORBs are derived from depleted and degassed mantle, in contrast to the derivation of OIBs, of hot-spot or plume provenance in less-depleted and less-degassed mantle, we might expect that OIBs would contain greater concentrations of noble gases than MORBs. On the basis of the Paris laboratory data, Allègre, Staudacher, and Sarda (1987) reached just this conclusion, but this generalization is not strongly supported by the broader database.

At least the volcanic rocks are derived from the mantle by extraction of partial melting. In view of the generalization that the noble gases are incompatible elements in crystal-liquid partitioning (i.e., they are preferentially partitioned into the melt; Section 2.4), we might expect that absolute noble gas concentrations in volcanic rocks will be higher than the concentrations in their mantle source rocks. In this light, it is further noteworthy that noble gas concentrations in mantle-derived rocks are comparatively low, specifically they are low compared to "atmospheric" abundances (Figure 6.3). At face value, this relation indicates that the total inventory of primordial noble gases in the mantle is less, perhaps much less, than the atmospheric inventory (i.e., that the Earth is already largely degassed). This is a plausible conclusion and can be reached by other lines of argument.

Figure 6.4 illustrates *patterns* of noble gas elemental abundances by normalizing the absolute concentrations to Ar and to atmospheric composition. The range of variation in elemental ratios is smaller than the range in absolute concentrations, mostly

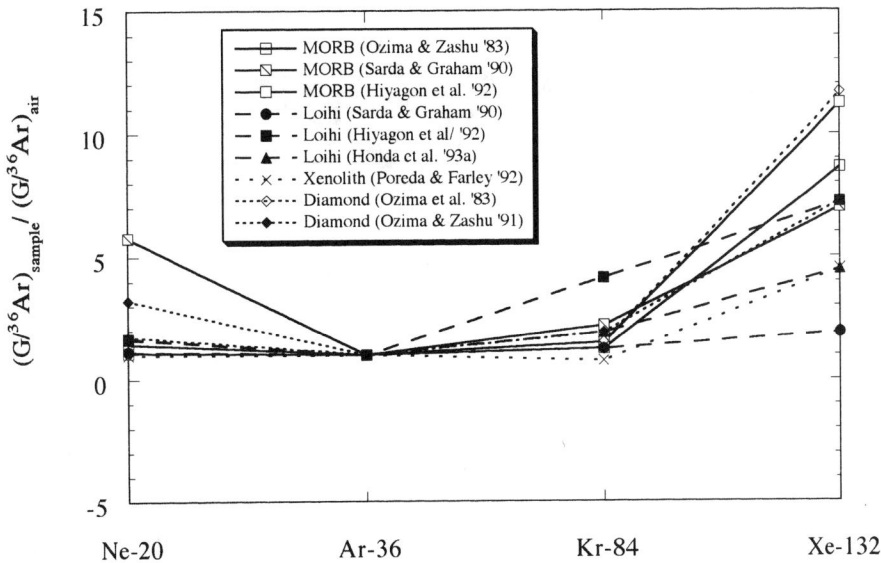

Figure 6.4 Elemental abundance patterns for noble gases in mantle-derived samples (cf. Figure 6.3 and Table 6.1). The ordinate is $(G/^{36}Ar)_s/(G/^{36}Ar)_a$, where subscripts s and a designate sample and atmosphere, respectively, and G represents any noble gas isotope.

less than one order of magnitude rather than two. Again in elemental abundance patterns (Figure 6.4), there is not a striking systematic difference between various types of mantle-derived samples. There *is* a characteristic enrichment in the relative abundances of the heavier gases Kr and Xe, relative to Ar, but the same is true of the lighter gas Ne. These features might reflect simply the different responses of the various noble gases to the processes of extraction from the mantle, but they also might represent real differences in elemental compositions.

A good deal of caution is required in the transition from observations to conclusions based on absolute abundances or even relative elemental abundance patterns. We have no in situ sample of the mantle, only natural samples variously derived from the mantle. The derivations involve chemical and physical processes whose effects on noble gas abundances are not well understood quantitatively, and we might suffer from even qualitative misconceptions about how noble gases behave under real mantle conditions. Also, the efficiency with which magmatic gases are retained when rocks appear on the surface is not well studied. Moreover, there are surely biases. Glassy margins of submarine volcanic rocks, for example, are certainly greatly overrepresented in published data collections.

Although it would be difficult to resolve a noble gas inventory in the mantle solely on the basis of these analytical data, correlation among noble gas contents in mantle-derived samples offer interesting information about mantle noble gas evolution in the mantle.

In Figure 6.5 we plotted He, Ne, Kr, and Xe contents against Ar content in MORB and OIB. Except for He and Ne in MORB, we see a rough correlation among these

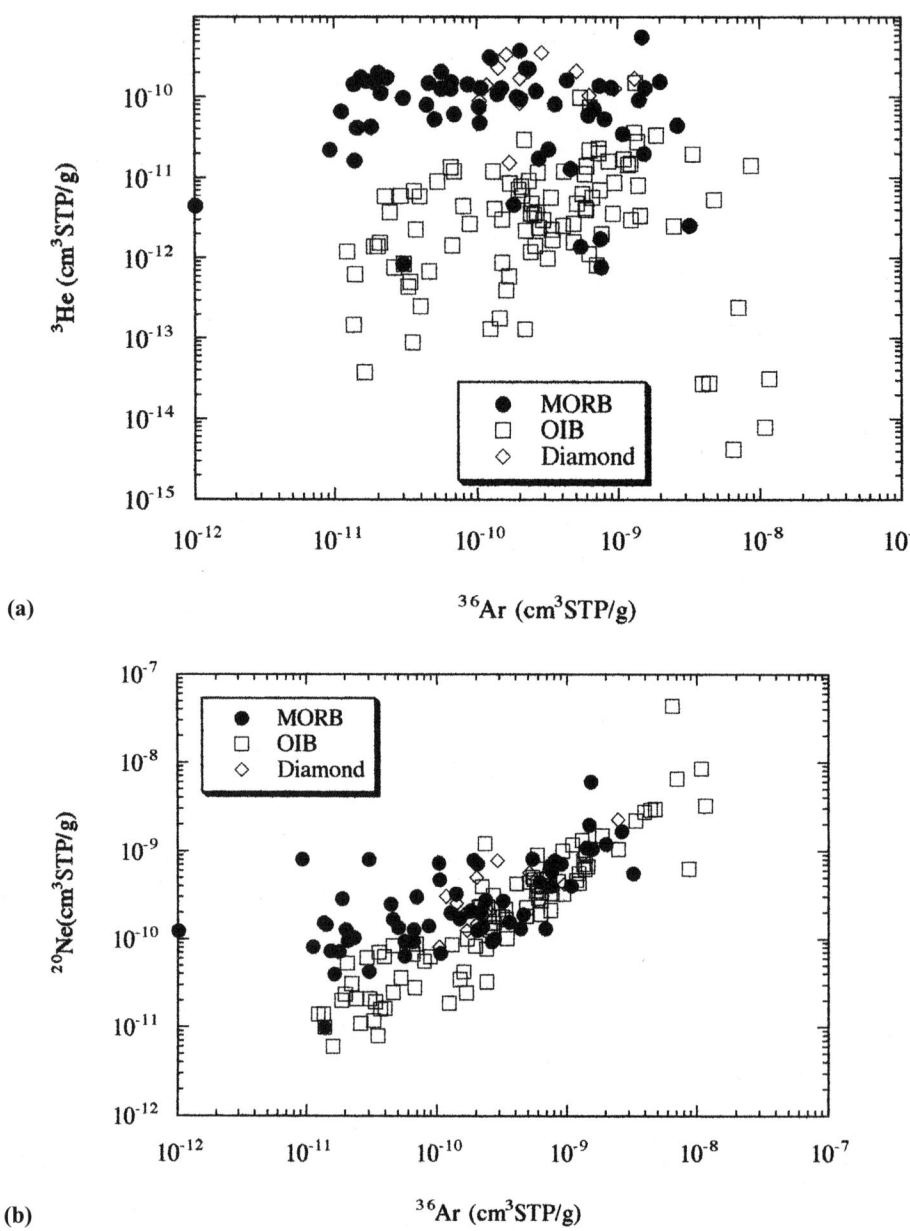

Figure 6.5 Correlation plots between ^{36}Ar abundance and ^{3}He, ^{20}Ne, ^{84}Kr, and ^{132}Xe abundances (cf. Table 6.1 for data source). Note that even though there are good correlations among the heavier noble gases (Ne in OIB; Ar, Kr, and Xe in MORB and OIB), correlations between He and the heavier noble gases are totally lacking.

noble gases. A poor correlation in the case of Xe may primarily be attributed to difficulty in the measurement because of its small sample size. It is remarkable that Kr and Ar contents are well correlated over three orders of magnitude. Also, note that there appears to be no distinct difference in this correlation between MORB and OIB.

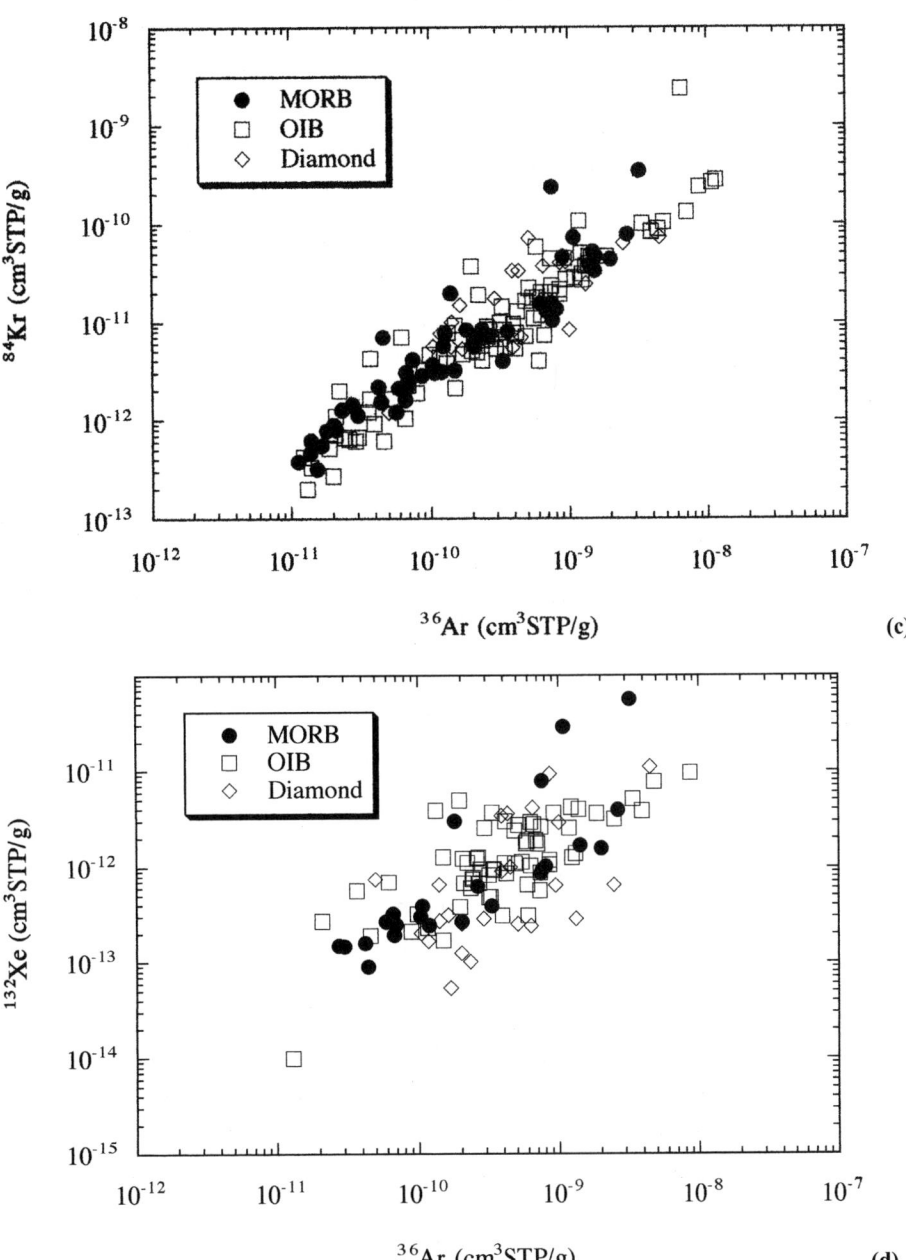

Figure 6.5 (*continued*)

These observations suggest that the heavier noble gases, Ne (except for in MORB), Ar, Kr, and Xe had behaved concomitantly throughout the whole geological processes until they were finally trapped in the glassy margins of the OIB and MORB samples.

On the other hand, He and Ne in MORB behave quite differently from the heavier noble gases. As seen in Figure 6.5, there is no correlation between He-content (and

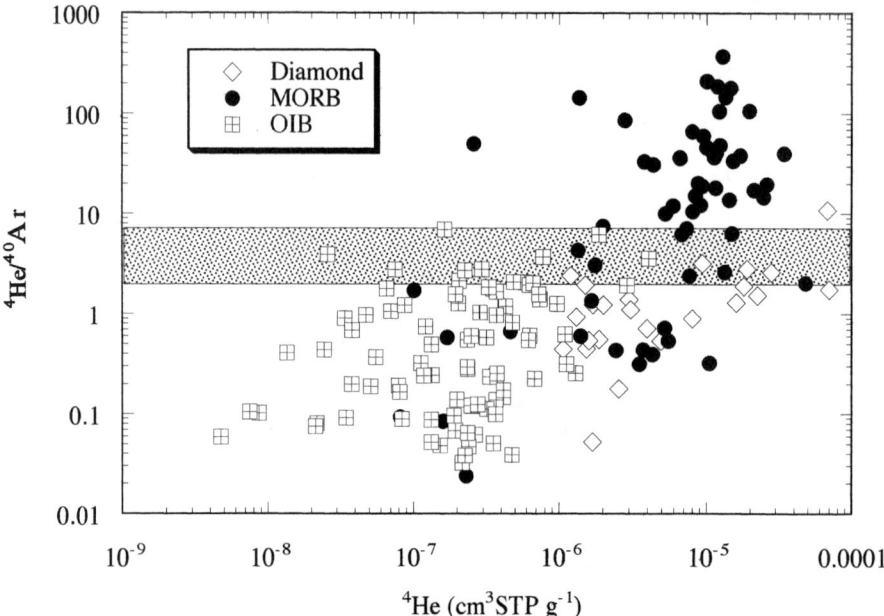

Figure 6.6 ^{4}He/^{40}Ar is plotted against ^{4}He. The shaded zone corresponds to a closed system for ^{4}He and ^{40}Ar (see text). Note that even though the majority of MORB data are in the "excess ^{4}He" zone, OIB and diamonds are essentially in the "^{4}He-deficient" zone. Reproduced from Ozima and Igarashi (2000).

Ne-content in MORB) and the heavier noble gas content both in the MORB and OIB samples. These results suggest that He (and Ne in MORB) are essentially decoupled from the heavier noble gases in the MORB and OIB samples or in their mantle sources. Of further interest is the fact that both ^{3}He and ^{4}He amounts are distinctly larger than those in OIB. This is contrary to what we would expect from a common assumption of "a degassed MORB-source mantle and less degassed OIB-source mantle." These observation may suggest that either He was imported into the MORB or their mantle sources, or it was removed from the OIB or their mantle sources. This further implies that one of the mantle sources for MORB and OIB or both of them have not remained as a closed system, at least for He (and Ne in MORB). This can be more clearly seen in the plot of ^{4}He/^{40}Ar and ^{21}Ne*/^{4}He data in the samples. Because of geochemical coherence of U and K in the case of the ^{4}He/^{40}Ar ratio (see Section 5.7) and of a constant production ratio of ^{21}Ne* to ^{4}He, both of which are derived from U, these ratios are constant in a closed system and, hence, can be used to assess a closed system condition.

As seen in Figure 6.6, even though the majority of MORB data are above the closed system range (shown by a shadow), OIB data are essentially located below the zone. A straightforward interpretation would be that ^{4}He was added into the MORB samples or its mantle source, whereas ^{4}He was depleted from OIB or its mantle source.

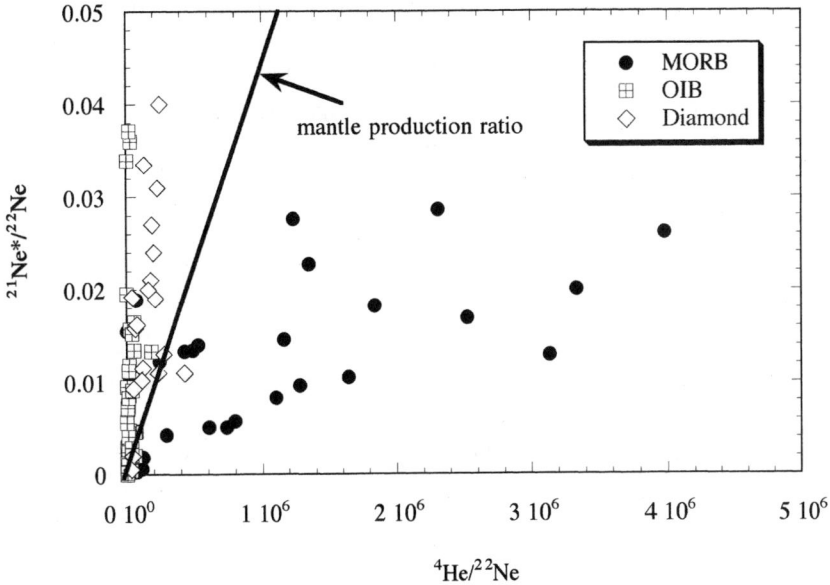

Figure 6.7 Nucleogenic $^{21}Ne^*$ (normalized to ^{22}Ne) derived from the decays of U, Th are plotted against ^4He (normalized to ^{22}Ne). The straight line corresponds to a constant production ratio of the nucleogenic $^{21}Ne^*$ to ^4He in the mantle (e.g., Yatsevitch & Honda, 1997). Note that even though MORB data are in the "excess ^4He zone," OIB and diamonds are in the "^4He-deficient" zone, in accordance with the result shown in Figure 6.6. Reproduced from Ozima and Igarashi (2000).

The same conclusion can also be seen in Figure 6.7, in which $^{21}Ne^*$ indicates a nucleogenic isotope produced from a reaction of mantle materials with α particles from U, Th decays and is calculated as $^{21}Ne^* = {}^{22}Ne \{({}^{21}Ne/{}^{22}Ne)_{obs} - ({}^{21}Ne/{}^{22}Ne)_{air}\}$. [This equation is an approximation. We assumed that the nucleogenic contribution is much larger than the difference between the air ($^{21}Ne/^{22}Ne$) and the primordial mantle ($^{21}Ne/^{22}Ne$).] The solid line corresponds to a constant $^{21}Ne^*/^4He$ production ratio estimated for the mantle (e.g., Yatsevich & Honda, 1997). If the mantle has remained a closed system or there has been no Ne/He elemental fractionation, data should lie on a production line. Most of the MORB data lie on the right side of the production line, whereas all the OIB and diamonds (except for one) are located on the left side. The result is in perfect accordance with the previous observation (Figure 6.6) that ^4He is deficient in the OIB but enriched in the MORB samples.

These results then lead us to some important questions, which are directly related to a fundamental issue in mantle geochemical dynamics. When and how did the failure of the open system condition take place? Did the loss or the gain of He occur in the samples or during magma production-transportation processes? Were they pertinent in the mantle sources? We may safely rule out the first possibility because there seems to be no plausible process to input a significant amount of He into a solid material. Unlike solid materials, both addition and loss of He would be possible in melt. However, it is very difficult to understand why the failure of a closed system

Table 6.2. *Ranges of noble gas isotopic compositions^a in mantle-derived materials*

Sample	^{3}He/^{4}He (10^{-5})	^{20}Ne/^{22}Ne	^{21}Ne/^{22}Ne	^{40}Ar/^{36}Ar	^{129}Xe/^{130}Xe	^{136}Xe/^{130}Xe
MORB	0.9–1.4	9.8–13	0.029–0.06	10,000–30,000	6.5–7.0	2.1–2.4
Loihi basalt	0.7–5	9.6–12	0.026–0.04	400–4,000	6.3–6.7	2.1–2.2
Samoan xenolith	1.2–4	9.8–12	0.029–0.042	350–12,000	6.7–7.0	2.2–2.4
Diamond	0.01–20	11.4–12	0.04–0.06	>300	6.7–7.5	2.3–2.6

^{a} Data sources are the same as in Table 6.1.

condition occurred totally in opposite directions, loss of He in the OIB but gain in the MORB, in spite of the similar formation process of these samples, namely, production of submarine volcanic glasses. This condition would suggest that the characteristic distinction between the MORB and OIB samples is rooted to respective mantle sources.

6.7 Mantle Noble Gas (Isotopic Ratios)

Unlike elemental concentrations, isotopic compositions are only affected a little by chemical differentiation processes. Mass-dependent isotopic fractionations can arise in chemical partitioning (cf. Section 2.9), of course, but on the scale of interest in the present context, plausible fractionation effects are small, especially at the high temperatures prevalent in the mantle. We can thus be much more confident that a noble gas isotopic composition measured in a mantle-derived sample is indeed characteristic of its mantle source. Representative mantle ranges for selected isotopic ratios are presented in Table 6.2.

^{3}He/^{4}He ratios in natural samples are highly variable, spanning some four orders of magnitude. Figure 6.8 shows a classic work by Polak et al. (1975) who first reported a distinct bimodality in the distribution of He compositions, as is quite evident in Figure 6.8. (Note that atmospheric composition lies between the mode.) With more data available now, the characteristic feature in the ^{3}He/^{4}He isotopic ratio distribution in terrestrial samples essentially remains much the same. The lower mode, near and below 10^{-7} in Figure 6.8, is clearly radiogenic He (see Section 5.6, now also called crustal), Polak et al. (1975) attributed the upper mode, near and above 10^{-5}, to the mantle, an attribution amply confirmed by subsequent investigations. This characteristic "signature" (high ^{3}He/^{4}He ratio) has become very widely used as a tracer to diagnose the presence of a mantle component in many geologic environments. (But caution is, as always, in order: He is highly mobile, and the presence of mantle-derived He in, say, a methane gas well, does not necessarily mean that the methane is also mantle derived; cf. Gold & Soter, 1982).

Mantle He is also sometimes termed primordial, but note that even in mantle He the ^{3}He/^{4}He ratio is in most samples significantly lower than the primordial compo-

6.7 Mantle Noble Gas (Isotopic Ratios)

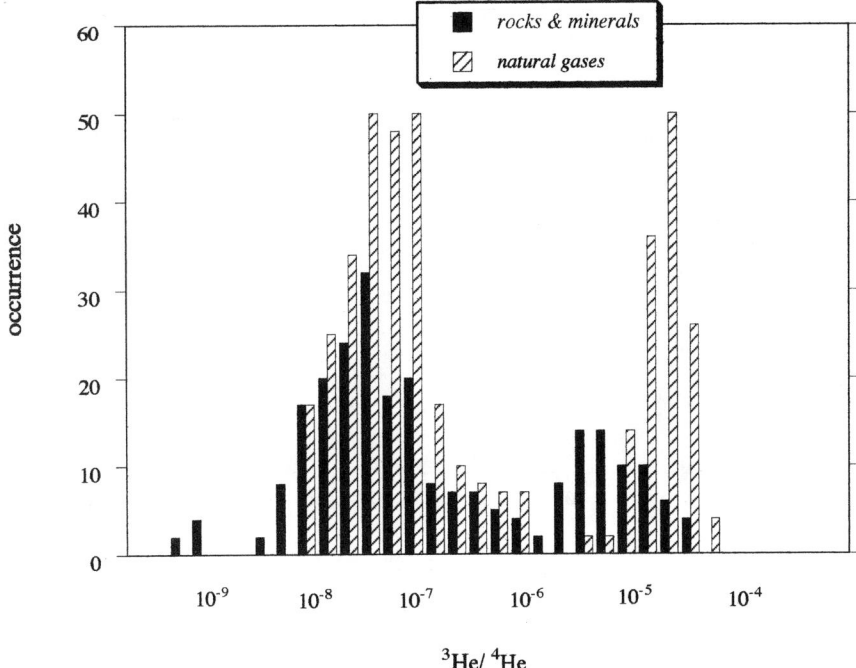

Figure 6.8 ^{3}He/^{4}He ratios in natural samples from the former Soviet Union and Iceland. Redrawn from Polak et al. (1975).

sitions known from study of extraterrestrial materials (Section 3.2); thus, for mantle He it is indeed true that essentially all the ^{3}He is primordial (rather than radiogenic) but in most cases most of the ^{4}He is radiogenic (rather than primordial).

Figure 6.9 shows the range of ^{3}He/^{4}He ratios in MORB and OIB, respectively. It is evident that, in the case of He composition, there is a distinct difference according to provenance. MORB He composition clusters tightly around a characteristic value ^{3}He/^{4}He = 1×10^{-5} [$R/R_a = 8$: It is common to quote ^{3}He/^{4}He ratios as multiples of R_a, the atmospheric ratio (1.4×10^{-6})]. OIB (hot-spot or plume) He is much more variable, but on average ^{3}He/^{4}He is significantly greater than the MORB value. The near uniformity of MORB He suggests that the MORB source reservoir, commonly taken to be the upper mantle, is rather well mixed on a global scale.

He composition is one of the ways in which noble gas data support the widely held view that the MORB-source reservoir is more depleted and more degassed than OIB-source reservoirs, at least those with relatively high R/R_a. In this context, it is helpful to keep in mind that, to a good approximation, the ^{3}He/^{4}He ratio is the ratio of primordial ^{3}He to radiogenic ^{4}He, which in turn is essentially the integrated ratio of He/U over geologic time. Thus, the MORB-source reservoir, relative to average OIB-source reservoir, has had low He/U for a long time. With plausible assumptions about initial uniformity, this translates to more extensive degassing of He relative to U, in the MORB reservoir.

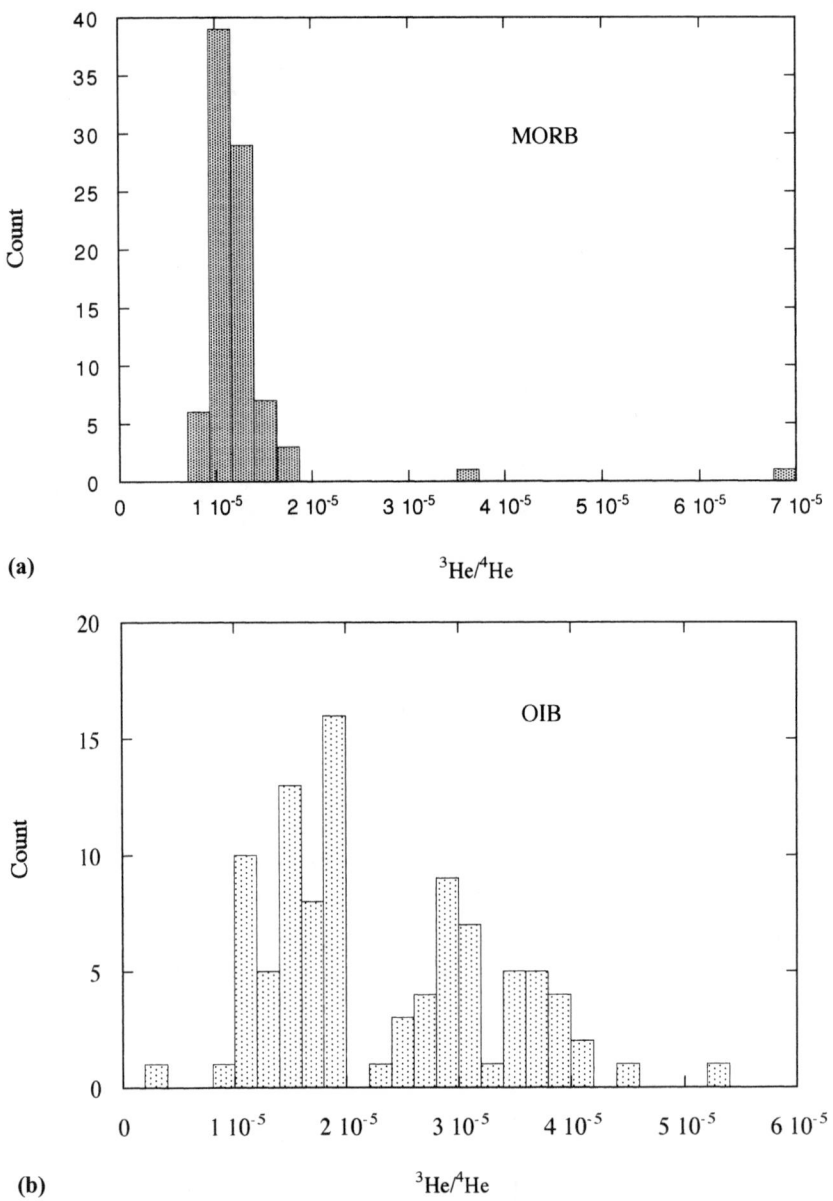

Figure 6.9 ^{3}He/^{4}He ratios in (a) MORB and (b) OIB. Data are compiled by Ozima and Igarashi (2000) and references are therein.

In Figure 6.10, MORB, OIB and diamond data are plotted in a ^{20}Ne/^{22}Ne–^{21}Ne/^{22}Ne diagram. A crude positive linear trend may be attributed to mixing of the air Ne with neons having higher ^{20}Ne/^{22}Ne and ^{21}Ne/^{22}Ne ratios in mantle sources. The nonatmospheric ^{20}Ne/^{22}Ne ratio in the latter components can be assigned to the primordial terrestrial Ne, but ^{21}Ne/^{22}Ne contains varied amount of nucleogenic ^{21}Ne* derived from U, Th decays.

6.7 Mantle Noble Gas (Isotopic Ratios)

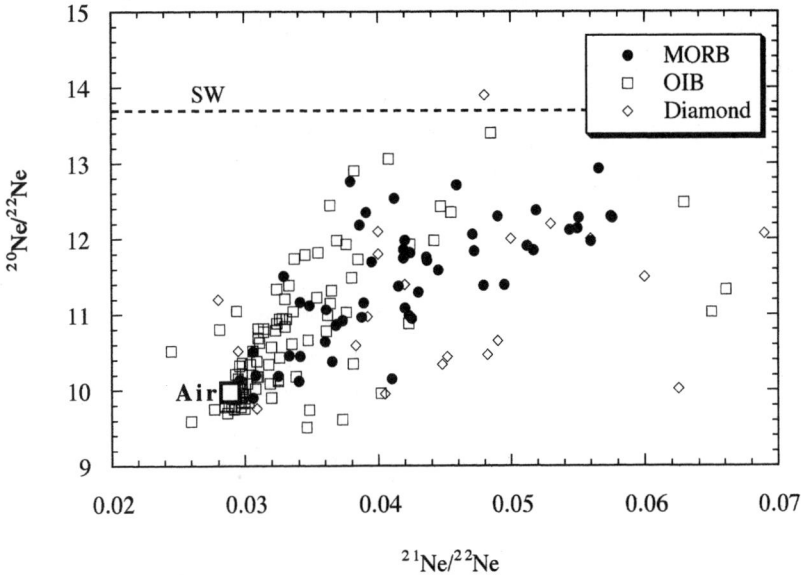

Figure 6.10 Ne isotopic data are plotted in a three-isotope plot. SW: solar Ne ($^{20}Ne/^{22}Ne = 13.7$). Rough positive linear trend suggests a mixing between atmospheric Ne and mantle Ne which has higher $^{20}Ne/^{22}Ne$ (primordial) and $^{21}Ne/^{22}Ne$ (addition of radiogenic ^{21}Ne) ratios than atmospheric Ne. MORB: Sarda et al. (1988), Hiyagon et al. (1992), Honda et al. (1993b), Moreira et al. (1995), Niedermann et al. (1997). OIB: Sarda et al. (1988), Staudacher et al. (1990), Hiyagon et al. (1992), Poreda and Farley (1992), Honda et al. (1993a), Rocholl et al. (1997), Valbracht et al. (1996), Valbracht et al. (1997). Diamond: Ozima and Zashu, 1991. SW: Benkert et al. (1993), Wieler and Bauer (1994).

Since both ^{21}Ne and ^{4}He are produced in the mantle by the U, Th decays with an approximately constant production ratio ($P_{21/4} \approx 10^{-8}$), Honda et al. (1991a, 1991b, 1993a, 1993b) first pointed out that $^{21}Ne/^{22}Ne$ in the mantle reservoirs should be correlated with their $^{4}He/^{3}He$ ratios. This relation can be expressed as

$$^{21}Ne*/^{22}Ne = (^{21}Ne*/^{4}He*)(^{4}He*/^{3}He)(^{3}He/^{22}Ne)$$
$$= P_{21/4}[(^{4}He/^{3}He)_{obs} - (^{4}He/^{3}He)_{0}]\,^{3}He/^{22}Ne \qquad (6.4)$$

where suffix 0 denotes the primordial component. Since $^{20}Ne/^{22}Ne$ can be assumed to be the same in the mantle, the spread of the data in the $^{20}Ne/^{22}Ne$–$^{21}Ne/^{22}Ne$ plot can be attributed to variations in $^{21}Ne/^{22}Ne$ or in $^{4}He/^{3}He$ in respective mantle sources. Therefore, samples with the same $^{4}He/^{3}He$ ratios should have the same $^{21}Ne*/^{22}Ne$ ratios and hence lie on a same mantle Ne–air Ne mixing line (an implicit assumption here is that there is no air He contamination in the sample). Since approximate linear trends (mixing lines) observed for the Loihi samples by Honda et al. (1991a) and MORB samples by Sarda, Staudacher, and Allègre (1988) are consistent with what is expected from Equation (6.4) with the solar values assumed for $^{20}Ne/^{22}Ne$, $^{3}He/^{4}He$, and $^{3}He/^{22}Ne$, Honda et al. (1991a) suggested that the primordial $^{20}Ne/^{22}Ne$ and $^{3}He/^{22}Ne$ values in the Earth are solar. However, be aware that the choices of the free parameters, $^{20}Ne/^{22}Ne$, $^{3}He/^{4}He$, and $^{3}He/^{22}Ne$, are not unique, and alternative choices of some nonsolar ratios are also consistent with Equation (6.4). Moreover, the Honda et al. model can also be questioned for its implicit assumption of a closed system evolution for both $^{21}Ne*$ and $^{4}He*$. If the mantle sources have been open either for $^{21}Ne*$ or $^{4}He*$ or for both, Equation (6.4) does not hold. On the basis of the larger database, Ozima and Igarashi (2000) found that there was no correlation between $^{21}Ne*/^{22}Ne$ and $[(^{4}He/^{3}He)_{obs} - ((^{4}He/^{3}He)_{0}]$, contrary to what would be expected from Equation (6.4). They, therefore, concluded that the mantle reservoir has not remained a closed system at least for He and Ne, and the solar mantle Ne may not be warranted. The failure of the closed system condition for the MORB and OIB mantle sources were also discussed in Section 6.6. As will be discussed in Chapter 7, Ne isotopic data on mantle-derived materials also suggest that mantle Ne is distinct from the solar Ne.

The $^{40}Ar/^{36}Ar$ ratio in MORBs is very high, perhaps 20,000 or up to 40,000 (cf. Burnard et al., 1997), high even in comparison to the air ratio and high enough to pose an experimental challenge to its measurement; in many cases, it may be questionable whether the observed ^{36}Ar is actually from the mantle rather than blank or atmospheric contamination (e.g., Burnard et al., 1997). The high ratio is another indication of extensive degassing of MORB-source mantle (i.e., high K/Ar leading to high $^{40}Ar/^{36}Ar$). The $^{40}Ar/^{36}Ar$ ratio in OIB-source mantle is considerably less than in MORB-source mantle, again a manifestation that OIB-source mantle is less extensively degassed than MORB-source mantle.

Although there is little doubt that $^{40}Ar/^{36}Ar$ is lower in the OIB-source mantle than in the MORB-source mantle, there is no consensus about just how much lower. Some observations in OIB volcanics yield $^{40}Ar/^{36}Ar$ ratios around 300, hardly more than the air ratio, and it can be argued that this is the best representation for the less-degassed mantle (e.g., Hart, Hogan & Dymond, 1985; Allègre et al., 1987). It may also be argued, however, that Ar that looks like modern air actually *is* modern air (e.g., Fisher, 1985a; Patterson, Honda & McDougall, 1990), and that OIB-source (lower mantle) $^{40}Ar/^{36}Ar$ is actually around a few thousand (e.g., Hiyagon et al., 1992; Poreda & Farley, 1992; Matsuda & Marty, 1995), manyfold higher than the air ratio but still manyfold lower than the MORB ratio. The counterargument, of course, is that the higher values may reflect contamination with MORB Ar. Formal mathematical models permit either result.

There is believed to be no significant nucleogenic production of either ^{36}Ar and ^{38}Ar in the Earth's interior, so that the present mantle $^{38}Ar/^{36}Ar$ ratio should reflect the primordial ratio. Available data (Figure 6.11 and also Section 7.3) present no strong evidence for variability in mantle $^{38}Ar/^{36}Ar$, nor for any difference from the atmospheric ratio, at least within about 1%. The issue is nontrivial because of the apparent difference between atmospheric and solar $^{38}Ar/^{36}Ar$ (Figure 7.3). To help assess the possibility that airlike $^{38}Ar/^{36}Ar$ is just air contamination, Figure 6.12 displays $^{38}Ar/^{36}Ar$ in relation to $^{40}Ar/^{36}Ar$ and $^{20}Ne/^{22}Ne$. Even for the most extreme known nonatmospheric $^{40}Ar/^{36}Ar$ (about 40,000; Burnard et al., 1997) and $^{20}Ne/^{22}Ne$

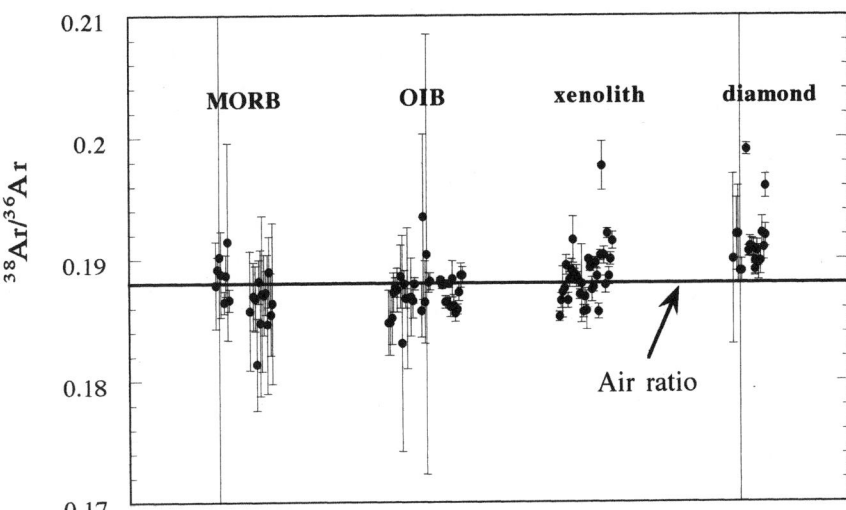

Figure 6.11 $^{38}Ar/^{36}Ar$ isotopic ratios in MORB, OIB, mantle xenolith, and diamonds. Except for diamonds, other data show $^{38}Ar/^{36}Ar$ ratios indistinguishable from the air ratio. Diamond Ar may contain nucleogenic ^{38}Ar (see text). MORB: Hiyagon et al. (1992), Honda et al. (1993b), Niedermann et al. (1997). OIB: Hiyagon et al. (1992). Xenolith: Poreda and Farley (1992), Rocholl et al. (1997), Valbracht et al. (1996). Diamond: Ozima and Zashu (1991), Wada and Matsuda (1998).

(details aside, solar composition presumably sets at least an upper limit to mantle $^{20}Ne/^{22}Ne$), there is no evident trend of $^{38}Ar/^{36}Ar$ covariation. The latter is quite contrary to the conclusion of solar-type mantle Ar, which was reached by Pepin (1998) on the basis of rather limited mantle Ar data.

Assertion of negligible nuclear production of ^{36}Ar and ^{38}Ar in the Earth as a whole does not preclude significant production in specific samples. There is *some* evidence for elevated $^{38}Ar/^{36}Ar$ in mantle-derived samples [e.g., in Zaire cubic diamonds (cf. Figure 6.11)]. Wada and Matsuda (1998) argued that even though this might reflect indigenous variation, it might also reflect nuclear production by $^{35}Cl\,(\alpha, p)^{38}Ar$. Fluid inclusions in the diamond are known to be relatively rich (a few percent) in Cl, so some nuclear production might not be surprising, but overall the issue remains open.

Kr compositions in mantle-derived samples are illustrated in Figure 6.13. Data for ^{78}Kr are comparatively imprecise, both because of its low abundance and because of ubiquitous hydrocarbon (C_6H_6) interference in the analysis. Within errors, there is no evidence for variation among the major isotopes (^{82}Kr, ^{83}Kr, ^{84}Kr, ^{86}Kr), nor for any difference between mantle and air Kr. Mean $^{80}Kr/^{84}Kr$ in MORBs is nominally about 3% above the air ratio; in view of the relatively large analytical errors for this ratio, and the absence of known specific-isotope effects for ^{80}Kr [save possibly $^{79}Br(n, p)$], the significance of this nominal difference is unclear.

In absolute terms, Xe is the least abundant of the noble gases in terrestrial materials (cf. Figure 6.3); precise isotopic analysis is correspondingly relatively difficult

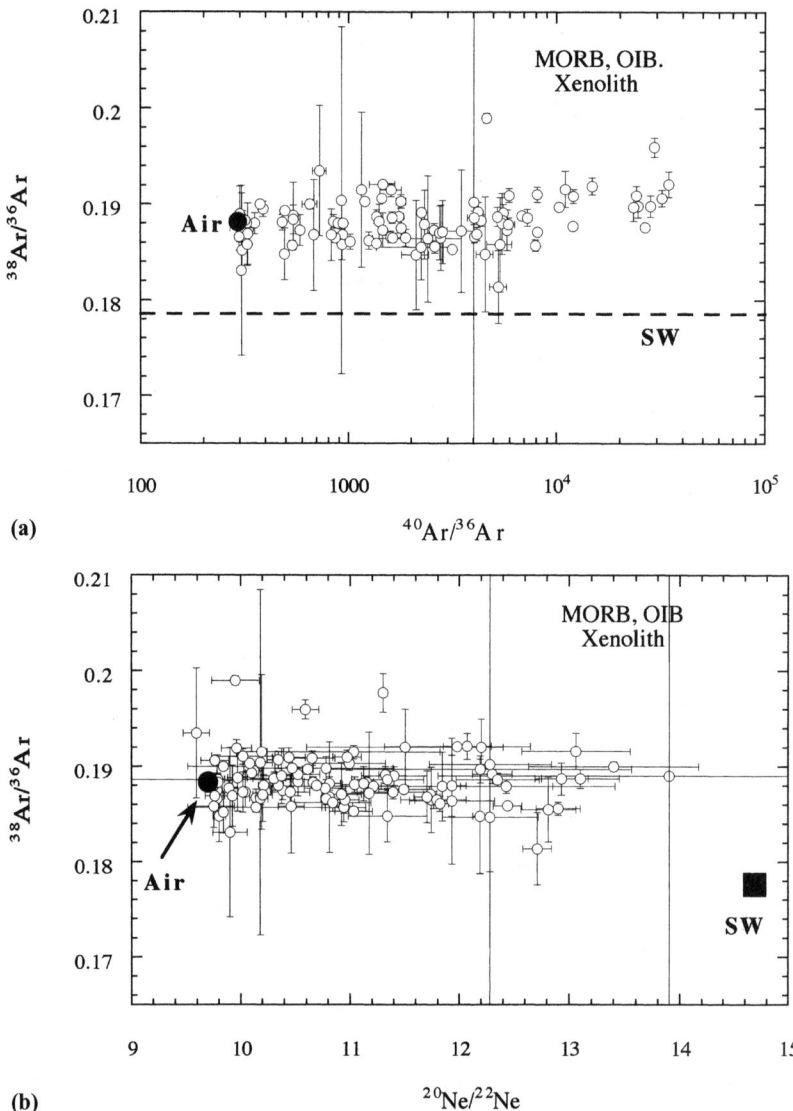

Figure 6.12 $^{38}Ar/^{36}Ar$ ratios in MORB, OIB and mantle xenolith are plotted against (a) $^{40}Ar/^{36}Ar$ and (b) $^{20}Ne/^{22}Ne$. Solar noble gas component (SW) is also shown. The mantle Ar is indistinguishable from the atmospheric Ar but differs from the solar component (see text for further discussion). Data are the same as in Figure 6.11 and SW from Table 3.3(a).

(as also for Kr) and the database of published analyses is sparse compared to that for He, Ne, and Ar. Available data are displayed in Figure 6.14. Data for the scarce isotopes ^{124}Xe and ^{126}Xe are not very precise and are not shown in this figure. Relative to atmospheric Xe, mantle Xe characteristically exhibits excesses of the heaviest isotopes ^{134}Xe and ^{136}Xe, which may be seen more clearly in Figure 6.15 than in Figure 6.14. The heavy isotope excesses are generally attributed to a significant component

6.7 Mantle Noble Gas (Isotopic Ratios)

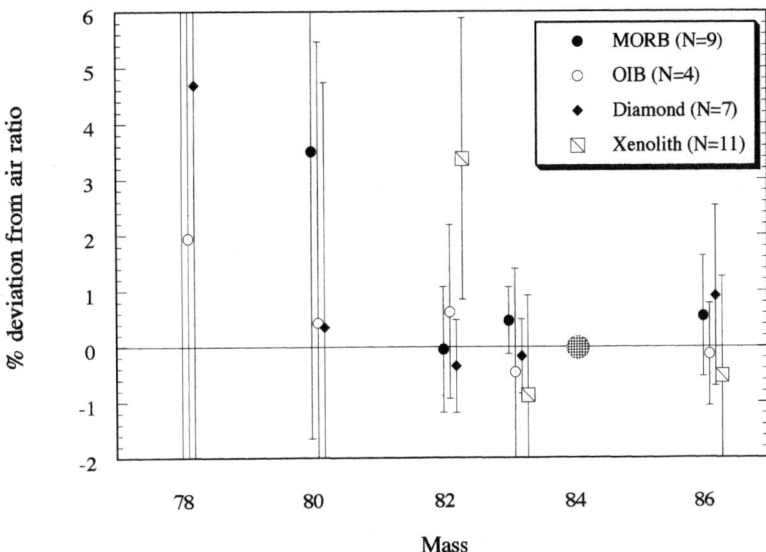

Figure 6.13 Kr isotopic ratios in MORB, OIB, diamond, and mantle xenolith are plotted as percentage deviation from the air Kr isotopic composition. MORB: Hiyagon et al. (1992). OIB: Hiyagon et al. (1992). Diamond: Ozima and Zashu, 1991. Xenolith: Poreda and Farley (1992). N: number of data analyzed.

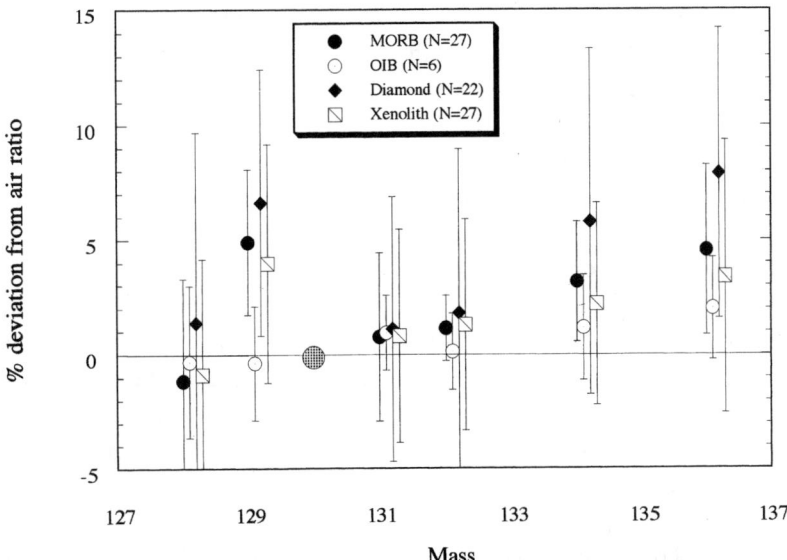

Figure 6.14 Xe isotopic ratios in MORB, OIB, diamond, and mantle xenolith are plotted as percentage deviation from the air Xe isotopic composition. MORB: Hiyagon et al. (1992), Kunz et al. (1998). OIB: Hiyagon et al. (1992). Xenolith: Poreda and Farley (1992), Rocholl et al. (1996). Diamond: Ozima and Zashu, 1991. N: number of data analyzed.

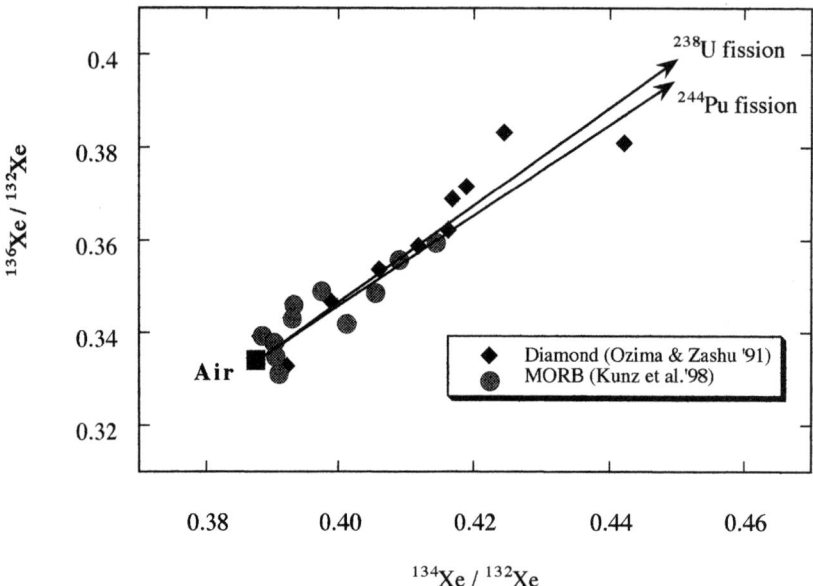

Figure 6.15 Xe isotopic data are plotted in a $^{136}Xe/^{132}Xe$-$^{134}Xe/^{132}Xe$ diagram. The linear trend indicates that the data represent mixing between the atmospheric Xe and fission-derived Xe, either from ^{244}Pu or ^{238}U or from both.

of spontaneous fission Xe. Both ^{238}U and ^{244}Pu are plausible candidates for the responsible fissioning nuclide. It is not possible to distinguish between these two candidates just from the data exhibited in Figure 6.15, but the distinction is important and is considered in more detail in Section 6.11. There are also significant excesses of ^{129}Xe in mantle-derived materials, which may be more clearly seen in Figure 6.17. Excess ^{129}Xe is generally attributed to decay of the short-lived radionuclide ^{129}I; its presence is also of considerable consequence in a variety of contexts, as discussed separately elsewhere (e.g., Section 6.11).

On average, the excesses in ^{129}Xe and the heavy Xe isotopes are larger in diamonds than in the other classes of mantle-derived materials. This is most simply understood in terms of a lesser degree of air contamination, with attendant degree of dilution of the isotopic effects, in the diamonds. If so, the diamond data set a lower limit to the true isotopic excesses in mantle Xe. There is nominally a slight excess of ^{128}Xe in the diamonds, but it is difficult to know how much credence to place in this result, and there is no comparable effect in the other mantle-derived samples. Although the Loihi data show heavy-isotope effects comparable to the other samples, they show considerably smaller ^{129}Xe excesses (relative to air) than MORB (Rocholl et al., 1996; Trieloff et al., 2000); again, it is difficult to judge whether this distinction is real.

Even besides the specific isotope nuclear components (fission Xe at the heavy isotopes, radiogenic ^{129}Xe from ^{129}I), atmospheric Xe differs significantly from Xe known in other solar system materials [i.e., solar, planetary, or Q Xe (cf. Section 7.3)]. This

Table 6.3. *Suggested representative noble gas isotopic ratios in the mantle*

Reservoir[a]	^{3}He/^{4}He (10^{-5})	^{20}Ne/^{22}Ne	^{21}Ne/^{22}Ne	^{40}Ar/^{36}Ar	^{129}Xe/^{130}Xe	^{136}Xe/^{130}Xe
MORB-source mantle	≈1.2	≈12	>0.06	>30,000	>7.0	>2.2
OIB-source mantle	≈4	≈12	>0.04	>3,000	>6.7	>2.2
Air	0.14	9.80	0.029	295.5	6.496	2.176

[a] "Upper" and "lower" are common associations but cannot be inferred from noble gas data; these designations may more justifiably be termed more degassed (MORB-source) and less degassed (OIB-source).

is one of the oldest and most intractable issues in noble gas geo- and cosmochemistry. An important issue in this regard is whether primordial Xe as preserved in the mantle is really like air (aside from the nuclear components) or whether it might be more like solar or meteoritic Xe. This issue is rather complicated and is discussed separately in Chapter 7. In the present context, we may simply note that the available Xe data for mantle-derived samples (Figure 6.13) do not provide any strong indication that the base Xe composition (other than the nuclear components) is meaningfully different from atmospheric Xe composition.

Table 6.3 summarizes our best (subjective) evaluation of the noble isotopic state of the mantle. In this table we distinguish between MORB-source and OIB-source mantle, following the common view that these represent the upper and lower mantle, respectively. The noble gas data themselves provide no information about the *location* (upper or lower) of their respective mantle sources, of course. They do, however, allow a clear and consistent distinction between a more- and less-degassed (upper and lower, respectively) mantle; the degassed distinction is also quite compatible with (although more extreme) the depleted distinction, which emerges from isotopic data for Pb, Sr, and Nd, for example.

6.8 Degassed, Less-Degassed, and Regassed Mantle

It is a truism, almost a cliche, that the mantle is not homogeneous. Describing the nature and distribution of the heterogeneities is a difficult task, however, as much or more so for the noble gases for other kind of observations. The data surveyed in the previous sections (Section 6.6 and 6.7) indicate a nonuniformity of mantle noble gases, and characterization of this nonuniformity in terms of the kinds of noble gases to be found in the different mantle regimes is one of the major problems of noble gas geochemistry.

A simple model, which is often advanced, is that there are basically two kinds of mantle, or two prominent poles in a spectrum of mantle characteristics. One kind is depleted mantle, material that has undergone at least one episode, or probably more, of partial melting and extraction of magma with the consequent diminution in the residue of a number of incompatible accessory and trace elements, those that are preferentially partitioned into the melt. In this model, partial melting of the depleted mantle produces the magma responsible for normal ocean floor volcanic rocks, erupting predominantly but not exclusively at oceanic ridges. The second kind of mantle, described by such terms as *undepleted*, *pristine*, and *less depleted*, has higher concentration of these incompatible elements, presumably because it has undergone less or no such differentiation. The inference is that undepleted mantle is closer to an original composition, retained since the formation of the Earth; the depleted mantle is the residue of the original mantle whose initial complement of incompatible elements is now mostly concentrated in the crust. The undepleted mantle is invoked as the source, by partial melting, of other kinds of magmas, in particular the alkalic magmatism frequently seen in ocean islands such as Hawaii or otherwise associated with hot spots (for comprehensive discussions of hot spots, see Anderson, 1991). A hot spot was first recognized by Morgan (1972) as a surface expression of deep convective plumes in the morphology of chain islands like Hawaii. Geochemical observations, including both elemental abundance and radiogenic isotope systematics, have played a central role in the development of this model, and recently seismic tomography appears to give a strong support. The model is undoubtedly oversimplified, but it serves as a very useful model nonetheless.

Depleted and undepleted mantle are also frequently referred to as upper and lower mantle, respectively. This is essentially a separate additional hypothesis, and the basic geochemical distinction between depleted and undepleted mantle may be valid even if their spatial assignments are not (e.g., Kellog, Hager & van der Hilst, 1999; see Section 6.13). It is possible, for example, that undepleted mantle is dispersed as relatively small pods in depleted mantle, or vice versa. The basic structural distinctions might also be lateral rather than vertical (e.g., if undepleted mantle is undepleted because it is mostly stagnant in the interior of global scale of convection cells or if depleted/undepleted reservoirs mysteriously remain associated with ocean/continent structures in spite of continental drift). In general, it is difficult to specify the physical location of a magma source. In the common view of depleted mantle and undepleted lower mantle, hot-spot magmatism is considered to be caused by upwelling of a deep convection plume from the lower mantle.

In the last decade, one of the most remarkable observations on mantle-derived noble gases is that MORB is characterized by an extremely high $^{40}Ar/^{36}Ar$ ratio, well above the air Ar ratio ($^{40}Ar/^{36}Ar = 295.5$) by more than two orders of magnitude, whereas hot-spot volcanics show considerably lower values. Similar distinct contrast is also seen in other radiogenic isotopic ratios such as $^{3}He/^{4}He$ and $^{129}Xe/^{130}Xe$ between MORB and hot-spot volcanics (Section 6.7). The difference can be reasonably attributed to early and very extensive degassing of the upper mantle or the source of MORB in comparison with the source for hot-spot volcanics; contribution of radiogenic isotopes such as ^{40}Ar, ^{4}He, and ^{129}Xe becomes more significant relative to the stable isotopes such as ^{3}He, ^{36}Ar, and ^{130}Xe, which were degassed. Hence, the noble gas observation is in good accordance with the commonly assumed chemically layered mantle model. The persistent emanation of the ^{3}He from the mantle also favors the chemically layered mantle. Although mantle noble gas data are in harmony with the chemically layered mantle model, mantle convection and plate motion appear to favor a physically layered mantle caused by phase transitions under high pressures. For example, a down-going slab, if it penetrated into the deeper mantle, would tend to mix the materials in both mantles and diminish the chemical distinction of the two layers. The involvement of the subducted slab would also introduce atmospheric noble gases into the mantle, since a major noble gas component in the subducting slab must be atmospheric noble gases that existed in pore water in oceanic sediments and was trapped in hydrous minerals via hydrous water. We discuss the contamination of atmospheric noble gases in mantle-derived materials or in the mantle later.

As discussed earlier, less prominent excesses in the radiogenic isotopic ratios such as ^{4}He/^{3}He, ^{40}Ar/^{36}Ar, and ^{129}Xe/^{130}Xe (Section 6.7) in hot-spot volcanics are generally attributed to the lesser degree of the degassing in the mantle source. However, since air contamination always reduces theses excesses (this effect is not important for He because of its very low concentration in air), it is important to see whether the difference in the radiogenic isotopic ratios between MORB and hot-spot volcanics is truly a reflection of the mantle dichotomy or is due to air contamination. It is also important to estimate the degree of the involvement of the subducted air noble gases in a MORB source mantle to elucidate the indigenous upper mantle component. Laboratory air contamination may be reduced to a tolerable level with modern ultra-high vacuum technology in a mass spectrometer and in a gas-handling system. Air contamination acquired by a sample after its emplacement on the surface may also be reduced by preheating in vacuum and by a step-heating technique. However, it would be difficult to discriminate indigenous air contamination, if a sample inherited it from a mantle source. Some authors argue that subducting oceanic crust has contributed a substantial amount of atmospheric noble gases into the mantle, at least into the upper mantle (e.g., Sarda, Moreira & Staudacher, 1999).

It is now well accepted that the oceanic crusts are subducting into the mantle along the island arcs and passive plate margins. Recent seismic tomography further suggests that some fraction of subducted oceanic slab, if not all of them, may penetrate into the lower mantle. Although the latter suggestion is still controversial, it is at least safe to assume that the subducting slab would introduce a nonnegligible amount of crustal materials into the upper mantle. Trace element geochemistry such as Nd, Sr, and Pb systematics revealed that there are several mantle source regions where radiogenic Sr, Nd, and Pb isotopic ratios are anomalous. These anomalous mantle regions are generaly interpreted as the result of assimilation of crustal materials (e.g., see Hofmann, 1997, for a recent review of this subject). However, Galer and O'Nions (1985) pointed out that Pb isotopic ratios in most of MORBs are consistent with a nearly closed system evolution over the age of the Earth. The inference of this observation is that recycling of incompatible trace elements from the crust into the upper mantle has only limited importance.

Some authors suggested that a subducting slab also introduced a significant amount of noble gases into mantle (e.g., Azbel & Tolstikhin, 1993). Porcelli and Wasserburg (1995a, 1995b) concluded that even though subduction of atmospheric He and Ne is not significant, a considerable amount of atmospheric Xe could be subducted into the upper mantle. However, these conclusions are model-dependent, and an alternative explanation is entirely possible.

There are a number of geochemical observations relevant to the estimation of a mixing degree of a subducted oceanic slab in the mantle. For this purpose, B and ^{10}Be have been successfully employed as a tracer (e.g., Morris, Leemann & Tera, 1990). Since B and ^{10}Be are highly enriched in the subducted components such as oceanic sediments, altered oceanic crust, or melt/fluid derived therefrom, the inventory of these elements is essentially determined by the amount of subducted materials but is insensitive either to chemical variations in the mantle or to the involvement of continental crust. Hence, B and ^{10}Be can be used to identify and ultimately quantify the subducted material in the mantle. To assess subduction recycling of noble gases, B would be a useful analogue because of their similarly highly incompatible nature in various geochemical cycles. Chaussidon and Jambon (1994) concluded from B-δ^{11}B systematics that B derived from the subducted oceanic slab amounts to less than 2% (by mass) of B in the mantle, and most of the B must be lost to the surface during subduction. Since noble gases would be more easily stripped off from the subducting slab than B, the estimation of 2% would be likely to be a maximum estimation of subducted noble gases in the mantle. Morris et al. (1990) came to a similar conclusion, "the sub-arc mantle has no memory of B enrichment imposed during earlier subduction." This conclusion is also in accordance with the persistent difference in ^{20}Ne/^{22}Ne and in radiogenic isotopic ratios ^{40}Ar/^{36}Ar and ^{129}Xe/^{130}Xe between the air and the mantle noble gases (see Section 6.3). Although the present state of our understanding of this problem is still far from deriving a definite conclusion, it would be fair to say that currently available noble gas data can reasonably be explained without invoking significant subduction of atmospheric noble gases into the upper mantle.

6.9 Mantle Degassing Mechanism

There is a general consensus that the terrestrial atmosphere was derived from the solid Earth or from the Earth-accreting solid materials during Earth accretion. A question then arises. How were volatiles (i.e., atmospheric constituents including noble gases) separated from a solid planetary body? At a first glance, it may seem to be an easy question, at least in the case of noble gases, because noble gases are not chemically bound with any known Earth-constituting minerals (cf. Section 2.1) and are highly diffusive, and hence must be easily separated from solid materials. However, if we consider the problem in a planetary scale, we realize that such intuitive explanation does not work. For example, we can easily see that a diffusion process, the most common gas-loss process in nature, is too slow to account for the major mantle degassing.

Diffusion coefficients of noble gases through solid materials are very low (cf. Section 2.7). Even for the lightest noble gas He, which has the largest diffusion coefficient among the five noble gases, the diffusion coefficient in common rock-forming minerals is likely to be less than 10^{-7} cm^2/s at a plausible average upper mantle temperature (1400°C) (Trull, Kurz & Jenkins, 1991). This corresponds to a diffusive length of $\sqrt{2 \times D \times t}$, which is less than 2 km for a lifetime of the Earth ($t = 1.4 \times 10^{17}$ s). The diffusive distance is too short to be of any importance in major degassing of the mantle. Diffusion coefficients for heavier noble gases are much lower than that of He, and it would be even less important for heavier noble gases in the mantle degassing. Therefore, we conclude that solid diffusion of noble gases through mantle can hardly play an important role in mantle degassing. For effective and large-scale noble gas degassing, we need some other processes than solid diffusion. Therefore, many authors assumed that magma generation and its transport would account for an effective degassing mechanism. Because of a small partition coefficient of noble gases between solid and melt, noble gases will be partitioned essentially in the melt (magma) phase. Once in magma, noble gases will easily be transported with ascending magma to the Earth's surface. It is then taken for granted that noble gases were immediately released into the atmosphere from magma when magma reached to the surface. As we mentioned in Section 6.7 and will discuss in more detail in Section 6.10, the existence of the excess in ^{129}Xe/^{130}Xe in mantle-derived materials relative to the air ratio indicates that the separation of the atmosphere from the solid Earth must have occurred within a few half-lives of the parent element ^{129}I, that is within a few ten million years. Hence, this unavoidable conclusion of the early degassing rules out a once-popular degassing scenario of a continuous degassing via volcanic activity throughout geological time, which was originally advocated by Rubey (1951).

Moreover, this conventional notion of noble gas degassing through magma generation and transportation has the following difficulty as an effective degassing mechanism. In the preceding scenario, it is assumed that the noble gas partition between melt and crystal favors overwhelming partition of noble gases into the melt. In contrast to this assumption, recent laboratory experimental results, although there are only three experimental results because of formidable experimental difficulty, suggest that noble gases, especially Xe, may not be as incompatible as hitherto assumed (cf. Section 2.4). Hence, if these laboratory experiments were correct, Xe would not be effectively degassed from the mantle by magma production. Therefore, unless these experimental results are disproved, we should at least be cautious in assuming that noble gases, especially the heavier ones, are always partitioned overwhelmingly in melt rather than in crystal.

Even if we accept that noble gases are mostly partitioned in magma and then effectively transported to the surface, we must further examine whether or not noble gases can be readily released from magma into the atmosphere once magma reaches the surface. Effective noble gas degassing mechanism from magma may be due to bubble degassing discussed by Zhang and Zindler (1989). Noble gases are very effectively partitioned into bubbles (mostly CO_2), which formed in ascending magma. Zhang and Zindler (1989) derived the following expression for a noble gas concentration in a bubble equilibrated with silicate melt;

$$(C_i)/(C_i)_0 = \exp(-Q_i V_g / V_m) \tag{6.5}$$

where C_i denotes the concentration of noble gas i in melt, subscript 0 denotes the initial concentration, and V_g and V_m denote the volume of gas and melt. Q_i is expressed as $1/(K_i \rho_m R T_m)$, where K_i is a partition coefficient of a noble gas i between melt and vapor ($K_i = C_i/P_i$; C_i in mol/g and P_i in atm), ρ_m and T_m are the density and temperature of the melt, and R is a gas constant. Values of Q_i range from 110 (He) to 3555 (Xe) for a typical magma (Zhang & Zindler, 1989). V_g/V_m corresponds to vesicularity in the melt. Taking a value of $Q = 1205$ for Ne in a typical magma (Zhang & Zindler, 1989), Equation (6.5) gives $C_i/C_{i,0} \approx 0.3$ for $V_g/V_m = 0.001$ and 0.002 for $V_g/V_m = 0.005$.

Equation (6.5) was derived for a fractional degassing case in which infinitesimal fraction of vapor separating at a given instant is assumed to be in equilibrium with the melt. However, if the whole vapor phase is assumed to be in equilibrium with the entire melt, which seems to be less realistic in a large-scale magma degassing, Equation (6.5) is modified to

$$(C_i)/(C_i)_0 = 1/\{1 + Q_i V_g / V_m\} \tag{6.6}$$

and the dependence of concentration change on vesicle becomes less severe. In both cases, we see an effective partition of noble gases in the vapor phase.

The preceding numerical example shows that with increasing vesicularity, noble gas will be very extensively extracted from the melt into bubble. Once noble gases were in bubbles, the latter would very quickly ascend to the surface due to buoyancy force (e.g., Toramaru, 1989, suggested an average ascending velocity to be about $1 \, \text{m s}^{-1}$ in magma), and release noble gases to the atmosphere. Although bubble formation in magma may be confined to a very surface region of a few kilometers depths (e.g., Toramaru, 1989), it is very likely that turbulent motion of magma will bring the deeper part of magma to the surface region where bubbles formed and scavenge noble gases from melt to release them to the atmosphere. Therefore, consecutive noble gas partitions, first between melt and crystal in a mantle source and then between gas (bubble) and melt at the surface, would result in a very effective, large-scale mantle degassing.

A similarly very effective but entirely different degassing process was advocated by Lange and Ahrens (1982). During the last stage of Earth accretion, Earth-accreting planetesimals must have had considerable impact velocity because of the earth's increasing gravitational force. Impact shock of planetesimals would then release a considerable amount of volatiles. Ahrens and his colleagues (Boslough et al., 1982; Lange & Ahrens, 1982, 1986; Lange, Lambert & Ahrens, 1985) made a series of experiments to examine this possibility. They showed that high-velocity impact of a

6.9 Mantle Degassing Mechanism

projectile on mineral targets such as serpentine and calcite gave rise to liberation of considerable amount of volatiles; the minerals begin to lose H_2O and CO_2, respectively, at the initial shock pressure of 10–15 GPa. Complete devolatilization occurred at shock pressures of 30–40 GPa. Tyburczy, Frisch, and Ahrens (1986) found similar results of shock-induced H_2O loss from carbonaceous chondrites. They interpreted that the volatile release was caused by high temperature, which resulted from the passage of a shock wave generated on impact. On the basis of these experimental results, they suggested that volatile release from Earth-impacting planetesimals may be responsible for the formation of the atmosphere. This is very plausible. However, it remains to be seen how far the experimental results can be extrapolated to the actual planetesimal impacting; one might immediately wonder whether impacting planetesimals were imbedded beneath the surface, most likely deep in the magma ocean so that volatiles were also buried in the solid Earth without being released into the atmosphere.

The preceding impact release of volatiles suggests that noble gases were also released during the planetesimal impact. The response of noble gases to impact shock has been studied by several workers. Most of the earlier studies were, however, to study whether or not the shock impact gives any disturbance on K-Ar (e.g., Davis, 1977; Jessberger & Ostertag, 1982; Bogard et al., 1987) and on I-Xe (Caffee et al., 1982) systematics. Although some experiments showed a significant disturbance, including the loss of radiogenic ^{40}Ar, others showed that noble gases were implanted on the target materials when the experiments were conducted in the presence of noble gases. Fredriksson and DeCarli (1964) and Pepin et al. (1964) showed that severely shocking a sample in Ar atmosphere will result in the implantation of Ar in the sample. Davis (1977) found that basalt powder and chips subjected to shock pressure ranging from 6.5 to 27 GPa, even in a moderately evacuated chamber, resulted in the implantation of a considerable amount of air Ar into the samples, and the implanted amount increased with increasing pressure. Later shock experiments by Bogard et al. (1987) and Wiens and Pepin (1988) not only confirmed the previous results, namely, shock implantation of gases (including noble gases), but also demonstrated that a variety of gases can be implanted during impact without elemental and isotopic fractionation and that melting is not required for gas implantation. These experimental results gave the robust grounds for the Martian origin of SNC (Shergotite-Nakhlite-Chassignite) meteorites because gases trapped in some SNC meteorites closely resembled those in the Martian atmosphere (e.g., Pepin, 1991).

Determination of absolute loss of noble gases at various shock pressures was attempted by Gazias and Ahrens (1991). For this, they prepared Ar-rich samples; vitreous carbon (3-mm diameter cylinder) was equilibrated with Ar atmosphere of 0.25–1.5 kbar at 773–973°C for 13 days so that the samples thus prepared contained a few weight percent of Ar. They found that 28% of the total Ar was released by driving 4-GPa shocks into the Ar-sorbed carbon. Since pulverization of the same sample material to the similar grain size as the shocked one gave much less Ar loss, they concluded that the Ar loss was not a result of the impact-induced diminution of grain size. As the possible causes for the Ar degassing, they suggested several possibilities such as (i) shock-induced high temperature, (ii) enhancement in Ar diffusion coefficient due to annealing of a sample after shock heating, (iii) development of microcracks, and (iv) short-lived volume change (i.e., "squeezing effect" caused by the shock pressure). Although the first possibility seems to be most likely, we must admit that the mechanism of impact degassing is still not well understood.

It is important to note that in the experiments by Gazis and Ahrens, Ar content in the samples was enormously high – larger by many orders of magnitude than Ar contents in meteorites and in common terrestrial rocks. Hence, Ar loss must be an only observable impact effect on the noble gas. On the contrary, in the earlier experiments where natural samples such as meteorites and terrestrial rocks were used, the initial amount of noble gases in theses samples were much smaller; hence, the implantation effect may become more important than degassing. The net effect of an impact shock on noble gases in rocks and minerals is determined as the result of a competition between two mutually opposing effects – degassing and implantation. The balance must primarily depend on the noble gas pressure in the environment during impact; with increasing environmental noble gas pressure, the implantation of noble gas would become more important. Impact velocity and other factors may also be important. In this regard, we emphasize that it is by no means clear whether noble gas degassing or noble gas implantation was more important in the case of planetesimal impact. Since a considerable amount of the primitive atmosphere must have been built in the later stage of the Earth accretion, the implantation effect may not be totally ignored in later-arriving planetesimals.

To this point, we have discussed several noble gas degassing mechanisms. If any of these degassing processes were responsible for the origin of noble gases in the atmosphere, this should account for the observed characteristics of terrestrial noble gases (e.g., the enrichment of the heavier noble gases in the mantle relative to the atmospheric abundance; cf., Figure 6.4). It is easy to show that the noble gas degassing due to noble gas partitions, first between crystal and melt and subsequently between melt and bubble, satisfy the constraint. We may even quantitatively explain the observed noble gas abundance pattern in the mantle and atmosphere by choosing values for the degree of partial melting, partition coefficients, and other free parameters from experimentally permissible ranges (e.g., Ozima & Zahnle, 1993; Azbel & Tolsitkhin, 1993), although success in the modeling does not necessarily guarantee that the assumed degassing indeed occurred.

In the case of impact degassing, however, it is not immediately clear whether the impact degassing can explain the observed characteristic noble gas elemental abundance pattern. The effect of an impact shock may be quite different for noble gases from that for major volatiles; although major volatiles such as H_2O and CO_2 are degassed as dehydration or decarbonazation due to impact shock, the shock seems to be less effective in degassing noble gases than in degassing these major volatiles (e.g., Gazias & Ahrens, 1991). Shock-induced thermal breakdown of mineral structures in

6.9 Mantle Degassing Mechanism

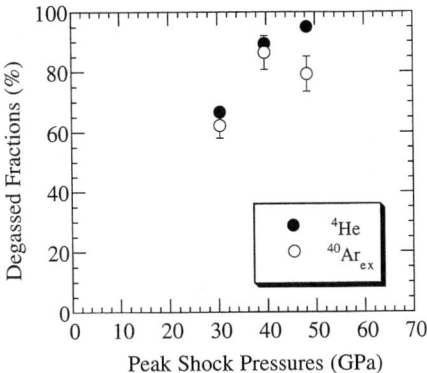

Figure 6.16 Noble gas degassing from olivine crystal powder due to shock. The olivine samples that were previously implanted by Ar and He were shot with a polycarbonate projectile in an evacuated (ca 10^{-2} Pa) chamber. The remaining noble gases in the target (olivine powder) after the shock treatment corresponds to the degassed fraction (ordinate). Note that the degassed noble gases showed little elemental fractionation. Reproduced from Azuma et al. (1994).

an impacting body seems to be a principal process of noble gas release in impact degassing and would be less fractionating than the equilibrium partition process. Hence, it is important to examine whether impact degassing can explain the observed systematic enrichment of the heavier noble gases in the degassed mantle. Experiments by Azuma et al. (1994) appear to show that there is little elemental fractionation in shock degassing of noble gases from olivine crystal. They measured He and Ar release from natural olivine crystals, which were previously implanted by He- and Ar-ion beams for the measurement. Although noble gas losses increased from about 60 to 90% and shock pressures increased from 30 to 50 GPa, they observed little difference in the degree of noble gas degassing between He and Ar (Figure 6.16). If the experimental results derived from the implanted noble gases also apply to noble gases trapped in natural rocks or minerals, this lack of significant elemental fractionation in shock degassing would be difficult to reconcile with the enrichment in heavier noble gases in mantle-derived materials relative to atmospheric noble gas abundance.

So far, we have discussed several plausible degassing mechanisms. However, the problem is still far from conclusive. The major separation of noble gases from the solid Earth may be due to an integrated process of the degassing processes already mentioned, or there might even be a new process yet unknown. We next ask what geological process(es) was responsible for the major noble gas segregation. There is no conclusive answer to this problem. For deepening our understanding of this problem, however, it may be useful to discuss some plausible scenarios that are at least in accord with major observational facts on terrestrial noble gases.

It is very likely that at a later stage of Earth accretion, energies deposited in the Earth by impacting planetesimals were large enough to melt a substantial fraction of the upper layer of the Earth (e.g., Abe & Matsui, 1986). That a large area of molten surface existed on the early Earth was independently suggested from geological evidence such as the widespread occurrence of komatiite in early Precambrian geological settings and also the analogue in the moon, namely moon oceans. All these considerations strongly argue for the existence of a "magma ocean" in the very early history of the Earth. Magma ocean seems to offer a most plausible geological event that was responsible for the major degassing of noble gases. Magma ocean formed as the total melt of the upper layer of the Earth eventually cooled down to form crystals. During the crystal settling, noble gases were expelled from the crystals because of the small partition coefficients, and they were enriched in melt. Owing to putative vigorous convective motion in the magma ocean, the major fraction of the magma ocean was eventually exposed to the surface, and subsequent bubbling resulted in very effective release of noble gases from the magma into the air. The bubbling of the magma ocean must have accompanied degassing of other major volatiles. As already discussed, however, the impact of degassing is likely to be more effective for major volatiles such as H_2O and CO_2 than for noble gases. Therefore, we should also keep in mind the possibility that the separation of the major volatiles or the evolution of the atmosphere was independent of the noble gas degassing.

Finally, let us make a rough estimation of how much noble gas was degassed from the solid Earth. Several estimates are based on a specific Earth degassing model. However, the conclusions are alike; all the degassing models concluded extensive noble gas degassing, some arguing even more than 99% degassing. Although these numerical values are crucially dependent on the choice of a specific mantle evolution model, the conclusion of extensive degassing can be shown quite independently of a mantle evolution model for the case of Ar from a simple inventory consideration. K content in the Earth can be fairly well constrained from the heat flow and cosmochemical elemental abundance considerations. Because of the very refractory nature of U, it may be safe to assume that its relative abundance in the Earth is similar to that in chondrites. Hence, we take U ≈ 20 ppb. We take a commonly assumed value of K/U (weight ratio) = 12,700 from the consideration of geochemical coherence of K and U. Therefore, we have K ≈ 250 ppm for a whole Earth. This much of K then produces about 150×10^{18} g of ^{40}Ar in 4.5 Ga. This is about twice as much ^{40}Ar as is in the present atmosphere. If atmospheric ^{40}Ar were derived from the solid Earth, degassing would then be about 50%. If degassing were confined only to the upper mantle, this would correspond to almost complete degassing of the upper mantle. Considering that stable noble gases had already existed in the Earth's interior long before the production of most of the radiogenic ^{40}Ar, we expect even higher degassing for stable noble isotopes such as ^{36}Ar and ^{38}Ar. Because of higher diffusivity of lighter noble gases, the degree of degassing for He and Ne would be higher than the heavier noble gases including Ar. However, in view of the recent experimental results on the high partition coefficient (Section 2.4) and some unexpected behavior of Xe at high pressures (Section 2.1), we should be careful in concluding the similarly extensive degassing for Xe.

6.10 Mantle Degassing Chronology

In Section 6.9, we discussed the mechanism of noble gas degassing from the mantle. In this section, we discuss mantle degassing chronology. This problem is directly related to the evolution of the atmosphere, since a large-scale noble gas degassing from the mantle may have accompanied degassing of other volatiles, which then constituted the major component of the present atmosphere. The degassing chronology has been discussed mainly on the basis of radiogenic noble gases and their isotopic ratios such as ^{4}He/^{3}He, ^{40}Ar/^{36}Ar, and ^{129}Xe/^{130}Xe, where numerators are radiogenic isotopes and the denominators are stable isotopes. These radiogenic isotopic ratios can serve as a very powerful time marker of the degassing event because degassing must have reset the storage of radiogenic noble gases resulting in triggering the "isotope clock."

Early attempts to use noble gases to study atmospheric evolution compared the amount of radiogenic noble gases in the atmosphere with the production in the Earth's interior (e.g., Damon & Kulp, 1957; Turekian, 1964). For a quantitative treatment, Turekien (1964) assumed that degassing of noble gas was represented by a first-order rate equation (i.e., the degassing rate is proportional to the amount in the mantle). ^{40}Ar evolution is then expressed as

$$\frac{(d\,^{40}\text{Ar})_\text{m}}{dt} = -k(^{40}\text{Ar})_\text{m} + \lambda_\text{e}(^{40}K)_\text{m}$$

$$\frac{(d\,^{40}\text{Ar})_\text{air}}{dt} = k(^{40}\text{Ar})_\text{m} \tag{6.7}$$

where subscript m represents the mantle, k denotes a degassing coefficient, and λ_e is an electron capture decay constant of ^{40}K. With the amount of ^{40}Ar in the present atmosphere as a constraint, Equation (6.7) yields a relation between a degassing coefficient k and the amount of ^{40}K in the present Earth. Since neither of the parameters is well known, it is not possible to derive a quantitative conclusion on mantle degassing from this degassing relation alone. To circumvent this difficulty, Ozima and Kudo (1972) proposed to use a maximum ^{40}Ar/^{36}Ar ratio observed in some submarine volcanics as a mantle representative value to constrain the rate equation. Simultaneous rate equations [Equation (6.7)] for ^{40}Ar as well as for ^{36}Ar were solved with the isotopic constraint for both the present k content in the mantle and the degassing coefficient, the results of which led them to conclude the early catastrophic argon degassing.

Here, it is important to note the basic premise of the rate process and its limitation in applying it to geological problems since this approach is still widely being used in discussing mantle degassing (e.g., Azbel & Tolstikhin, 1990; O'Nions & Tolstikhin, 1996). The rate equation implicitly assumes that noble gas concentration within an object instantaneously responds to degassing to keep the concentration uniform in the whole system. This requires that the homogenization proceeds in the system with a time constant much faster than the degassing rate from the reservoir. Degassing rate or residence time of noble gases in the mantle has been estimated by several authors. For example, Kellog and Wasserburg (1990) obtained 1.4 Ga for He, and O'Nions and Tolstikhin (1994, 1996) suggested 1 Ga for He, Ne, and Ar in the upper mantle. Therefore, for the rate equation to be a reasonable approximation to describe mantle degassing process, homogenization time of noble gases in the mantle must be required to be much less than 1 Ga. Such a very rapid mixing may be possible for fluids introduced with subducting materials (McKenzie, 1979). On the other hand, Davies (1984) concluded that clumps could persist in the mantle for 560 Ma and 2.4 Ga for upper- and whole-mantle convection, respectively. Therefore, if noble gases were firmly trapped in clumps, mixing time would be considerably longer than 100 Ma, and the use of the rate equation to describe noble gas mantle degassing must be questioned. However, if a reservoir is much smaller (e.g., a magma chamber), it would be reasonable to use the rate equation to discuss its degassing.

A seemingly more geologically feasible process of noble gas degassing may be by mass transfer (e.g., Kellog & Wasserburg, 1990; Porcelli & Wasserburg, 1995a, 1995b). For this process, noble gas degassing is expressed as

$$dN/dt = -C\, dM/dt$$

where N and C denote the amount and concentration of noble gas in the mantle, and dM/dt indicates the transfer rate of the mantle material. If the transfer of the mantle material proceeded as magma generation and its uprising, a case generally assumed for upper mantle degassing, C should be multiplied by a fractionation factor taking a preferential partition of noble gases in magma into account. A simple case for such fractionation may be a batch melting for which C should be replaced by $C\{1 - (1-f)^{1/D}\}/f$, where D denotes a partition coefficient of noble gas between melt and solid and f denotes a degree of partial melting. In this mass transfer model, the mass transfer rate must be assumed quite independently of a model for noble gas degassing.

A number of mantle noble gas evolution models have been proposed. Here, the degassing process has been discussed mostly either on the basis of the first-order rate equation (Tolstikhin, 1975; Hamano & Ozima, 1978; Azbel & Tolstikhin, 1990; O'Nions & Tolstikhin, 1994, 1996) or on a mass transport equation (e.g., Kellog & Wasserburg, 1990; Porcelli & Wasserburg, 1995a, 1995b; Kamijo, Hashizume & Matsuda, 1998) with a specific mantle evolution model. Some models involve not only the degassing of noble gases, but also the subduction of the atmospheric and crustal components into the mantle (Allègre et al., 1987; Azbel & Tolstikhin, 1990,

1993; Porcelli & Wasserburg, 1995a, 1995b). A key constraint in all these degassing models is the observation that noble gases are highly radiogenic in the degassed mantle (represented by MORBs) than the counterpart in the atmosphere; for example, $^{40}Ar/^{36}Ar$ in MORB (>30,000) is almost two orders of magnitude larger than the atmospheric ratio (295.5). The difference is interpreted that noble gases, once liberated into the atmosphere, ceased to change their isotopic compositions, whereas radiogenic noble gases have continued to accumulate in the mantle. The earlier the degassing, the more difference in the isotopic ratios would be observed. In spite of considerable variation among these mantle degassing models, it is interesting to note that the conclusions are rather alike; that is, they all indicate a very early and catastrophic (major degassing at one time) degassing. Apart from this general conclusion of the very early separation of the atmosphere from the solid Earth, however, details of the degassing processes drawn from these degassing models should be regarded as tentative because of large uncertainties in the mantle evolution models on which any noble gas evolution model crucially depends.

However, it is important to note that quite apart from mathematical modeling, a very fundamental characteristic of noble gas degassing chronology can be deduced from a comparison of some radiogenic noble gas isotopic ratios between the mantle and the atmosphere. For example, the much higher $^{40}Ar/^{36}Ar$ ratio in the mantle than in the atmosphere is reasonably attributed to extensive degassing of a stable noble gas (^{36}Ar) from the mantle before significant accumulation of the radiogenic ^{40}Ar in the mantle, suggesting early sudden degassing. This "early degassing" can be concluded more quantitatively from the excess ^{129}Xe in the mantle relative to the atmospheric ratio, since the excess ^{129}Xe in the mantle indicates that the short-lived parent ^{129}I still existed when separation of the atmosphere from the mantle took place. The half-life of ^{129}I ($T_{1/2} = 16.7$ Ma) further imposes a quantitative constraint on the time of degassing. Degassing time must be comparable with, or shorter than, the half-life, again suggesting a very early (< a few ten million years) mantle degassing. This conclusion is quite independent of details of the mantle evolution model and may be regarded as one of few robust constraints on mantle evolution.

In the preceding mantle degassing models, degassing or the quantitative separation of noble gas from radioactive parent elements (i.e., Ar from K, Xe from I as well as from U and Th, and He from U and Th) was attributed to fractionation in magma sources. Noble gases will be essentially partitioned in melt because of their very incompatible nature, therefore giving rise to their very effective separation from the mantle. However, as we discussed in Section 6.9, this assumption may be challenged. If the radiogenic noble gases such as ^{40}Ar or ^{136}Xe are less incompatible than their respective radioactive parent elements K and U (cf. Section 2.4), magma extraction from the mantle would result in relative enrichment of noble gases in the mantle relative to their radioactive parents elements. We then expect less radiogenic noble gas (e.g., smaller $^{40}Ar/^{36}Ar$ ratio) in the degassed mantle than in the undegassed mantle, which is clearly contrary to observation (see Table 7.3). This contradiction may simply be due to the flaw in the partition experiments. However, if the experimental results are correct, we may have to conclude that magma generation and its transportation have not played a major role in atmospheric evolution. In the latter case, we must seek an alternative major degassing process.

Noble gas evolution models that have been proposed so far crucially depend on a choice of a specific mantle evolution model, and validity of these models must be proved independently in the first place. Among these models, Xe degassing chronology based on ^{129}I-^{129}Xe systematics can be discussed with less dependence on a specific mantle evolution model. This is because ^{129}Xe has a half-life much shorter than the age of the Earth, and hence decay of ^{129}I or the evolution of ^{129}I-^{129}Xe systematics had essentially completed long before the commencement of a major mantle evolution. Next, we show an example of Xe degassing chronology.

As we showed in Section 6.9, ^{129}Xe/^{130}Xe in mantle-derived materials is in general higher than the atmospheric ratio. The difference is reasonably attributed to very early degassing of Xe from the mantle. In mantle-derived materials, excess Xe isotopes relative to the atmospheric Xe can also be seen in $^{131-136}$Xe, which were derived from either ^{238}U or ^{244}Pu spontaneous fission. Here, we take ^{136}Xe as a proxy of all the fission $^{131-136}$Xe. Besides the excesses in ^{129}Xe and $^{131-136}$Xe, the most remarkable observation in mantle-derived Xe is that the mantle Xe data form a nice linear array in a ^{129}Xe/^{130}Xe-^{136}Xe/^{130}Xe diagram (Figure 6.17). The linear trend was first recognized as a characteristic feature of MORB Xe by Staudacher and Allègre (1982) (they used ^{134}Xe/^{130}Xe instead of ^{136}Xe/^{130}Xe) and was later confirmed by other researchers for other mantle-derived materials such as diamonds (Ozima & Zashu, 1991), MORB and Loihi (less conspicuous than in MORBs) basalts (Hiyagon et al., 1992), and mantle xenoliths (Farley & Poreda, 1993). Many authors now regard the linear trend as one of the key constraints to understanding the isotopic evolution of Xe in the mantle (e.g., Staudacher & Allègre, 1982; Ozima et al., 1985; Azbel & Tolstikhin, 1993; Porcelli & Wasserburg, 1995a).

Excesses in $^{131-136}$Xe can be attributed to spontaneous fission either from ^{244}Pu (Allègre et al., 1983; Ozima et al., 1991) or from ^{238}U (Fisher, 1986; Allègre et al., 1987) or to both (Azbel & Tolstikhin, 1993; Porcelli & Wasserburg, 1995a; Kunz, Staudacher & Allègre, 1998). Opinion is still divided among researchers (Section 6.11) as to whether the excess Xe is due to ^{244}Pu or ^{238}U. Later, for instruction's sake, we present a schematic model to discuss Xe isotopic evolution assuming that the excesses are entirely due to ^{244}Pu. We note that more sophisticated models such as those that consider not only Pu-fission Xe but also U-fission Xe can easily be constructed in a way similar to this basic approach. Let us suppose that a fraction f of Xe were degassed into the atmosphere at time t_d. Taking a stable isotope ^{130}Xe as a normalization isotope, Xe isotopic ratios in the atmosphere and the mantle are then expressed as

$$(^{136}Xe/^{130}Xe)_m = (^{136}Xe/^{130}Xe)_a + \frac{(^{244}Pu)_0}{f(^{130}Xe)_0} {}^{136}Y_{Pu} \exp(-\lambda_{244} t_d) \quad (6.8)$$

where subscripts a and m denote atmosphere and mantle, $^{136}Y_{Pu}$ a fission yield of ^{136}Xe, and λ_{244} are the decay constants of ^{244}Pu. We can set up similar equations for

6.10 Mantle Degassing Chronology

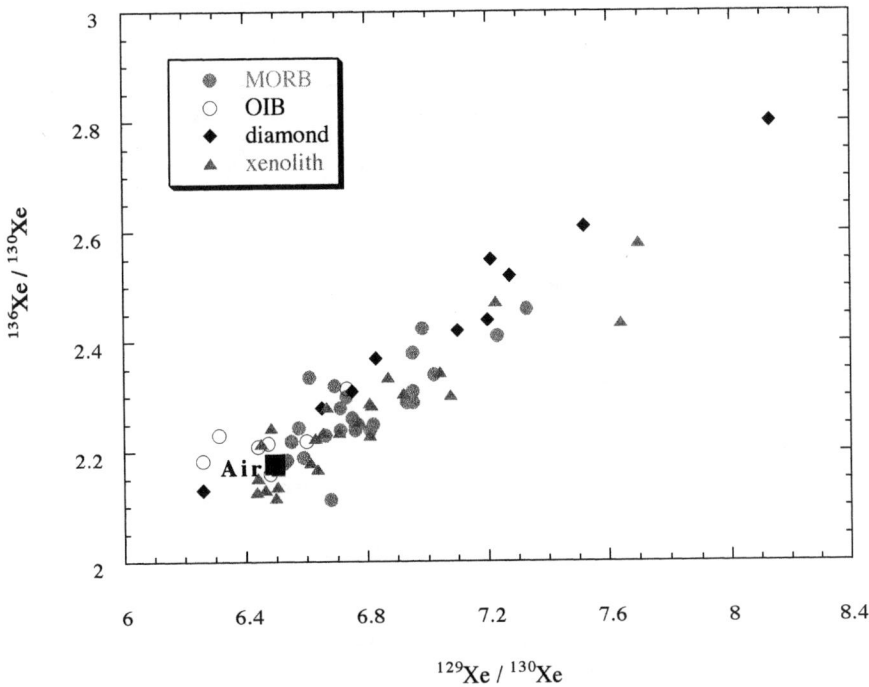

Figure 6.17 Xe in mantle-derived materials are plotted in a $^{136}Xe/^{130}Xe$-$^{129}Xe/^{130}Xe$ diagram. The observed linear trend can be attributed to the mixing between the atmospheric Xe (solid square) and a mantle Xe (not well known), which has excesses in ^{136}Xe and ^{129}Xe. The excess in ^{136}Xe is plausibly due to fission either from ^{244}Pu or from ^{238}U or from both, and that in ^{129}Xe is from radioactive decay of ^{129}I. MORB: Hiyagon et al. (1992), Kunz et al. (1998). OIB: Hiyagon et al. (1992). Xenolith: Poreda and Farley (1992), Rocholl et al. (1996). Diamond: Ozima and Zashu, 1991.

$^{129}Xe/^{130}Xe$ as Equation (6.8). We assume that the linear correlation in Figure 6.18 represents a mixing line between air Xe and mantle Xe. The slope of the line is then represented as

$$\text{slope} = \frac{(^{136}Xe/^{130}Xe)_m - (^{136}Xe/^{130}Xe)_a}{(^{129}Xe/^{130}Xe)_m - (^{129}Xe/^{130}Xe)_a} \tag{6.9}$$

which is further expressed with the use of Equation (6.8) and the equivalent equation for ^{129}I,

$$\text{slope} = \frac{(^{244}Pu)_0}{(^{129}I)_0} {}^{136}Y \frac{\exp(-\lambda_{244} t_d)}{\exp(-\lambda_{129} t_d)} \tag{6.10}$$

To solve t_d from Equation (6.10), we must know $(^{244}Pu/^{129}I)_0$ at the time of the Earth formation. The ratio is rewritten as $(^{244}Pu/^{129}I)_0 = (^{244}Pu/^{238}U)_0 (^{238}U/^{127}I)_0 (^{127}I/^{129}I)_0$.

For illustration's sake, we assume the solar abundance ratio of 0.01 for $^{238}U/^{127}I$ (Anders & Grevesse, 1989), the chondritic value of 0.006 for $(^{244}Pu/^{238}U)_0$ (Hudson

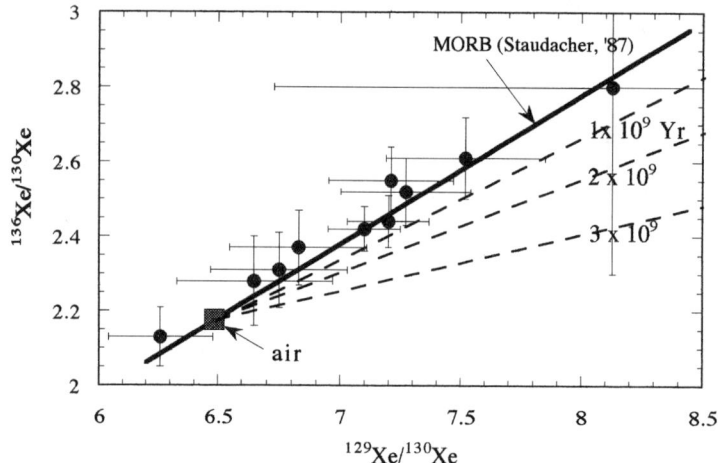

Figure 6.18 Xe in diamonds are plotted in a $^{136}Xe/^{130}Xe$-$^{129}Xe/^{130}Xe$ diagram (Ozima & Zashu, 1991). The solid line was proposed for MORB by Staudacher, 1987 (cf. Fig. 6.17). The dotted lines correspond to mixing lines between the atmospheric Xe and mantle Xe at various assigned ages (shown on the lines), of which excess ^{136}Xe is assumed to be derived only from ^{238}U fission. The almost identical linear trend both for the diamonds and MORB indicates either that the excess ^{136}Xe in the diamonds is essentially due to ^{244}Pu fission (if the ages are much older than 1 Ga) or that diamond ages are much younger than 1 Ga.

et al., 1989), and a commonly assumed chondritic initial ratio of 10^4 for $(^{127}I/^{129}I)_0$ (e.g., Swindle et al., 1991). Putting these values in Equation (6.10), we obtain $t_d \approx 2.7 \times 10^8$ years. Since $(^{127}I/^{129}I)_0$ in the Earth is not known, the degassing time thus obtained is only tentative. However, because of a logarithmic dependence of t_d on $(^{127}I/^{129}I)_0$, an order of magnitude change in the latter would result in a change only by a factor of 2.3. Therefore, we may assume that the conclusion of a very early degassing is valid. If we take $(^{127}I/^{129}I)_0 = 10^6$, which is inferred from observational data (see Section 7.7), t_d would be reduced to about 1.3×10^8 years.

In the preceding model, we assumed that the excesses in $^{131-136}Xe$ are due to ^{244}Pu. However, ^{238}U-derived fission Xe may also be important, especially after the decay of ^{244}Pu. Azbel and Tolstikhin (1993) extended the model in which they took into consideration both ^{244}Pu fission Xe and ^{238}U fission Xe. They also considered recycling of atmospheric Xe into the mantle. After choosing suitable values for a recycling efficiency parameter, a specific Xe isotopic composition in the present mantle, and other parameters, they showed that the model can simulate the observed linear trend reasonably well (Figure 6.17). Porcelli and Wasserburg (1995a) proposed a similar model. Although somehow ad hoc choices of several free parameters for some specific mantle evolution models may enable us to explain the observed linear trend (Figure 6.17), we need independent proof for the choices of these particular values for the parameters. Some of these problems are difficult to answer. Hence, it would not be very rewarding to pursue further a model calculation until we have solid constraints on these parameters. We emphasize, however, that quite apart from detailed model calculations, the observation of the excess ^{129}Xe in the mantle relative to the atmospheric Xe offers unrefutable evidence of the very early (within a few half-lives of ^{129}I) separation of the major fraction of Xe from the MORB source mantle.

6.11 Excess $^{131-136}$Xe: ^{244}Pu or ^{238}U Derived?

The problem of whether the excesses in $^{131-136}$Xe in mantle-derived materials and therefore in the mantle were derived from ^{238}U fission or ^{244}Pu fission is one of the fundamental problems in noble gas geochemistry. U content in the mantle seems to be sufficient to account for the fission Xe observed in mantle-derived materials (e.g., Fisher, 1985b; Allègre et al., 1987). On the other hand, the cosmic elemental abundance ratio (^{244}Pu/^{238}U = 0.006 ± 0.001 at 4.5 Ga ago; Hudson et al., 1989), if applied to the Earth, favors ^{244}Pu, since ^{244}Pu fission Xe would then be an order of magnitude abundant more than ^{238}U fission Xe in the Earth (Bernatowicz & Podosek, 1978). Direct measurements of Xe in mantle-derived materials would resolve this intriguing question. However, because of the extremely small amount of Xe and possible atmospheric Xe contamination in mantle-derived samples, it is still not possible to resolve unequivocally the parent element for the fission Xe (see Figure 6.15). In contrast to these samples, CO_2 well Xe, some of which are assumed to be derived from the mantle because of its association with typical mantle-type He and Ne, have been regarded to be useful in resolving this problem because a large amount of the sample can be available for mass spectrometry.

Since the discovery of excess ^{129}Xe in CO_2 well gas from Harding County, New Mexico, United States (Butler et al., 1963), which was the first known case of excess ^{129}Xe in any terrestrial sample and was subsequently confirmed by several researchers (e.g., Phinney, Tennyson & Frick, 1978; Smith & Reynolds, 1981), three CO_2 well gases, one from Australia and two from Colorado, have been reported to show excess ^{129}Xe (Caffee et al., 1999). In all cases, the CO_2 well Xe showed not only the excess in ^{129}Xe but also excesses in $^{131-136}$Xe. The latter excess was confirmed essentially as a result of ^{238}U spontaneous fission (Phinney et al., 1978; Caffee et al., 1999). Phinney et al. placed an upper limit of 20% on the possible ^{244}Pu fission Xe contribution in the Harding County well gas. Here are additional interesting observations: (i) these CO_2 well Xe were accompanied by nonatmospheric Ne and He (only for the New Mexico sample, but no He measurement for the other two samples), which were best interpreted to be mixtures of the mantle He and Ne (Ne in the Colorado samples may also have nucleogenic neon) with atmospheric ones, and (ii) CO_2 well Xe are enriched in the light isotopes ($^{124-128}$Xe) relative to the atmospheric Xe. From the first observation, Caffee et al. concluded that the CO_2 well Xe were also derived from the mantle. From the second observation, they suggested the possibility that the CO_2 well Xe contained a small fraction of solar Xe because solar Xe is also enriched in the light Xe isotopes relative to the atmospheric Xe (cf., Wieler & Bauer, 1994; also see Section 7.3).

Noting that the isotopic composition of the Harding County CO_2-Xe fell on the same linear trend as MORB in a $^{136}Xe/^{130}Xe$-$^{129}Xe/^{130}Xe$ diagram (Figure 6.18), Staudacher (1987) concluded that the CO_2 well Xe must also come from the same mantle source as MORB. However, except for the Australian sample, such relation is not clear in other CO_2 well Xe. If the CO_2 well Xe were proved to come from the mantle, this would give a definite evidence for the ^{238}U fission origin of the excesses $^{131-136}Xe$ in mantle Xe. However, Ozima, Podosek, and Igarashi (1985) argued from the following reasoning that the origin of the CO_2 well Xe cannot be attributed to the mantle reservoir from which the atmospheric Xe was derived. The excesses in ^{129}Xe and $^{131-136}Xe$ relative to the atmospheric Xe are reasonably attributed to the extinct nuclides – ^{129}I and ^{244}Pu – and to ^{238}U. Because half-life of ^{244}Pu is longer than that of ^{129}I, ^{244}Pu-derived excess $^{136}Xe_{Pu}$ relative to ^{129}I-derived $^{129}Xe*$ should be more prominent in the Earth's interior than in the atmosphere. Accordingly, we have the following inequality relationship between the mantle and the atmospheric Xe:

$$(^{136}Xe_{Pu}/^{129}Xe*)_{atm} < (^{136}Xe_{Pu}/^{129}Xe*)_m \qquad (6.11)$$

Here, note that this inequality relationship does not depend on a specific mantle evolution model; hence, the inequality relationship is rather general. Ozima et al. (1985) showed that if $^{136}Xe_{Pu}$ is indeed less than 20% of the excess ^{136}Xe observed in the CO_2 well Xe (Harding County), it violates the preceding inequality relationship. They, therefore, concluded that the CO_2 well Xe cannot be related to the mantle source from which the atmospheric Xe was derived. As a possible source for the CO_2 well Xe, Caffee et al. (1999) therefore suggested some region in the lower mantel that is slightly richer in iodine (this is to explain relatively high excess ^{129}Xe in CO_2 well Xe) relative to primordial Xe when compared to the reservoir from which the atmosphere was derived.

The problem of whether the excess $^{131-136}Xe$ in the mantle is due to ^{244}Pu or ^{238}U bears profound implications not only in the mantle dynamics but also in the planetary evolution. For example, if the excesses were proved to be due to ^{238}U fission, we must question a constant ratio of Pu/U for meteorites and planets, one of the basic premises in cosmochemistry, which was concluded from the very refractory nature of U and Pu and their chemical coherence. This would further evoke a question as to the validity of the elemental abundance in the Earth, which was deduced from cosmic elemental abundance based on meteorite analyses. A straightforward approach to this important problem is to improve experimental precision in Xe isotopic measurement, which includes reduction of atmospheric Xe contamination both in an experimental line and in a sample of mantle-derived materials with more abundant mantle Xe as well as improvement in mass spectrometry. As such, Kunz et al. (1998) analyzed a popping rock (Section 6.2) with the hope that the comparatively abundant trapped Xe in this sample would yield enough analytical precision to resolve this problem. From a graphic inspection of the data in a $^{134}Xe/^{130}Xe$-$^{136}Xe/^{130}Xe$ diagram, they suggested about 30% of ^{244}Pu fission Xe in the excess ^{136}Xe. However, the analytical data still show considerable scatter (see Figure 6.15), and it remains to be seen whether or not the claimed value (30%) is statistically significant.

Alternatively, Ozima and Zashu (1991) suggested the use of diamonds to resolve this problem. Diamonds are generally very old. Some are older than 2 Ga. If Xe in such an old diamond were shown to lie on the same linear trend in a ^{136}Xe/^{130}Xe-^{129}Xe/^{130}Xe diagram as MORBs, we must conclude that the parent element was ^{244}Pu because the same linear trend indicates no change in the ^{136}Xe excess in the last few billion years, which is only consistent if the parent element was a short-lived ^{244}Pu. Figure 6.18 shows a case for Zaire cubic diamonds (Ozima & Zashu, 1991). Both diamonds and MORB form the same linear trend. If the age of the diamonds were shown to be older than some billion years, the result would give an unequivocal support for ^{244}Pu as a progenitor. However, if the age turned out to be younger than 1 Ga, we may not resolve the progenitor, and either ^{244}Pu or ^{238}U could be the parent element. Unfortunately, except for the eruption age (kimberlite eruption age), which is Cretaceous, the crystallization age of the Zaire cubic diamond has not been determined yet.

Finally, it should be noted that noble gases in the mantle are very likely to constitute only a minor fraction in the terrestrial noble gas inventory (cf. Section 6.9); hence, excess Xe isotopes observed in mantle-derived materials are unlikely to be representative of the terrestrial noble gas. In the present solid Earth, excess Xe from ^{238}U fission would then become more conspicuous than ^{244}Pu-derived excess Xe because of the much longer half-life of ^{238}U. In support of this argument, we note that the air Xe, the proxy of the terrestrial noble gas, appears to have excess Xe, which is more likely to be attributable to ^{244}Pu fission than to ^{238}U fission (cf. Section 7.2 and Table 7.1).

6.12 The Mantle He Flux

There are many models for the evolution of the atmosphere and/or degassing of the solid Earth. Most of them conclude or assume that the major features of the atmosphere were established early, within a small fraction of the age of the Earth. Nevertheless, the mantle is still degassing and offering thereby an observational constraint on models for atmospheric evolution. Among the noble gases (indeed, among all the excess volatiles that constitute the generalized atmosphere in the geochemical sense), a quantifiable degassing flux is identified only for He. That identification is possible only because He does not accumulate in the atmosphere so that its background level in air and air-saturated water is much lower than it otherwise would be.

Direct observation of present-day mantle degassing was first reported by Clark et al. (1969) and Mamyrin et al. (1969) for the case of ^{3}He; a ^{3}He degassing rate was subsequently determined by Craig et al. (1975). The degassing rate reported by Craig et al. is still the best available estimate of this important geochemical parameter and has been very widely used in various discussions concerning mantle dynamics. It should accordingly be useful to outline here some of the details for how this value was obtained.

The flux estimate is actually based on ^{4}He supersaturation. The governing relation is

$$F_4 = \rho C^*(\Delta_E\text{He})w = 3 \times 10^5 \text{ atoms/cm}^2/\text{s} \qquad (6.12)$$

where F_4 is the flux of juvenile ^{4}He, ρ is the density of sea water (1.04 g/cm^3), C^* is the solubility equilibrium concentration of ^{4}He (4×10^{-8} cm^3 STP/g), w is the mean vertical advective velocity of the world ocean. There are several estimates of w including those based on heat balance, formation rate of bottom water, dissolved inorganic carbon, and carbon flux into the sea. All are consistent with a mean velocity of about 5 m/yr (Craig & Lupton, 1976). (Δ_EHe) is the fractional excess (supersaturation) of dissolved ^{4}He. Evaluation of (Δ_EHe) is an exacting process, involving precise experimental measurements and corrections for temperature and pressure effects on solubility and a correction for supersaturation due to bubble injection (see Section 4.3). Craig et al. (1975) inferred a global average supersaturation (Δ_EHe) of 2.5% due to injection of juvenile ^{4}He at the seafloor. The global average was based on five sites (North Atlantic, South Atlantic, Southwestern Pacific, North Pacific, East Pacific Rise); the excesses were not uniform, being highest at the East Pacific Rise and lowest (almost no excess) in the Atlantic. The numerical result in Equation (6.12) is the average flux over the whole Earth. The ^{3}He flux F_3 then follows:

$$F_3 = (1.6 \times 10^{-5})F_4 = 4.8 \text{ atoms/cm}^2/\text{s} \qquad (6.13)$$

Craig et al. (1975) cite minimal 20% uncertainties in both (Δ_EHe) and w and assign a 50% uncertainty to the overall result. With due allowance for thin geographic and temporal sampling, an error of a factor of two would not be surprising. This is an important geochemical parameter that clearly warrants more attention.

An independent estimation of the mantle He-flux would be possible if we know the He escape rate from the Earth's upper atmosphere (e.g., Kockart, 1973). The present atmosphere contains about 0.7 ppm of He. Because He is too light to be gravitationally bound to the Earth, the present He concentration in the atmosphere can be concluded to represent a stationary value in balance between the mantle He influx and its outflow from the upper atmosphere. Therefore, if we know the outflow flux, we can equate it to the mantle He flux, or vice versa. However, the former estimation is even more difficult, and the present best estimate of the He escape flux is still based on the mantle He flux.

Farley et al. (1995) recently applied a global circulation model (GCM) for the world ocean to the He flux problem, assuming a source function that injects juvenile He only along ridge axes at a rate proportional to the spreading rate. They iterated the Hamburg Large-Scale GCM (Meier-Reimer, Mikolajewicz & Hasselmann, 1993) until steady-state ^{3}He distribution was obtained and concluded that the reasonable

6.12 The Mantle He Flux

Table 6.4. ^3He and ^4He flux in oceanic and continental areas

Locality	^{3}He (10^4 atoms m^{-2} s^{-1})	^{4}He (10^{10} atoms m^{-2} s^{-1})
Oceanic area		
Mean[a]	4.8	0.3
Mean[b]	4	
Continental area		
Extensional crust[c]	10–11	1–0.1
Stable or loading crust[c]	<0.01	<0.001
Great Artesian Basin[d]	0.2	3.1
Paris basin[e]	0.03	0.4
Lake Nemrut (North Turkey)[f]	~6000	~600
Laacher See (Germany)[g]	7400	1000 ± 200
Island arc		
Lake Mashu (North Japan)[h]	870	92
Kanto Plain (Central Japan)[i]	0.16	1
Chinshui (Taiwan)[j]	3.9	2.7
Chuhuangkeng (Taiwan)[j]	7.2	2.4
Subaerial volcano		
Mean	0.37[k]	
	0.88–1.15[l]	

[a] Craig et al. (1975).
[b] Farley et al. (1995).
[c] O'Nions & Oxburgh (1988).
[d] Torgersen & Clark (1985, 1992).
[e] Marty et al. (1993).
[f] Kipfer et al. (1994); ^{4}He flux corresponds to the mantle flux, and ^{3}He flux was calculated from the measured value of ^{3}He/^{4}He = 1.03×10^{-5} by the present authors.
[g] Aeschbach-Hertig et al. (1996).
[h] Igarashi et al. (1992).
[i] Sano (1986).
[j] Sano et al. (1986).
[k] Marty and Le Cloarec (1992); estimated from ^{210}Po/^{3}He ratios.
[l] Allard (1992); estimated from CO_2/^{3}He ratios.

agreement between the model and measurements justified their assumption (sources other than ridge axes not important on a global scale). Their estimate of total mantle ^{3}He flux (over the last millennium or so) is about 10^3 mol/yr (≈ 4 atoms/cm^2/s), consistent with the Craig et al. (1975) estimate [Equation (6.13)].

Mantle He also emanates in continental and island arc regions, but flux estimates are more difficult than for the world ocean. Simply by comparing areas, we do not expect that continental or arc fluxes will make a major difference to the global flux calculated for the world ocean, but the subject is of keen interest because He serves as a tracer for mantle influences. It *is* expected, and observed, that flux varies according to tectonic regional setting. Summary data are exhibited in Table 6.4.

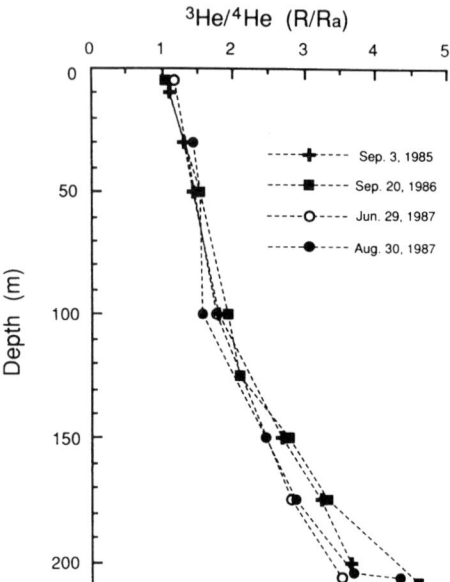

Figure 6.19 Vertical profile of ^{3}He/^{4}He ratio (relative to the air ratio) in Lake Mashu, northern Japan. Symbols correspond to respective sampling dates. Reproduced from Igarashi et al. (1992).

Juvenile He can be observed in lakes in a manner similar to that for the ocean. Fluxes also calculated with a similar approach, but with added complications such as seasonal overturn. Igarashi et al. (1992), for example, observed a systematic variation of ^{3}He/^{4}He ratio with depth in Lake Mashu (*ca.* 2–5 km, depth 200 m) in northern Japan (Figure 6.19). They attributed the depth variation to injection of the mantle He from the bottom and the seasonal variation to different accumulation times between twice-annual seasonal overturns. With an assumed circulation model, the depth profile can be translated to a flux with an eddy diffusion coefficient K_n of 0.03 cm^2/s. Igarashi et al. obtained a mantle ^{3}He flux of 8.7×10^2 atoms/cm^2/s, more than two orders of magnitude greater than the mean oceanic value. Such a high value presumably reflects regional magmatic activity and perhaps spatial concentration in the lake bottom. Interestingly, the He/heat ratio in Lake Mashu is close to the mean global ratio. Kipfer et al. (1994) also reported an unusually high mantle flux, about 6×10^8 atoms/cm^2/s (of ^{4}He), for Lake Nemrut, a crater lake in the Nemrut volcano, northern Turkey. They suggested that the very high mantle He is possibly related to local extensional tectonics and associated magmatic activity. In Maar lake in the East-Eifel volcanic field in Germany, Aeschbach-Hertig et al. (1996) observed a ^{3}He flux of 7.4×10^3 atoms/cm^2/s, more than three orders of magnitude higher than the mean oceanic flux. It seems that high mantle He flux is rather common in lakes in volcanic regions. Aeschbach-Hertig et al. noted, however, that because of the restricted area of volcanic lake sources their overall contribution to the global ^{3}He flux is evidently small.

Sano (1986) and Sano et al. (1986) found He isotopic variations with depth, ^{3}He/^{4}He decreasing toward the surface, in two natural gas wells in northern Taiwan. This relation is interpreted as a mantle flux to the bottom of the well, progressively diluted by radiogenic He released from the surrounding sediment as the gas migrates upward. With a simple mixing model, they obtained mantle He fluxes close to the mean oceanic value (Table 6.4), but the situation in a gas well is rather complicated, and it remains to be seen whether or not the coincidence with the oceanic value is accidental.

Over wider continental areas, He flux may be observed in groundwater aquifer systems, in which it is not uncommon to find nonatmospheric He. Typically, the ^{4}He is inferred to be of mostly (crustal) radiogenic origin, but mantle ^{3}He/^{4}He is so much higher than radiogenic ^{3}He/^{4}He (Section 6.7) that minor additions of mantle He are frequently detectable. Mantle ^{3}He flux can then be estimated from radiogenic ^{4}He flux. O'Nions and Oxburgh (1988) find that mantle He flux in continental areas can vary by orders of magnitude [e.g., from 10^5 to 10^6 atoms/cm^2/s (of ^{4}He) in the Pannonian Basin in central Europe (a region of crustal extension) to less than 10^3 atoms/cm^2/s in stable areas such as Britain or those undergoing depositional loading such as the Molasse and Po basins in central Europe]. Evidently upward migration of mantle He is impeded by overlying crust in tectonically stable regions but not in regions undergoing tectonic extension. The distribution of mantle He flux can provide a "map" of underlying tectonic/magmatic activity (Figure 6.20).

The He fluxes listed in Table 6.4 are total fluxes, not resolved into mantle versus crustal or primordial versus radiogenic origin. The oceanic flux can safely be taken to be all from the mantle, but in other regions the crustal component is evident. Note that even in the mantle most of the ^{4}He is radiogenic, rather than primordial, but also that, because ^{3}He/^{4}He is radiogenic, He is so low. Even in some cases where crustal ^{4}He flux dominates mantle ^{4}He flux, the ^{3}He is mostly primordial mantle ^{3}He. In some continental and arc settings, the juvenile flux density is high compared to the oceanic value, but these are rather localized areas characterized by recent magmatism and probably do not make a substantial contribution to the global flux. Overall, there is not enough geographic coverage to determine how nonoceanic flux, on average, compares to oceanic flux.

Because of the essentially primordial nature of ^{3}He and of its extremely high mobility, the absolute degassing rate of juvenile volatiles from the mantle has been most successfully estimated for ^{3}He, and the degassing rate thus determined is used to scale degassing rates of other volatiles such as CO_2, SO_4, and CH_4. Recently, however, increasing accumulation of data on CO_2 flux from subaerial volcanoes encouraged an independent estimate of its degassing rate. As such an attempt, Allard (1992) estimated the degassing rate of CO_2, from which he deduced a mantle ^{3}He flux. He found that thus estimated mantle ^{3}He flux is equivalent to 20–40% of the ^{3}He flux from midocean ridges and, hence, suggested that primordial ^{3}He degassing

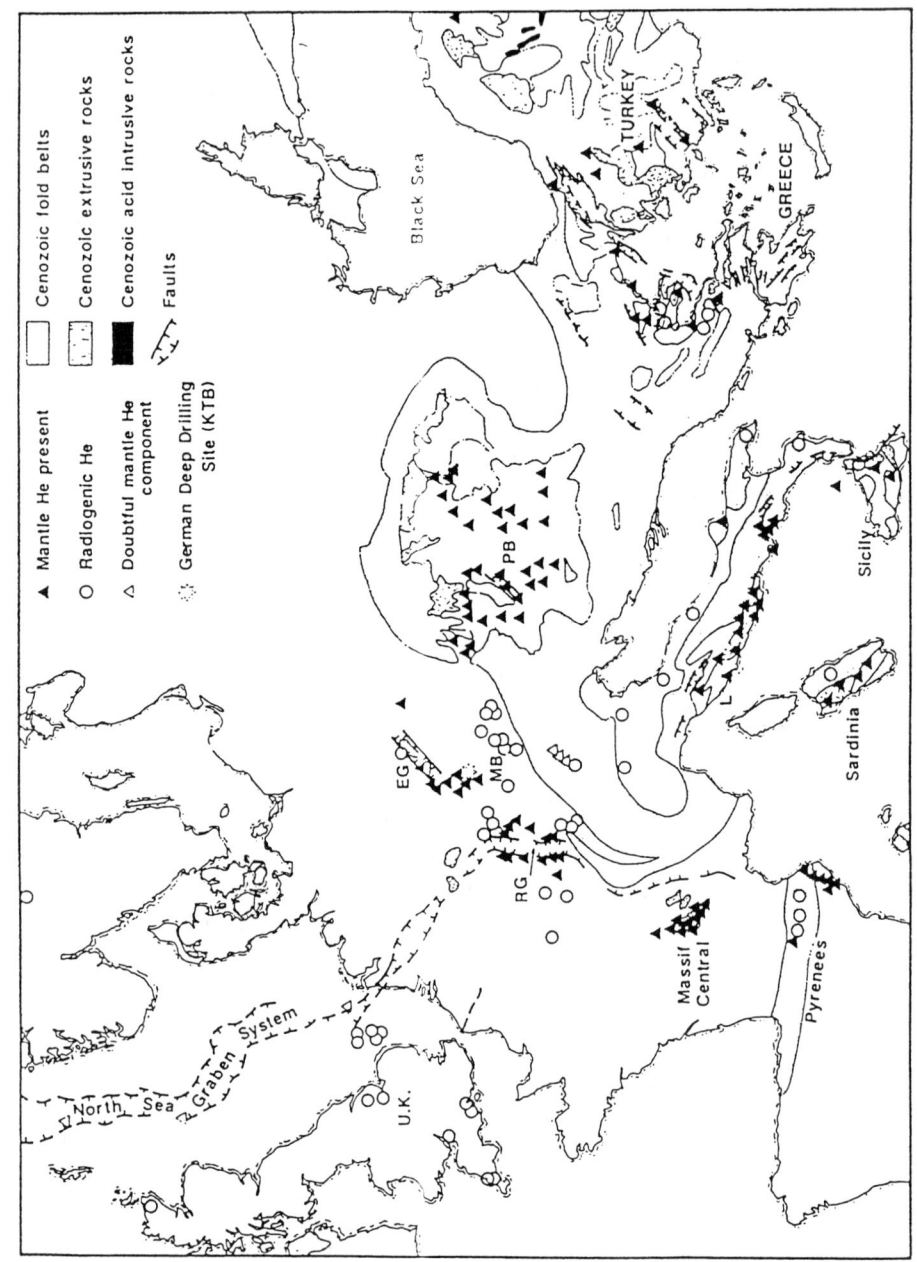

Figure 6.20 Distribution of helium isotopes in the crust of western Europe. KTB, the German deep-drilling site in Oberpfalz; MB, Molasse Basin; PB, Pannonian Basin; RG, Rheingraben; EG, Egergraben. Reproduced from Oxburgh and O'Nions (1987).

from subaerial volcanic regions may be larger than hitherto expected. In Table 6.4, we also show a preliminary estimation of a mantle ^{3}He by Marty and Le Cloarec (1992), who estimated the mantle ^{3}He flux from the observed ^{210}Po/^{3}He ratios and average ^{210}Po emission of 5×10^4 Ci a^{-1} (Lambert, Le Cloarec & Pennisi, 1988) from subaerial volcanoes.

Although there is still considerable uncertainty in the estimation of the absolute value of the mantle He flux, there is almost no direct estimate of the mantle degassing rate for other elements than He. In an early study of mantle degassing, Fanale (1971) suggested from inventory consideration that even though noble gases were extensively degassed from the mantle very early in the Earth's history, other major volatiles were less degassed. Because of large uncertainty in parameters used in his inventory consideration, the suggestion was not conclusive and required further examination. To examine this important problem, Marty and Jambon (1987) proposed to use the mantle He flux as a reference to estimate C flux, one of the major volatiles. They found that C/^{3}He ratios in MORB glasses collected throughout the world had a narrow range with an average value of 2×10^9. The value is also similar to those in hot-spot and arc volcanics. The fairly uniform values from the largely separated areas suggest that the ratio represents a large reservoir that is very likely to be the mantle. With the use of the mantle ^{3}He flux of 4×10^4 atoms m^{-2} s^{-1}, they obtained a C flux of 2×10^{12} mol a^{-1}. They also estimated the C/^{3}He ratio in the crust (air + crust; sometimes referred to as Rubey's exosphere) and found that the ratio is about two orders of magnitude smaller than the MORB ratio, the value being about 4×10^7. Marty and Jambon therefore concluded that carbon is still retained mostly in the mantle, possibly as graphite (or diamond) or dissolved in minerals. Trull et al. (1993) made quite extensive study on C-He systematics in hot-spot xenoliths from Loihi-Hawaii, Reunion, and Kerguelen hot spots. They found that CO_2 and He abundances vary by more than three orders of magnitude, but that they are strongly correlated (Figure 6.21).

The C/^{3}He ratios range from $(2\sim20) \times 10^9$, which largely overlap with MORB values, although they are somewhat more variable. The results are in good agreement with the earlier work by Marty and Jambon. In contrast to the rather similar C/^{3}He ratios between hot-spot xenolith and MORB, C/^{4}He ratios in the hot-spot samples [$(4\sim40) \times 10^4$] are by an order of magnitude larger than the values in MORB [$(0.5\sim7) \times 10^4$]. The difference is likely to reflect the upper mantle differentiation, which resulted in an increase of the U/He ratio or U/C ratio (He/C was nearly the same between MORB and the hot spot) in the MORB mantle source. Another implication of the results is that the upper (MORB source) and lower (hot-spot source) mantle reservoirs had similar initial C/^{3}He and ^{3}He/^{4}He ratios. Significant carbon recycling to the upper mantle is unlikely, but recycling of carbon to the lower mantle is consistent with C-He systematics and can account for both the similarity of ridge and hot-spot carbon isotopic compositions and the low atmospheric (air + crust) C inventory (Trull et al., 1993).

In the case of subducting area, however, Van Soest, Hilton, and Kreulen (1998) called attention to the importance of the arc crust carbon in the C/He flux; the crustal

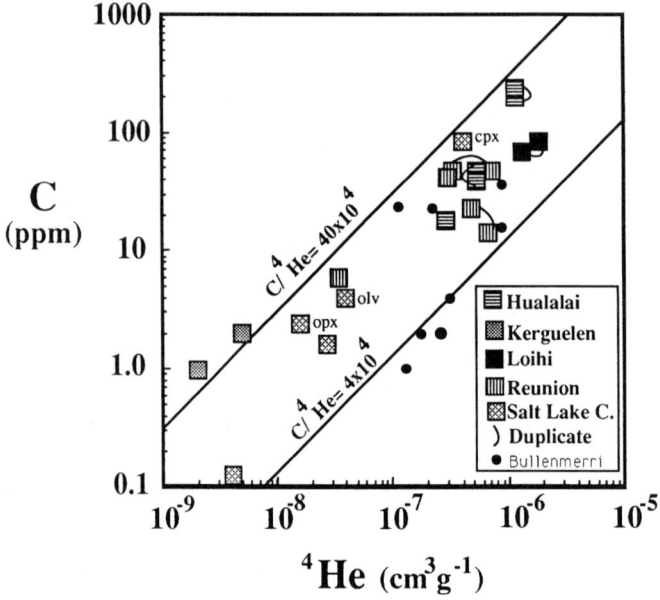

Figure 6.21 C and He abundances in fluid inclusions in hot-spot xenoliths. olv = olivine; opx = orthopyroxene; cpx = clinopyroxene. Reproduced from Trull et al. (1993).

carbon may contribute up to 20% of the total carbon output in the central Lesser Antilles. Also, Hilton, McMurty, and Goff (1998) cautioned that $CO_2/^3He$ would vary considerably in response to changes in magma chemistry.

6.13 ^{4}He/Heat and Mantle Dichotomy

When a ^{238}U atom decays to a final stable isotope ^{206}Pb, it yields eight atoms of ^{4}He and dissipates an energy of 7.6×10^{-5} erg. The energy will be stored as heat in the surrounding medium. For the decay of ^{238}U, we therefore have a ^{4}He/heat production ratio of 1.05×10^{12} atom/J. Taking account of ^{235}U and ^{232}Th, which also emit ^{4}He and heat, and of another important heat-generating element, ^{40}K, the ^{4}He/heat production ratio in the Earth (U/K $\equiv 10^4$, U/Th $\equiv 3.5$) is about 10^{12} atom/J.

Since atomic diffusion is so slow, partial melting of mantle material and the subsequent magma transportation would be the most effective process to extract helium from the mantle and to transport it to the surface. Magma transportation is also a very effective process for heat transport. Hence, we expect that the heat and helium generated in a magma source under oceanic ridges will essentially be removed by the ascending magma. Therefore, the emerging heat and helium at ridges can be expected to show a ^{4}He/heat ratio close to the theoretical production ratio. However, as seen in Figure 6.22, the majority of the data are considerably smaller than the theoretical value.

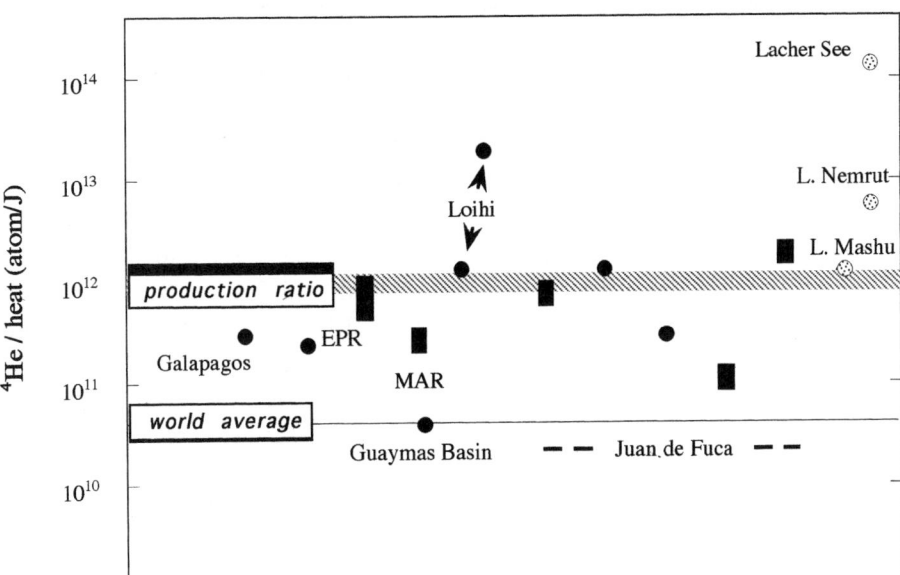

Figure 6.22 ^{4}He/heat (atom/J) measured at various localities. The production ratio is calculated from heat and ^{4}He produced from ^{238}U, ^{235}U, ^{232}Th, and ^{40}K (heat only). The world average is calculated from the average heat flow $0.082 \, \text{W/m}^2$ and the He flux $4 \times 10^9 \, \text{atoms} \, \text{m}^{-2}/\text{s}$ (Craig & Lupton, 1976). Note that both the world average and the observed values are significantly below the production ratio. L. Mashu, Igarashi et al. (1992); L. Nemrut, Kipfer et al. (1994); Laacher See, Aeschbach-Hertig et al. (1996); MAR, Jean-Baptiste et al. (1998). Other data are from the compilation by Lupton, Baker, and Massoth (1989).

The disagreement becomes even larger if we take into account the preferential extraction of ^{4}He relative to heat from a mantle source during magma generation. The extraction of trace elements such as noble gases into magma can be described as equilibrium partition between silicate melt and residual mantle crystals. Because of very low partition coefficient of He between melt and crystal, most of ^{4}He accumulated in the mantle source region that participated in the partial melting is extracted into the magma and transported to the surface. On the other hand, only the heat associated with the upwelling magma will be transported to the surface. This results in an enhancement of the ^{4}He/heat ratio in the magma; otherwise, the observed ^{4}He/heat ratio overrepresents the ratio in the mantle source. To show this quantitatively, we assume a batch-melting model for noble gas partition in magma. Noble gas will be enriched in magma by a factor of $(1/f)\{1 - (1 - f)^{1/D}\}$ relative to the initial concentration in the mantle, where f and D denote the degree of fractionation and a partition coefficient of noble gas between melt and crystal. The enrichment factor ranges from about 5 to 20 for a reasonable choice of $f = 0.2$–0.01 and $D = 0.05$ (cf. Table 2.4). We note that, because of a very small value of D, the enrichment factor is insensitive to the choice of a D value. Therefore, the ^{4}He/heat ratio in the source mantle may be by a factor of 5–20 as small as the observed values at the surface, and the imbalance would become even larger.

If we use world-averaged values of heat flow and ^{4}He flux instead of the direct observational values, the discrepancy between the averaged ratio and the theoretical production ratio also becomes significant; the ratio of the world average ^{4}He flux (4×10^9 atoms m^{-2} s^{-1}; Craig et al., 1975) to that of heat flow (0.082 W m^{-2}; Stacey, 1992) is smaller by about a factor of 30 than the theoretical production ratio.

The imbalance between heat flow and ^{4}He flux can also be seen from the consideration of a uranium inventory in the Earth. O'Nions and Oxburgh (1983) pointed out that even though a reasonable geochemical model with 5 ppb of U (K/U = 10^4, Th/U = 3.8) for the upper mantle can approximately account for the observed helium flux, it will yield only 3% of the observed heat flux at ridges. This result indicates that the remaining 97% of the heat flow must come from somewhere other than the upper mantle, namely either from the lower mantle or from the core or from both, whereas little extraneous ^{4}He flux is required. This led O'Nions and Oxburgh to conclude that ^{4}He flux from the lower mantle is essentially inhibited.

To explain the imbalance, O'Nions and Oxburgh (1983) and Oxburgh and O'Nions (1987) proposed that a barrier, which is suggested to exist between the upper and the lower mantle from seismic observation, has trapped helium in the lower mantle and retarded the heat transport from the lower mantle to the upper mantle. O'Nions et al. (1983) suggested, from a semiquantitative discussion, that delayed heat transfer from the lower mantle to the upper mantle with a time constant of about 2 Ga would enhance the present heat flow by a factor of two. McKenzie and Richter (1981) made numerical calculation on a two-layered mantle convection and showed that heat transfer from the lower mantle to the upper mantle is considerably retarded to give rise to an enhancement of the present surface heat flow up to a factor of two. If the thermal barrier not only retards the heat transfer and hence enhances the present surface heat flow but also essentially prevents the ^{4}He flux from the lower to the upper mantle, this would qualitatively explain the imbalance. If this indeed were the case, we would expect a large amount of ^{4}He accumulation in the lower mantle. However, it is difficult to conclude such a large accumulation of ^{4}He in the lower mantle from the currently available scarce noble gas data derived from mantle-derived materials.

The boundary between the lower and the upper mantle, characterized by a large jump in seismic velocity and known as a 670-km discontinuity, is the most distinct boundary in the mantle. The discontinuity has been argued to be due either to the first-order phase transition in an isochemical medium consisting of olivine and other mantle-forming minerals or to a compositional difference or to both. If the boundary were due to the compositional difference, its persistent existence argues for the inhibition of a mantle-wide convection, and the layer may act as the effective He barrier. In this case, except for mass flow from the lower mantle in plumes, helium movement through the boundary layer would essentially be governed by solid diffusion. The diffusion time through the boundary layer of a thickness d is $d^2/2D$. Taking

$D = 10^{-8}\,\text{cm}^2/\text{s}$, which is the value for olivine at 1300°C at an atmospheric pressure (Hart, 1984; Trull et al., 1991) and $d = 30\,\text{km}$, we have $t \approx 10^{13}$ years. Hence, we may safely assume that diffusive helium transportation from the lower mantle is negligible even if integrated over the age of the Earth.

Allègre, Hofmann, and O'Nions (1996), from a potassium (and hence radiogenic ^{40}Ar) inventory consideration, concluded that the mantle is likely to be two-layered with an upper mantle outgassed in ^{40}Ar and a lower mantle relatively undegassed in ^{40}Ar. They pointed out that K contents in the whole Earth ($\approx 250\,\text{ppm}$), which was estimated from the generally assumed values of U (20 ppb) and U/K (12,700) for the whole Earth, is much higher than K content in the upper mantle ($\approx 50\,\text{ppm}$) estimated from average K content in MORB (500 ppm) and the potassium partition in the MORB magma (about ten times enrichment relative to the mantle source). They, therefore, concluded that the marked difference in the K content between the upper and the whole Earth must indicate a chemically layered mantle. Such a compositionally layered mantle is also required for the He-barrier model proposed by O'Nions and Oxburgh; the boundary required by these models is not necessarily the same as the 670-km seismic discontinuity.

The mantle barrier or the layered mantle, if it exists, would also explain the long-standing problem of the persistent outgassing of the primordial helium from the mantle; the persistent ^{3}He flow shows a storage of a considerable amount of the primordial ^{3}He in the Earth's interior in spite of its extreme mobility. A commonly assumed hiding place for the primordial helium is the core. However, recent experiments on noble gas partition between silicate and iron melt at high pressures suggest that only a negligible amount of noble gas may be partitioned in iron melt (Matsuda et al., 1993). The experimental results seem to argue against the core, leaving the lower mantle as the plausible hiding place for the primordial helium. We then interpret that the present ^{3}He outgassing is due to continuous leaking from the deeper mantle via channels such as hot spots and by some diffuse entrainment process at the transition interface between lower and upper mantle throughout geological time (Oxburgh, 1991). The present outgassing rate of the primordial ^{3}He is estimated to be about 4 atoms/cm^2/s (Craig & Lupton, 1976). If this rate lasted for 4.5 Ga, the total amount of ^{3}He outgassed from the Earth would be about $10^{16}\,\text{cm}^3$ STP, which is much smaller than the ^{3}He amount in the lower mantle inferred from the amount in OIB samples. Although the assumption that the outgassing rate has remained constant may be questioned, this simple calculation suggests that the primordial ^{3}He stored in the mantle is more than enough to sustain a continuous helium outgassing at the present rate throughout the whole history of the Earth.

The mode of the mantle convection, layered or mantle-wide, is one of the most fundamental problems in current earth science. As we discussed earlier, ^{4}He-heat systematics appears to suggest that the lower mantle (apart from the exact locale) is essentially isolated from the upper mantle by a barrier that impedes He migration between the layers. Other noble gas characteristics, for example much higher ^{40}Ar/^{36}Ar and ^{129}Xe/^{130}Xe in the upper mantle than in the lower mantle, also appear to be consistent with the layered mantle. On the contrary, recent seismic tomography appears to show that some subducting slabs are penetrating into the lower mantle or an accumulation of downwelling slabs above the 670 km discontinuity resulting in frequent avalanches of upper-mantle material into the lower mantle (e.g., Van der Hilst, Widiyantoro & Engdahl, 1997, and references therein). It is then important to examine whether such deep penetration of a subducting slab can be reconciled with the chemically distinct two-layered mantle favored by noble gas geochemistry. The problem is essentially related to a time constant for homogenization of imported material in the lower mantle and also to the degree of subduction input such as its time duration and spatial scale. Recently, Kellog et al. (1999) proposed a numerical mantle convection model that may reconcile the noble gas isotopic distinction between the upper and the lower mantle with the substantial input of subducting material into the lower mantle. With choices for acceptable values for viscosity and other free parameters and the assumption that the deep layer above the CMB (core mantle boundary) is enriched in heat-producing elements (25.6 ppb of U in the deep layer, but 7 ppb in the upper mantle), they showed that compositionally distinct layers separated at about 1600 km (assumed from the tomographic results by Van der Hilst & Kárason, 1999) can remain over the age of the Earth while a considerable fraction of the subducting slab reaches the CMB. The model also offers an explanation for the "missing heat" pointed out by O'Nions and Oxburgh (1983); the assumed high potassium content at the CMB would provide necessary heat. Alternatively, Davies (1999) suggested that either of the following three possibilities or some combination of them would resolve the apparent contradiction between the mantle dichotomy suggested by noble gas systematics and geophysical evidences for large mass flow between the upper and lower mantle: (1) the Earth has 50% less K than is usually estimated, (2) ^{40}Ar is sequestered in the core, or (3) ^{40}Ar has been lost from the Earth entirely. Among the three possibilities, Davies favors the first one, pointing out that the Earth's K/U ratio is only about 6000, compared with the values of 10,000–13,000 commonly accepted. These recent studies at least show that the contradiction between the geochemical inferences and geophysical observations on the mantle dichotomy may not be as insurmountable as it might appear.

Chapter 7

Noble Gases in the Earth

7.1 Overview

In previous chapters, we have discussed noble gas characteristics in major geological divisions in the Earth, such as ocean, crust, and mantle. Differences in noble gas isotopic ratios and in relative elemental abundances among these divisions are primarily due to the addition of radio- and nucleogenic isotopes and to elemental fractionation in the course of evolution of the respective regions. An implicit assumption underlying these discussions is that the Earth was endowed with a primordial noble gas common to all the regions. Although we still do not have direct evidence to prove (or disprove) this, discussions in previous chapters and also in this chapter seem to support this assumption. A question then arises: what would be the primordial noble gas in the Earth? In this regard, we emphasize that the primordial noble gas in the Earth, especially its isotopic compositions, is a key constraint as a reference in any Earth evolution model including noble gases. Through various discussions of noble gas characteristics given in previous chapters, we have learned that the primordial noble gas in the Earth was likely to be derived from solar noble gas by some mass-dependent fractionation. In this chapter, especially in Sections 7.3 and 7.7, we will discuss these questions in more detail.

As already mentioned, current noble gases in the Earth have been modified by addition of radio- and nucleogenic isotopes and by elemental fractionation due to various geological processes. Therefore, to resolve the primordial noble gas in the Earth, we must in the first place asses these secondary effects. This can be most conveniently done by comparing noble gases among major distinct regions; differences in noble gas characteristics would reflect those in chemical compositions and geological processes in respective divisions, which we may independently assess. This will be discussed in Section 7.2. As we will discuss in Section 7.3, comparison of the terrestrial noble gas with major extraterrestrial componenets such as the solar noble gas (SW-noble gas) is also crucial in resolving the primordial component in the Earth.

Among the five noble gases, Ne and Xe deserve special attention because their isotopic compositions are unique (as far as we know) to the Earth, suggesting that their evolution processes are fundamentally related to some specific processes of Earth evolution. In Section 7.4, we will discuss Ne in that a key issue is to understand the distinct difference between mantle neon and atmospheric neon isotopic compositions. In Section 7.5, we discuss a long-standing "missing Xe" problem.

A recent development in the study of extraterrestrial noble gas components such as the solar noble gas and meteorite noble gas has shed a new light in understanding the origin of the terrestrial noble gases, including missing Xe. Another unique feature of Xe is that it has isotopes produced from short-lived nuclide $^{129}Xe^*$ from ^{129}I and $^{131-136}Xe_{Pu}$ derived from ^{244}Pu spontaneous fission. In meteoritics, these extant Xe isotopes have been studied extensively to resolve early meteorite chronology, but little has been done for the Earth. In Sections 7.6 and 7.7, we discuss concomitantly both $^{129}Xe^*$ and fission $^{136}Xe_{Pu}$ in a framework of a $^{129}I/^{127}I$—$^{244}Pu/U$ systematics, which will be shown to be useful in resolving the early evolutionary history of the Earth. Finally, in Section 7.8, we address a question regarding how the terrestrial noble gases were acquired by the Earth. This question is relevant not only to the origin of the terrestrial noble gas but also to Earth evolution.

7.2 Primordial Noble Gas in the Earth

From the preceding chapters, it should be clear that there are two distinct major noble gas reservoirs in the Earth; one is the atmosphere, and the other is the solid Earth. Even though the atmospheric component is homogeneous in its isotopic composition and in its relative elemental abundance, noble gases contained in rocks, minerals, and fluids or gases (e.g., methane or CO_2 well gases) show considerable variations not only in relative elemental abundances but also in isotopic compositions. However, the latter variation is essentially due to radiogenic and nucleogenic components. As we discussed in Chapter 6, apart from these secondary components, noble gases in mantle-derived materials show a fairly uniform isotopic composition, and we may discuss mantle noble gas as a distinct component. Noble gases in crustal materials, including those dissolved in sea water, are predominantly atmospheric. Since the amount of noble gases in the crust is much less than that in the atmosphere, it is sufficient to consider the atmospheric and mantle components in the discussion of the bulk Earth noble gas. As we discussed in Chapter 6, it seems reasonable to assume that isotopic composition of nonradiogenic noble gas in the mantle is indistinguishable from the atmospheric noble gas except for $^{20}Ne/^{22}Ne$. A further implication of these empirical observations is that the atmospheric noble gases were degassed from the solid Earth while fractionation of Ne took place in the atmosphere after the degassing. We discard He in the present discussion, since He is not gravitationally conservative in the atmosphere.

Contrary to isotopic composition, it is difficult to infer an elemental composition representative of the Earth's interior from empirical noble gas data in mantle-derived materials because of likely elemental fractionation. We have discussed this difficulty in Chapter 6. However, there are reasonable grounds to suppose that the major fraction of terrestrial noble gases now resides in the atmosphere. For example, from a simple consideration of ^{40}Ar budget in the Earth, we have shown that more than half of radiogenic ^{40}Ar has been degassed from mantle to the atmosphere (Section 6.6). A nonradiogenic component or the primordial terrestrial noble gas must have been subjected to even higher degree of degassing, since the latter component existed long before most of radiogenic ^{40}Ar had been accumulated in the mantle. Hence, it would be reasonable to assume that the amount of atmospheric noble gas gives a rough estimate of a noble gas inventory in the whole Earth. In comparison with the noble gas inventory, relative elemental abundances would be better inferred from atmospheric data, since noble gases were essentially partitioned into gaseous phase on degassing because of their geochemical imcompatibility (Section 2.4), preserving approximately their original abundance ratio in the atmosphere. Therefore, it would be reasonable to assume that apart from argon, which is mostly radiogenic ^{40}Ar, relative elemental abundance of atmospheric noble gas approximates that of the primordial terrestrial noble gas. However, caution is due in the case of Ne and Xe. Difference in Ne isotopic composition (i.e., ^{20}Ne/^{22}Ne between mantle and atmosphere) is likely to be due to neon isotopic fractionation caused by neon escape from the primitive atmosphere. As we discuss in Section 7.4, to fractionate neon by the observed amount (i.e., about 30% reduction in ^{20}Ne/^{22}Ne), a few ten percent of the initial amount of Ne is required to escape from the primitive atmosphere (Zahnle et al., 1990b). If this were the case, the relative abundance of the primordial Ne would be underrepresented by the atmospheric Ne. Underrepresentation of primordial component is also suspected in the case of Xe as we will discuss in Section 7.5.

Contrary to this argument, Owen, Bar-Nun, and Kleinfeld (1992) suggested that heavy noble gases were supplied to the Earth by impacting icy planetesimals (comets). Porcelli and Wasserburg (1995a, 1995b) argued from their modeling calculations of noble gas evolution in the Earth that most of noble gases in the atmosphere were supplied by late accreted veneer materials such as comets, and that only a few percent of the atmospheric noble gas was derived from the Earth's interior. However, if this were the case, we would have to make an ad hoc assumption that the extraterrestrial materials happened to have the same noble gas isotopic compositions as those observed in the Earth's interior. This seems to be too fortuitous, since as we discuss in Section 7.3, isotopic compositions of any known extraterrestrial noble gas components are distinctly different from the terrestrial noble gases. Therefore, it seems to be more reasonable to assume that atmospheric noble gases were essentially

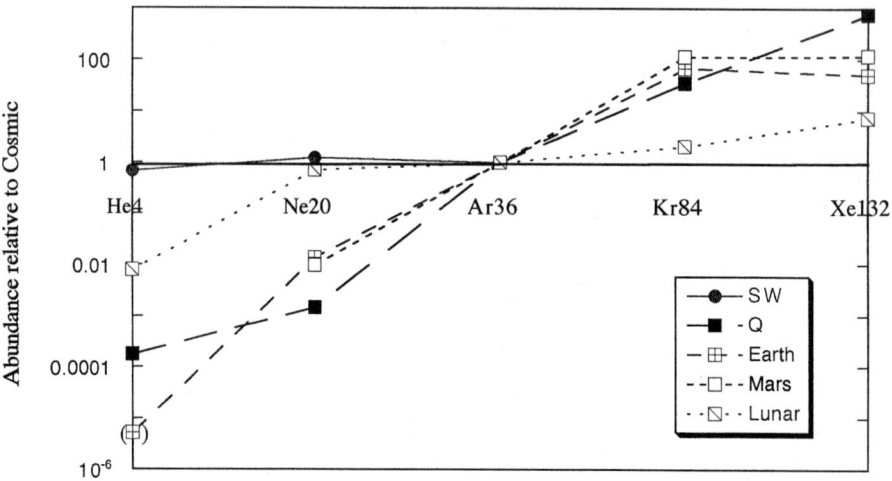

Figure 7.1 Elemental abundance of noble gases relative to cosmic abundance (Anders & Grevesse, 1989). Data for Earth (atmosphere), SW (solar wind implanted on Al foils on the moon), Lunar (solar wind implanted on lunar soils), Q (chondrites), and Mars are from Table 3.2.

derived from the solid Earth, and except for Ne, nonradiogenic noble gas isotopic compositions in the bulk Earth are well represented by the atmospheric ones.

To resolve the primordial terrestrial noble gas, it would be useful to examine major noble gas reservoirs in the early solar system, which could have supplied noble gases to the Earth. As we discussed in Chapter 3, two major noble gas components occur very widely in the solar system and can be a potential source for the terrestrial noble gas. They are solar noble gas (representative of the sun), which is generally assumed to be best represented by solar wind noble gas implanted on Al-foil target plates on the moon (elemental ratio) and on lunar breccia (isotopic ratio) (e.g., Ozima et al., 1998), and Q phase noble gas (see Wieler, 1994, for a review), which occurs very widely in various chondrites. Next we will compare the bulk Earth noble gas, which we assume to be represented by atmospheric noble gas with these two major noble gas components in the solar system.

Figure 7.1 shows a noble gas elemental abundance relative to ^{36}Ar for the Earth atmosphere, Q, SW, and lunar soils [cf. Table 3.2, 3.3(a), and 3.3(b)]. We also included the supposed Martian atmospheric noble gas (e.g., Pepin, 1991). The abundances are normalized to the solar (cosmic) abundance.

In Figure 7.1, noble gases in solid objects [Earth, Mars, and meteorites (Q)] show large enrichment in the heavier noble gases relative to the solar abundance, whereas the lunar soils show a slight fractionation and the SW shows almost no fractionation (no measurement of Kr and Xe because of their extremely low abundance in SW). Such a fractionation trend is what we would expect if noble gases were captured by solid objects from gaseous solar nebula. It is also worth noting that even though the fractionation is almost identical for Earth and Mars, the Q component shows a little

7.2 Primordial Noble Gas in the Earth

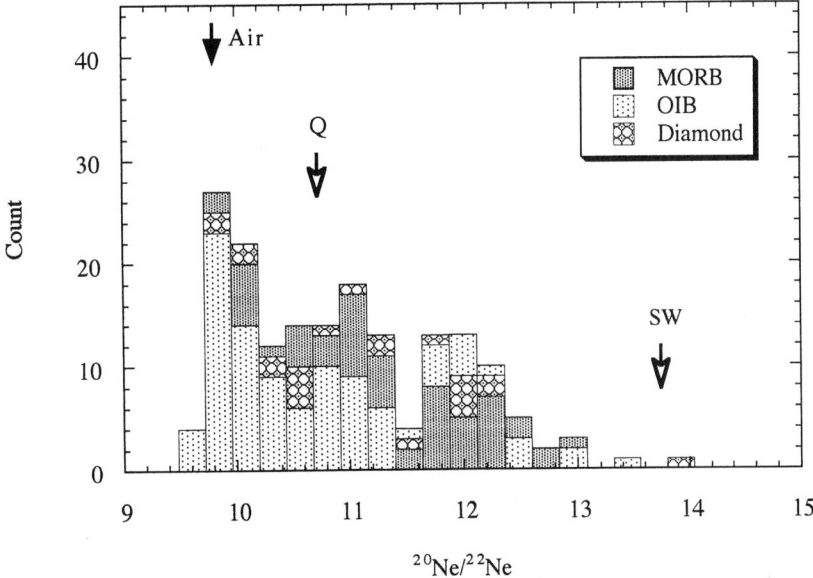

Figure 7.2 ^{20}Ne/^{22}Ne ratios in mantle-derived materials are compared with those of air, SW (solar wind), and Q. MORB: Sarda and Graham (1990), Hiyagon et al. (1991), Moreira et al. (1995), Niedermann (1997). OIB: Sarda et al. (1988), Staudacher et al. (1990), Hiyagon et al. (1992), Poreda and Farley (1992), Honda et al. (1993a, 1993b), Valbracht et al. (1996, 1997). Diamond: Ozima & Zashu (1991). Data for Q and SW are from Table 3.3.

different fractionation trend, especially in Xe and Ne. Not just the Earth and Mars but also the meteorite noble gases Q are suggested to be related to the solar noble gas by a simple mass-dependent fractionation (Section 7.3). This relation might indicate that noble gases in the terrestrial planets are more closely related to each other than to meteorites; the terrestrial planets, for example, might have captured noble gases through a similar process.

In Figures 7.2–7.5, we show histograms for ^{20}Ne/^{22}Ne, ^{38}Ar/^{36}Ar, ^{86}Kr/^{84}Kr, and ^{128}Xe/^{130}Xe for MORB, OIB, and diamonds. These isotopic ratios are least modified by later addition of nucelogenic components in the Earth, and hence are more representative of the primordial ratios. SW values in the figures (arrow mark) indicate isotopic ratios measured on lunar breccia, which we assume to represent the solar noble gas isotopic composition.

^{20}Ne/^{22}Ne: Many researchers have suggested that ^{20}Ne/^{22}Ne ratios are distinctly different between the air and mantle (e.g., Craig & Lupton, 1976; Sarda et al., 1988; Ozima & Zashu, 1988; Honda et al., 1991a, 1991b; Farley & Poreda, 1993). ^{20}Ne/^{22}Ne ratios observed in mantle-derived materials generally range from the air ratio (9.80) up to values close to the SW ratio (13.7). These observations have led some authors to assume that the primordial terrestrial Ne is solar and the spread in ^{20}Ne/^{22}Ne ratios is due to mixing between the air Ne and the solar Ne in the mantle (e.g., Pepin, 1991;

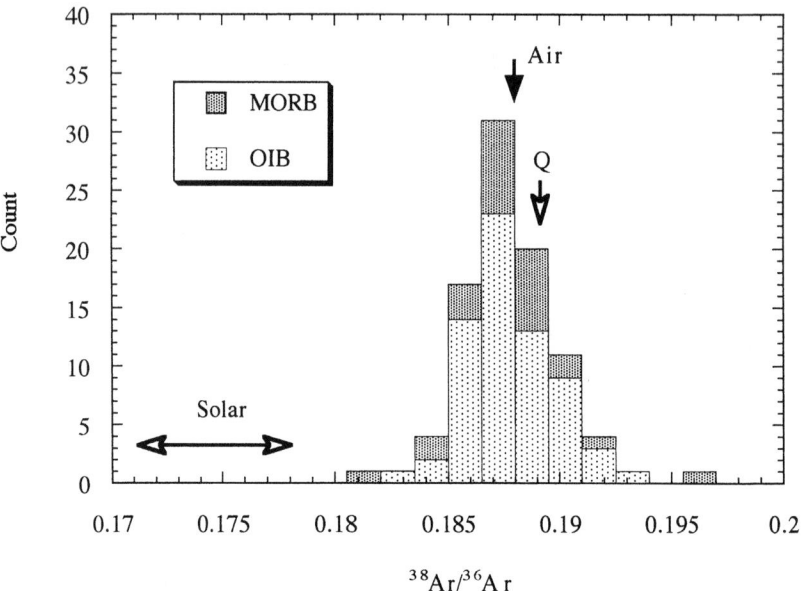

Figure 7.3 $^{38}Ar/^{36}Ar$ ratios in mantle-derived materials compared to air, solar (SW), and Q. Data sources are the same as in Figure 7.2. Solar: Pepin (1998). Diamond data are not included because they appear to have some nucleogenic ^{38}Ar (see Section 5.3).

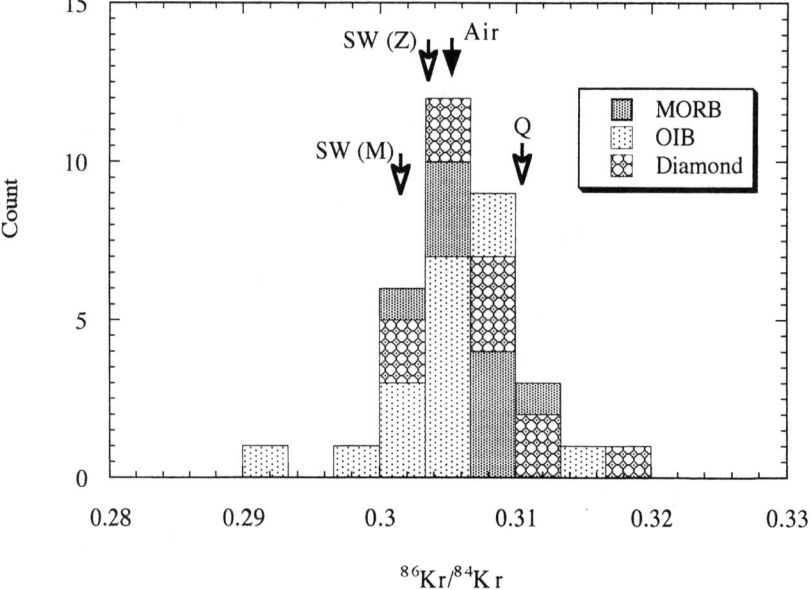

Figure 7.4 $^{82}Kr/^{84}Kr$ ratios in mantle-derived materials compared to air, SW, and Q. Data sources are the same as in Figure 7.2. Z: Zurich (Wieler & Baur, 1994); M: Minnesota (Pepin et al., 1995).

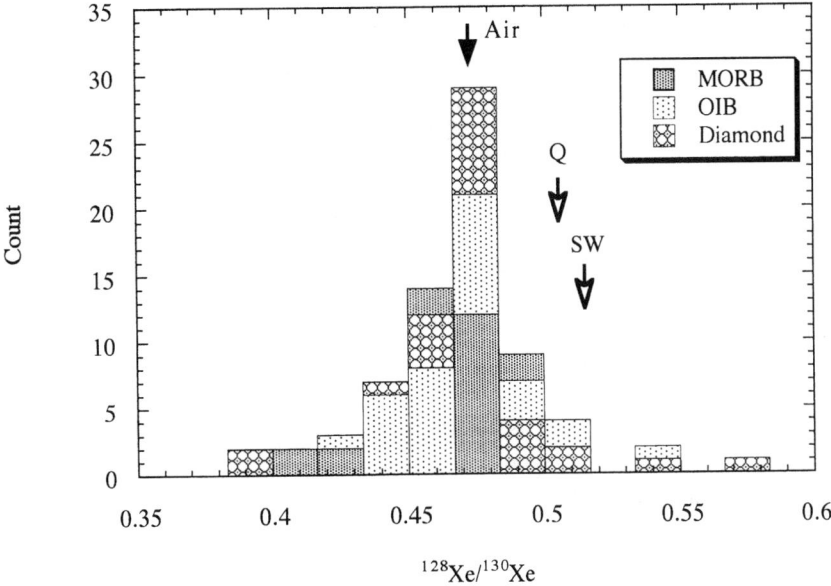

Figure 7.5 $^{128}Xe/^{130}Xe$ ratios in mantle-derived materials compared to Air, SW and Q. Data sources are the same as in Figure 7.2.

Honda et al., 1991a, 1991b; Valbracht et al., 1997; Moreira et al., 1998). However, some favor significantly lower mantle ratios than the solar value (e.g., Ozima & Zashu, 1991; Niedermann et al., 1997; Ozima & Igarashi, 2000). Although the histogram does suggest the existence of a nonatmospheric component (or components), which has a higher $^{20}Ne/^{22}Ne$ ratio than the air ratio, it is not possible to conclude from Figure 7.2 alone whether or not this second component is solar Ne. From a statistical test on the distribution (Figure 7.2), Ozima and Igarashi (2000) suggested that the distribution may result from mixing of three independent Ne components (air Ne and two independent mantle components, both of which differs from solar Ne). However, this remains to be confirmed by more empirical data on mantle-derived materials. As far as the currently available observational data are concerned, however, we may at least argue that the histogram (Figure 7.2) hardly confirms a common assumption that the distribution resulted from a two-component mixing (i.e., air Ne and solar Ne).

$^{38}Ar/^{36}Ar$: The $^{38}Ar/^{36}Ar$ distribution histogram (Figure 7.3) fits well to a Gaussian distribution curve with a mean value indistinguishable from the air ratio (Ozima & Igarashi, 2000). The well-defined Gaussian distribution for the MORB and OIB samples shows without any reasonable doubt that Ar observed in the mantle-derived samples consists essentially of a single component. The corollary to this conclusion is that the pristine mantle Ar was identical with the air Ar, unless a perfect homogenization of Ar between the atmosphere and the mantle was realized. However, such a perfect homogenization can easily be ruled out from the observed distinct difference in $^{40}Ar/^{36}Ar$ and other radiogenic isotopic ratios such as $^{21}Ne/^{22}Ne$ and $^{129}Xe/^{130}Xe$ between the atmospheric noble gases and the mantle components. The identical isotopic composition between the mantle and the atmosphere is also in accordance with a general consensus that the atmospheric noble gases were degassed from the mantle. Although suggested values for the solar Ar still have a considerable range (Benkert et al. 1993; Becker et al. 1998), they are distinctly outside the observed distributions. It is difficult, however, to conclude whether the primordial terrestrial Ar meaningfully differs from Q-Ar. In summary, it seems to be safe to conclude that the primordial Ar in the Earth differs from the solar Ar (cf. Section 6.7). A claim for solar Mantle Ar (Pepin, 1998) seems to be biased by a small number of data used in this discussion (Kunz, 1999).

$^{86}Kr/^{84}Kr$: A distribution for $^{86}Kr/^{84}Kr$ ratios in MORB, OIB, and diamonds (Figure 7.4) is fairly well represented by a normal distribution with a mean value identical to the atmospheric ratio. This observation is consistent with the previous conclusion for Ar isotopic ratio, and we may conclude that the atmospheric Kr was derived from the Earth's interior or that the primordial Kr is the same as the atmospheric Kr. There is still some disagreement about a solar $^{86}Kr/^{84}Kr$ ratio among researchers (e.g., Wieler & Bauer, 1994; Pepin, Becker & Rider, 1995), but both ratios are very close to the air ratio. There is also little difference between the SW and Q. Hence, it is difficult to conclude from the empirical data alone whether the terrestrial Kr differs meaningfully from these extraterrestrial components.

$^{128}Xe/^{130}Xe$: The distribution of $^{128}Xe/^{130}Xe$ (Figure 7.5) is approximately Gaussian around the air ratio, again supporting the conclusion that the atmospheric component was derived from the Earth's interior. The terrestrial components seem to be significantly different from SW-Xe and Q-Xe.

Apart from fission and radiogenic components, it appears that the principal underlying relationship between terrestrial Xe and solar Xe is a strong (about 3.5%/amu) mass-dependent isotopic fractionation. This is shown in Figure 7.6. To a good approximation, atmospheric Xe is related to SUCOR Xe, a solar Xe composition calculated to be *su*rface-*cor*related Xe in a lunar mare soil. However, it is also clear that slight deviation from a linear trend indicates that atmospheric Xe and solar Xe cannot be related solely by fractionation.

Assuming that both characteristic meteorite Xe (often referred to as planetary Xe; see Chapter 3) and terrestrial Xe were derived from a common primordial Xe (not assuming the solar Xe), Pepin and Phinney (1978) attempted to resolve the relationship among these Xe components by means of a multidimensional correlation technique. They concluded that atmospheric Xe contains about 4.6% of fission Xe either from ^{244}Pu or ^{238}U (because of similar fission yields, it was not possible to resolve a precursor element, although ^{244}Pu was slightly preferred). Monoisotopic radiogenic $^{129}Xe^*$ is unlike fission Xe, which has four isotopes with known fission yields and hence can be analyzed by the correlation technique, in that it is not possible to apply

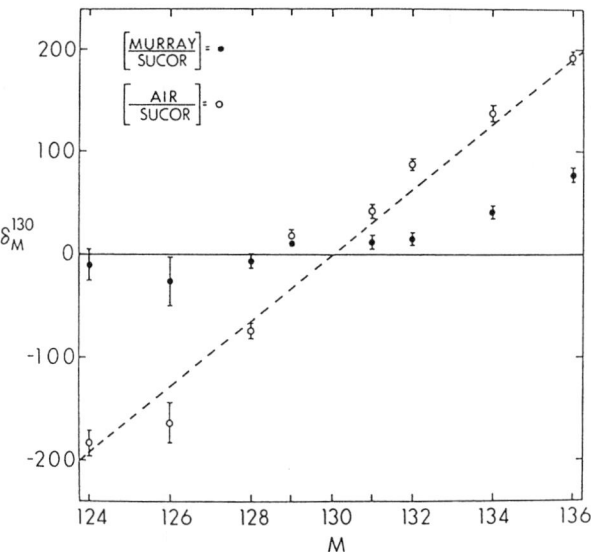

Figure 7.6 Isotopic compositions of meteorite (Murray), solar, and terrestrial Xe, displayed as per mil variation of observed isotope ratios (normalized to ^{130}Xe) in air and the carbonaceous chondrite Murray from the corresponding ratios SUCOR, a solar Xe composition calculated to be surface-correlated Xe in a lunar mare soil. The dashed line, illustrating linear fractionation, is primarily for reference. Reproduced from Podosek (1978).

this technique. Hence, radiogenic ^{129}Xe* was identified from comparison with the least radiogenic ^{129}Xe found in the Novo Urei meteorite, and they concluded that 6.8% of atmospheric ^{129}Xe is radiogenic ^{129}Xe*. The correlation analyses also enabled Pepin and Phinney (1978) to estimate the primordial Xe, which they concluded to be slightly different from solar Xe. Pepin and Phinney (1978) thus resolved primordial Xe as U-Xe.

The U-Xe thus calculated is very similar to the "Primitive Xe" proposed by Takaoka (1972) to represent the primordial Xe component in the solar system from a visual correlation of chondrite and achondrite Xe isotopic data. However, with a refined statistical technique and expanded database for meteorite noble gases, Igarashi (1995) reexamined the problem and concluded that atmospheric ^{136}Xe contains 2.8 ± 1.3% of fission Xe either from ^{244}Pu or ^{238}U with slight preference toward ^{244}Pu (Table 7.1). Contrary to the conclusions about U-Xe by Pepin and Phinney, Igarashi concluded that there was no meaningful difference within the statistical significance between thus determined primordial Xe and solar Xe (or more precisely SW-Xe). Therefore, Igarashi concluded that it was not necessary to invoke a primordial Xe component such as U-Xe, which differs from the solar Xe. Although the number of meteorite Xe isotopic data used by Igarashi is more than five times as many as those used by Pepin and Phinney, the statistical analysis by Igarashi is still subject to

Table 7.1. *Primordial Xe in Earth and in solar system*

	^{124}Xe	^{126}Xe	^{128}Xe	^{129}Xe	^{130}Xe	^{131}Xe	^{132}Xe	^{134}Xe	^{136}Xe
Atmosphere (present)	2.337	2.18	47.15	649.6	≡100	521.3	660.7	256.3	217.6
Primordial Earth (1) (nonradiogenic)	2.337 ± 0.007	2.180 ± 0.011	47.146 ± 0.047		≡100	516.5 ± 0.8	651.5 ± 0.9	249.3 ± 1.9	211.5 ± 3.0
Primordial Xe (1)	2.803 ± 0.084	2.461 ± 0.065	50.09 ± 0.47		≡100	501.0 ± 1.6	613.2 ± 1.1	220.8 ± 1.7	176.4 ± 2.5
U-Xe (2)	2.947	2.541	50.873	628.7	≡100	499.58	604.79	212.88	166.34
SW-Xe (3)			51.27	636.47	≡100	498.49	603.86	221.01	179.35

Table 7.1 References: 1, Igarashi (1995); 2, Pepin & Phinney (1978); 3, renormalized to ^{130}Xe from the data given by Wieler & Baur (1994).

considerable uncertainty due to the small number of data. Hence, it is obvious that we need more data to resolve definitely the question of whether a fundamental component such as the U-Xe, which differs meaningfully from the solar Xe, ever existed. However, as will be discussed in Section 7.3, empirical data strongly suggest that heavier noble gases Ne, Ar, Kr, and Xe in the Earth were related to the solar noble gas by a simple mass-dependent fractionation. This suggests that the fundamental noble gas component (including Xe) is the solar component. Therefore, unless further experiments show otherwise, the solar Xe as the primordial component in the solar system would be a useful paradigm in discussing terrestrial and meteorite Xe.

In Figure 7.7, atmospheric (present) Xe, SW-Xe, and the primordial Earth Xe (nonradiogenic terrestrial Xe) are shown relative to the primordial Xe proposed by Igarashi (1995). Here, the primordial terrestrial Xe was also deduced from the multivariational correlation analyses of Xe isotopic data, which included both the atmospheric Xe and meteoritic Xe. Results are given in Table 7.1. We note that SW-Xe is indistinguishable from the estimated primordial Xe at $^{124-130}$Xe. Slight excesses in atmospheric Xe relative to the primordial Xe at $^{131-136}$Xe are reasonably attributed to addition of fission Xe either from ^{244}Pu or from ^{238}U or from both.

The primordial terrestrial Xe composition is unique in the solar system and, hence, is likely to reflect some specific Earth evolution processes. It has been suggested that the fractionation of the primordial terrestrial Xe relative to the solar Xe could have resulted through gravitational attraction in earth-accreting planetesimals (e.g., Ozima & Nakazawa, 1980; Zahnle, Pollack & Kasting, 1990a), while Pepin (1997) argued that the fractionation took place during atmospheric loss from the Earth caused by a giant moon-forming impact. In a later section, we will discuss in more detail the Xe isotopic fractionation process.

In summary, we may reasonably conclude that except for Ne, and apart from radio- and nucleogenic components, the isotopic composition of atmospheric noble gases well represent the bulk Earth and can be assumed to approximate a primordial terrestrial noble gas component. Thus defined, primordial terrestrial noble gases are

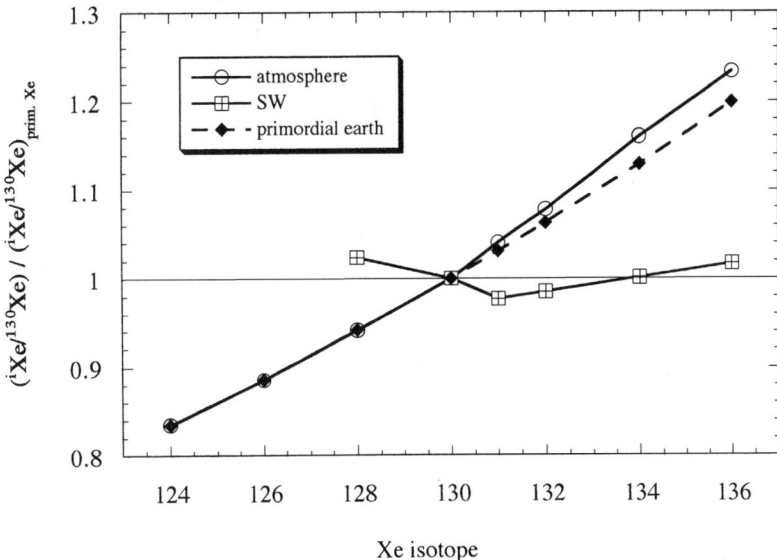

Figure 7.7 Isotopic compositions of atmospheric (present) Xe, SW-Xe, and calculated primordial Earth Xe (see text) are shown relative to the calculated primordial (solar system) Xe (Igarashi, 1995). Data are given in Table 7.1. Note that SW-Xe is indistinguishable from the calculated primordial (solar system) Xe. Slight excesses in the atmospheric $^{131-136}$Xe relative to the primordial Earth Xe are due to the addition of fission Xe either from ^{244}Pu or from ^{238}U or from both, which were produced in the Earth.

distinct from the two widely occurring noble gas components in the solar system (i.e., Q component and SW component).

7.3 SW, Q, and Terrestrial Noble Gases

More than a half century ago, Suess (1949) first proposed the modern approach of the compilation of cosmic elemental abundance (including noble gases) on the basis of nucleosynthesis systematics. With the aid of this cosmic elemental abundance, Signer and Suess (1963) noted that there appeared to be a simple mass-dependent fractionation relationship between the cosmic elemental abundance ratios of noble gases and those in the Earth's atmosphere. They further suggested that the terrestrial noble gas was captured from the solar nebula (which they regarded to have the same elemental composition as the cosmic abundance) by Earth's gravitational force, resulting in a mass-dependent relative elemental abundance of the terrestrial noble gas. Two decades later, Pepin and Phinney (1978) readdressed this fundamental issue with the use of then available SW noble gas data. The latter data were first made available from the analyses of the *Apollo* lunar samples (lunar soils and Al foils). They showed that the terrestrial Xe isotopic composition (nonradiogenic) was related to the solar Xe by a systematic mass-dependent fractionation, which is close to the fractionation that results from diffusion loss or Rayleigh distillation. Such a genetic relation between the solar noble gases and the terrestrial noble gases is in harmony with a general consensus that the sun and planets formed from the same starting material (i.e., the primitive solar nebula), and that even though this original composition is preserved in the sun, only negligibly small fractions were partitioned into planets in that a partition between gas and solid resulted in the mass-dependent elemental fractionation.

Figure 7.8 displays noble gas depletion factors in relation to mass in a Rayleigh fractionation diagram for multiple components (see Section 2.9 for Rayleigh fractionation). In this plot, if terrestrial noble gases were derived from solar noble gases by Rayleigh distillation, the data points for the elemental ratios should form a straight line passing through the origin. As seen in Figure 7.8(a), two distinct linear trends appear. While all the data points off the Rayleigh distillation line are attributable to the ratios involving Xe, other ratios lie very nicely on the Rayleigh fractionation line. Figure 7.8(b) displays the same data except that the atmospheric Xe abundance is multiplied by a factor of 7. Hence, the almost perfect linear array of points is difficult to dismiss as coincidence. Ozima and Podosek (1999), therefore, suggested that the Earth's noble gas abundance pattern, relative to solar/cosmic composition, could have been established by mass-dependent Rayleigh distillation. As will be discussed in Section 7.5, this corrected amount of Xe can be interpreted to correspond to "missing Xe."

Ozima et al. (1998) also suggested that Q noble gases in meteorites are likely to be related to the solar noble gases by Rayleigh fractionation, as shown in Figure 7.9. In this figure, however, isotopic ratios were compared because, unlike the terrestrial atmosphere, elemental ratios are not well defined in meteorites. In the Rayleigh fractionation diagram, the slope of the linear trend is proportional to the degree of distillation (Section 2.9). Therefore, the steeper slope for the atmospheric noble gases than for Q may suggest more severe distillation in the former.

The same fractionation relationship between the terrestrial noble gases or Q and the solar noble gases suggests that the fractionation occurred in the early solar nebula, which seems to be the only likely locale for the observed severe and large-scale fractionation. This relation, in turn, suggests that the terrestrial noble gases were already fractionated before Earth formation. The latter inference is in perfect accordance with the fractionated solar primordial noble gases in the Earth, the conclusion discussed in Section 7.2.

In concluding this section, we note that we have used the terms *solar noble gas* and *SW noble gas* somewhat loosely. Strictly speaking, the term *solar noble gas* should apply to the average noble gas composition in the sun, and the term *SW noble gas* to those observed in solar wind (see Section 7.1 and Chapter 3). The latter is very likely to be fractionated from the former. Even among SW noble gases, they slightly differ not only in elemental composition but also in isotopic composition, depending on how SW noble gas is observed (e.g., in lunar soil or in Al foil target in a spacecraft). However, because the solar noble gas composition, either isotopic or

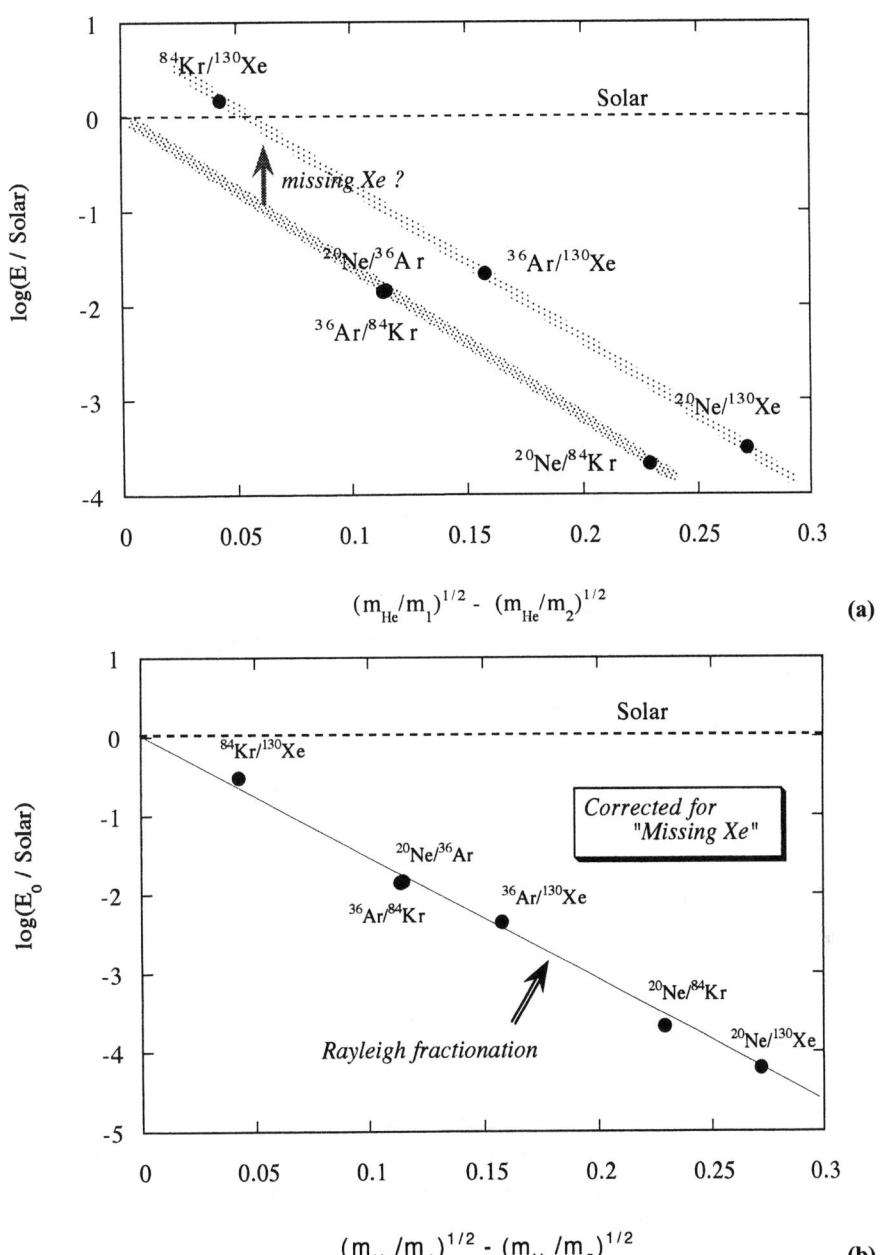

Figure 7.8 Display of noble gas (Earth atmosphere) abundance ratios (ordinate), normalized to the corresponding solar ratios (Anders & Grevesse, 1989), against a measure of mass difference (abscissa). If the abundance pattern were established by mass-dependent Rayleigh distillation of solar noble gases, the points should fall on a straight line passing through the origin (Ozima et al., 1998). (a) The abundance pattern for the terrestrial atmosphere. (b) The same data except that the abundance of Xe is multiplied by 7 (corresponding to about 90% of missing Xe). After Ozima and Podosek (1999).

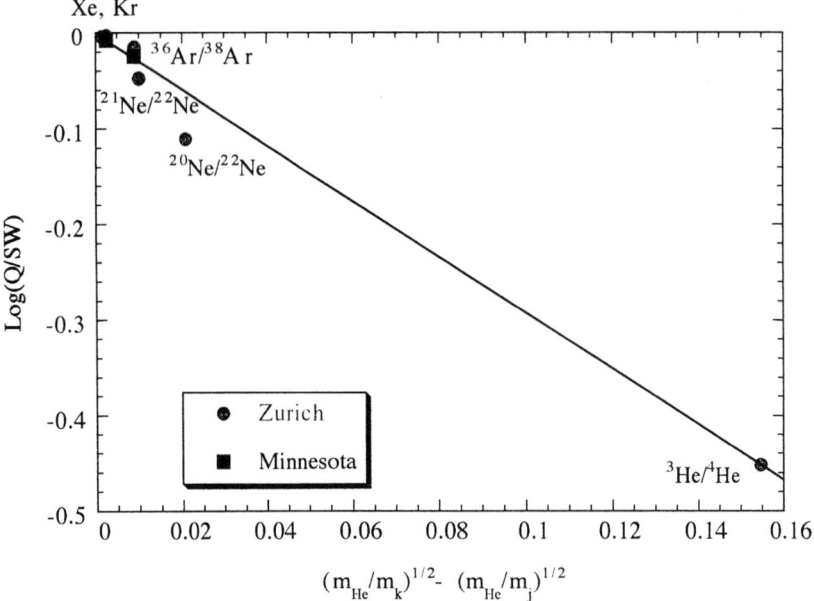

Figure 7.9 Isotopic ratios of Q noble gases were plotted in the same diagram as Figure 7.8. Since elemental ratios of the Q are not well defined, isotopic ratios are used. After Ozima et al. (1998). Minnesota: Pepin et al. (1995), Becker et al. (1996). Zürich: Benkert et al. (1993), Wieler and Bauer (1994).

elemental, is not well known, it is conventional to use *SW noble gas* for *solar noble gas* as a proxy. Theoretically, estimated cosmic elemental and isotopic abundances (e.g., Anders & Grevesse, 1989) may be the better proxy for the average compositions (elemental and isotopic) of noble gases in the sun and, hence, in the early solar nebula.

7.4 Ne Fractionation

In contrast to almost identical isotopic (nonradiogenic) compositions of Ar, Kr, and Xe between the mantle and the atmosphere, there is a large difference in Ne isotopic composition ($^{20}Ne/^{22}Ne$) between them: $^{20}Ne/^{22}Ne$ in the atmosphere is 9.80, whereas ratios in mantle-derived materials generally are larger than 10. The difference in the isotopic ratio appears to be too large to be attributed to common mass-dependent fractionation processes such as Rayleigh distillation. It has been suggested that atmospheric Ne isotopic composition resulted from mixture of solar Ne and some extraterrestrial component. For example, Marty (1989) suggested that the atmospheric Ne may be a mixture between SW-Ne ($^{20}Ne/^{22}Ne$ = 13.7) and planetary Ne ($^{20}Ne/^{22}Ne$ ≈ 8.7), which was once believed to be representative of Ne in meteorites. However, it is now apparent that "planetary" Ne is a composite consisting of some extrasolar Ne and a characteristic meteoritic Ne (Q-Ne) (Huss et al., 1996), and its isotopic com-

7.4 Ne Fractionation

position is unlikely to represent a pertinent reservoir in the early solar system. Moreover, no terrestrial sample has been found to show a $^{20}Ne/^{22}Ne$ smaller than atmospheric ratio, casting a doubt on this suggestion.

A promising process proposed to explain the Ne isotopic fractionation appeals to the mass-dependent escape of lighter trace elements, such as Ne, that accompany massive hydrodynamic flow of H_2 from the Earth's atmosphere (Zahnle & Kasting, 1986; Hunten, Pepin & Walker, 1987; Sasaki and Nakazawa, 1988; Pepin, 1991). A principal premise in these models is that a primitive atmosphere was essentially comprised of water molecules (steam atmosphere – Abe and Matsui, 1985; Matsui and Abe, 1986) and that a substantial amount of hydrogen was generated in the upper atmosphere by photodissociation of the water. The assumption seems to be reasonable, since a H-C-N-O system (major volatiles in the solar system), which was in thermodynamic equilibrium with molten silicate (supposed magma ocean) would form a predominantly H_2O atmosphere with minor amounts of N_2, CO, and CO_2 (Abe & Matsui, 1985). Intense levels of solar EUV (extreme ultraviolet) radiation, which is likely to have occurred during the early stage of solar evolution, would have resulted in dissociation of water to form hydrogen molecules in the upper atmosphere. At the same time, the EUV radiation would have supplied thermal energy into the atmosphere which gave rise to large hydrodynamic escape flow of hydrogen from the Earth (Sekiya, Nakazawa & Hayashi, 1980). Hunten et al. (1987) showed that a sufficiently vigorous hydrodynamic flow of hydrogen would drag other lighter constituents. Because the lighter elements and isotopes will be more easily dragged by the hydrodynamic flow, noble gases remaining in the atmosphere will become more enriched in the heavier isotopes, resulting in mass-dependent fractionation.

Zahnle, Kasting & Pollack (1990b) refined the model by taking into account a second major constituent such as N_2, CO, or CO_2 in addition to hydrogen. This may be a more realistic picture of the primitive atmosphere. In discussing a two-major-component atmosphere, they proposed a "diffusion-limited" hydrogen escape model in which only molecules that are lighter than the "limiting" second abundant heavy constituent are assumed to escape from the atmosphere. Hence, if we assume that the limiting heavy constituent is CO or N_2, only helium and neon could have escaped. However, if we assume that CO_2 is the limiting constituent, we must conclude that some argon escaped from the atmosphere.

Following Zahnle et al. (1990b), an escape factor, x_i, for a trace constituent i is defined as

$$x_i = f_i/(F_i f_H)$$

where $F_i = N_i/N_H$ is a mixing ratio of an escaping constituent i to a major constituent H (hydrogen) in the atmosphere, and f_i, f_H denote their escaping fluxes. With escaping fluxes thus defined, the following expression for an isotopic fractionation factor $F_{ij}[\equiv (N_i/N_j)/(N_i/N_j)_0]$ between isotope i and isotope j can be derived:

$$F_{ij} = \exp\{(x_i - x_j)/(N_H f_H)\Delta t\} \tag{7.1}$$

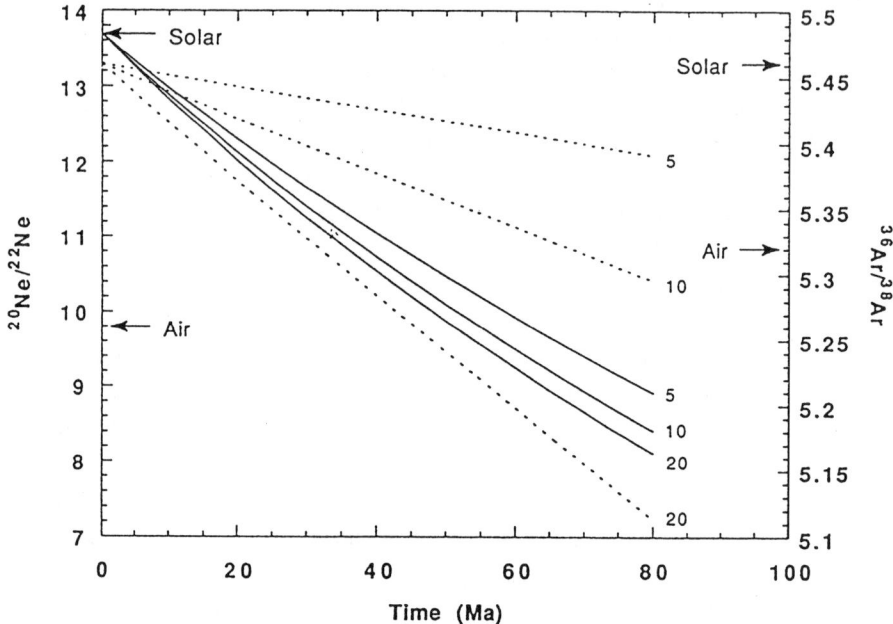

Figure 7.10 Neon (solid lines) and argon (dashed) isotopic fractionations produced by diffusion-limited hydrogen escape from representative terrestrial atmospheres, as a function of its lifetime Δt. Three examples that differ in the amount of CO_2 present are shown. Each assumes 270 bars of water vapor (i.e., the mass of the present ocean). Carbon dioxide contents are 5, 10, and 20 bars. After Ozima and Zahnle (1993).

where Δt is the duration of the steam atmosphere existence and N_H represents the column density of hydrogen; for a 100-bar steam atmosphere, $N_H = 3.4 \times 10^{27}$ molecules cm^{-2}. f_H is the escaping flux of hydrogen molecules (molecules cm^{-2} s^{-1}). x_i and x_j are determined as a function of f_H. The hydrogen flux f_H is a function of temperature at the base of the exosphere, the intensity of solar insulation, and other physical parameters defining the structure of the steam atmosphere. However, if we specify the limiting heavy constituent such as CO or CO_2, this will impose a constraint on a minimum intensity of f_H. The extent of neon and argon isotopic fractionation predicted by Equation (7.1) as a function of Δt is shown in Figure 7.10 for three representative CO_2—H_2O atmospheres, in which CO_2 is assumed to be the limiting constituent.

From Figure 7.10, we can see that if intense EUV radiation had lasted more than 70 Ma, this would account for the observed Ne isotopic fractionation in the atmosphere relative to solar (SW) Ne, regardless of the amount of the CO_2 content. Although not shown in Figure 7.10, a fractionation in a pure steam atmosphere is little different from those with the second constituent. Figure 7.10 also shows that the existence of the second major constituent CO_2 in the steam atmosphere would result in a much larger fractionation effect in the argon isotopic ratio than in the neon isotopic ratio. This calculation is made for an assumed steam atmosphere of 270 bars, which corresponds to the mass of the present ocean. Note that the amount of the primitive steam atmosphere deduced from the thermodynamic calculation by Abe and Matsui (1985) is close to the present ocean. Hence, the model predicts that upon cooling, the steam atmosphere formed the ocean.

In this scenario, it is assumed that the early solar EUV flux was intense enough – at least a few hundred times as intense as the present flux – to heat up the atmosphere to cause a massive hydrogen flow. Recent observations on some T Tauri stars, a proxy of the early sun, seem to suggest the existence of such intense EUV radiation (e.g., Pepin, 1991). Although the hydrodynamic hydrogen flow seems to give a reasonable explanation for the observed large Ne isotopic fractionation (~20%/amu) between the atmosphere and the mantle (cf. Section 6.7), the observed nearly identical isotopic compositions in Ar, Kr, and Xe between the atmosphere and the mantle indicate that the hydrodynamic escape of hydrogen has not affected these heavier noble gases.

7.5 Missing Xe

It has been long recognized that the relative abundance of Xe in atmospheric noble gas is considerably lower than that in a characteristic noble gas component in meteorites, widely referred to as a planetary noble gas component. This is shown in Figure 7.11.

The ratios of ^{130}Xe/^{36}Ar and ^{130}Xe/^{84}Kr in Earth's atmosphere are a factor of 15 lower than in the "planetary component." Because of its uniformity in elemental and isotopic compositions, it has been assumed that planetary noble gas represents a fundamental noble gas component in the solar system pertinent not only to meteorites but also to planetary bodies, including terrestrial planets (see Chapter 3). Therefore, the relative deficiency of Xe in the Earth has been regarded to reflect some fundamental process characteristic to Earth's evolution and has attracted the interest of researchers as the missing Xe problem. However, in spite of much effort in searching for the missing Xe in the Earth, for example in sediments (Podosek et al., 1981a, 1981b; Bernatowicz et al., 1984), in glacier ices (Wacker & Anders, 1984; Bernatowicz, Kennedy & Podosek, 1985) and in silica (Matsuda & Matsubara, 1989), hiding Xe, which can account quantitatively for the missing Xe, has not been found. Also, as we discussed in Section 7.4, the planetary noble gas may no longer be considered to be a meaningful proxy of the primordial terrestrial noble gas. Hence, there seems to be a general feeling that missing Xe, a problem that initially stemmed from the comparison of relative abundance of Xe between atmosphere and meteorites, has lost its original appeal and is no longer a problem. However, developments in our understanding of extraterrestrial noble gas components, especially elucidation of Q as a fundamental noble gas component in meteorites and of the solar noble gas as a representative of the solar system, have shed a new light on this problem.

Comparing Q noble gas and solar noble gas (Section 7.3, and also Sections 3.2 and 3.3), we showed that they may be related to each other by Rayleigh distillation.

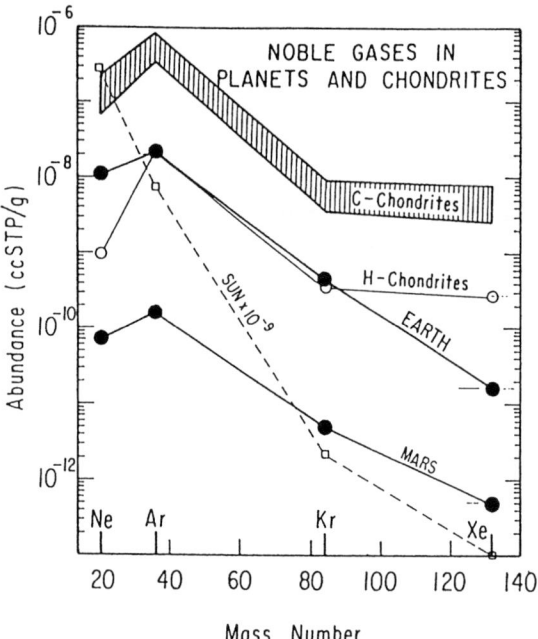

Figure 7.11 Noble gases in planets and chondrites. Noble gases in chondrites and planetary atmospheres are greatly underabundant relative to the Sun (10^{-5}–10^{-12} ×) and show similar strong fractionation, with the lighter gases preferentially depleted. After Wacker and Anders (1984).

If corrected for missing Xe of about 90% of the primordial Xe, the terrestrial Xe is also related to the solar noble gas by Rayleigh distillation. This consistent relationship among the major noble gas components in the early solar system would give strong support for the assumed missing Xe. However, note that although it is difficult to dismiss the missing Xe as an artifact, there is still no convincing explanation how, where, and when the missing Xe event took place.

Tolstikhin and O'Nions (1994) emphasized the importance of gas-melt partition of noble gas in mantle degassing and suggested that this may explain the missing Xe. Because of lower solubility of heavier noble gases in silicate melts, Xe would be preferentially partitioned in the gas phase, resulting in relative depletion of Xe in a mantle reservoir or enrichment in the atmosphere. Subsequently, most of the degassed Xe in the atmosphere was lost by an assumed hydrodynamic hydrogen flow, and continuing mantle degassing of lighter noble gases has finally established the present noble gas inventory in the atmosphere. Because of some not well-constrained free parameters involved in this scenario, it is difficult to assess. Perhaps, a more likely explanation for the missing Xe would be to assume that Xe is hiding in some Earth interior. In this regard, recent development in high-pressure experiments on noble gas is worth noting.

There has been a suggestion that under extremely high pressures Xe might become metallic and be incorporated in an iron core (e.g. Stevenson, 1985). However, Caldwell et al. (1997) have shown experimentally that Xe does not alloy with iron-melt at core-mantle boundary pressure. Matsuda et al. (1993) studied noble gas partition between silicate melts and iron melt under high pressures up to 10 GPa. They found that with increased pressure, all noble gases showed a rapid solubility decrease in iron melt (Figure 2.7). Therefore, they concluded that Xe is unlikely to be incorporated into the iron core. However, under much higher pressures (say >100 GPa), which prevail in the lower mantle and in the core, behavior of noble gas could be drastically changed. Jephcoat (1998) speculated that solidification of Xe under high pressures (Jephcoat & Besedin, 1996), which are easily attainable in the mantle, might facilitate migration of Xe into the deeper part in the Earth, since the density of solid Xe well exceeds that of the surrounding mantle. However, an obvious difficulty in this speculation is in attaining a finite size of solid Xe for effective gravitational segregation. Because of an extremely low concentration of Xe in the mantle ($\leq 10^{-11}$ cm^3/g-mantle), a solid phase of Xe, if formed, would be too small to be controlled by gravity; 10^{-11} cm^3/g of Xe forms a particle a few nanometers in size in 1 cm^3 of mantle material. However, Jephcoat argued that because of the high density of solid Xe, these particles, once formed, could be embedded in iron melt and be entrained into the core.

The possibility of accommodating a substantial amount of Xe in the core is still a matter of debate. This problem not only depends on physical characteristics of Xe under high pressure but also is fundamentally related to the core formation process – the physical (e,g., pressure and temperature) and chemical conditions (e.g., oxygen fugacity) under which iron-silicate melt segregation took place. Conversely, missing Xe, which is now difficult to dismiss, would impose a constraint in resolving the core formation process. Moreover, if the primordial Xe in the Earth were indeed more than an order of magnitude larger than the present estimate, this would impose a severe constraint on Earth evolution. In Section 7.7, we will discuss possible implications of missing Xe on Earth evolution.

7.6 Radiogenic ^{129}Xe* and Fission Xe

Terrestrial Xe, after correcting for mass-dependent fractionation (see discussion that follows) shows excesses in ^{129}Xe/^{130}Xe and $^{131-136}$Xe/^{130}Xe relative to solar Xe. It is now generally accepted that the former decayed from an extinct nuclide ^{129}I ($T_{1/2}$ = 15.7 Ma) and the latter derived by spontaneous fissions from both ^{244}Pu and ^{238}U. In the latter fission-produced Xe isotopes, ^{136}Xe has the highest fission yield; hence, the biggest fractional effect occurs at ^{136}Xe, which is therefore commonly used as a proxy for the entire fission Xe isotopes. These excess Xe are very widely observed in meteorites and have been used to discuss early chronology of meteorite evolution because of their relatively short half-lives (e.g., Whetherill, 1975; Podosek, 1999; Podosek &

Ozima, 1999). Next we will discuss these isotopic ratios in the Earth and their implications for the early chronology of Earth evolution.

^{129}I/^{127}I: Present atmospheric Xe shows about 2% excess in ^{129}Xe/^{130}Xe ratio relative to that of solar Xe (Table 7.1). If we make a correction for the assumed fractionation between solar Xe and primordial terrestrial Xe of about 3.5%/amu (see Table 7.1 and Figures 7.6 and 7.7), a net excess in ^{129}Xe/^{130}Xe relative to the solar Xe would become about 5.5%. The most reasonable explanation so far proposed for this excess is the addition of radiogenic ^{129}Xe* produced from ^{129}I, which was trapped within the Earth. Once a substantial size of a proto Earth was formed, in situ decayed ^{129}Xe* would be essentially retained in the proto Earth. The Xe closure could have even predated the formation of a proto Earth, provided that Earth-accreted planetesimals (a few to a few ten kilometers in size according to a standard planet formation model, e.g., Levy & Lunine, 1993) had essentially retained Xe, and that the Xe was then totally accommodated in the Earth. Alternatively, it also would be possible to suppose that the final Xe closure was not realized until well after the formation of a full-sized Earth. For example, a catastrophic event such as an often assumed moon-forming giant impact would have dissipated a substantial fraction of Xe from the Earth's atmosphere. However, this seems to be unlikely. Some simulation models of a giant impact (e.g., Ida, Canup & Stewart, 1997) indicate that mass was essentially conserved in the Earth after the moon-forming giant impact. Without specifying any event corresponding to the Xe-closure time, we hereafter designate Xe closure as the time of formation of the Earth. A total amount of ^{129}I trapped in the Earth can then be estimated from the amount of radiogenic ^{129}Xe*, which we may assume to be 0.055 × (total terrestrial ^{129}Xe) from the preceding argument.

The preceding estimation of radiogenic ^{129}Xe* in atmospheric Xe was made with reference to SW-Xe (^{129}Xe/^{130}Xe = 6.3647) trapped in lunar soils. An implicit assumption in this argument is that SW-Xe is totally free of radiogenic ^{129}Xe*. However, this assumption is difficult to ascertain, since lunar breccias are likely to contain excess radiogenic ^{129}Xe (e.g., Wieler & Bauer, 1994). Pepin and Phinney (1978) used Xe from Novo Urei for the reference, that has the least radiogenic ^{129}Xe/^{130}Xe ratio (^{129}Xe/^{130}Xe = 6.287 ± 0.029; Marti 1967) known in the solar system, and they concluded that the radiogenic fraction of atmospheric ^{129}Xe is 0.068 ± 0.005.

The initial isotopic ratio of iodine [i.e., $(^{129}I/^{127}I)_0$] yields crucial information on the early chronology of meteorites and planets and has been extensively discussed in the literature (e.g., Whetherill, 1975; Swindle & Podosek, 1988; Podosek & Ozima, 1999). We can easily see that the earlier the formation time of meteorites or planets (or more precisely the Xe closure time in these bodies), the larger the ratio. Here, we make a common assumption that the proto-solar nebula was endowed with a uniform ^{129}I/^{127}I ratio when it was isolated from an interstellar molecular cloud; hence, difference in the initial ratio $(^{129}I/^{127}I)_0$ reflects only difference in Xe closure time in respective planetary bodies. Therefore, from the comparison of the initial ratios, we may infer a relative time interval in their formation, which is commonly named as a

7.6 Radiogenic ^{129}Xe* and Fission Xe

formation time interval and denoted as Δt. It is a common practice to adopt Bjurböle meteorite as a reference time origin so that a ratio $(^{129}I/^{127}I)_{sample}/(^{129}I/^{127}I)_{Bjurböle}$ can be directly related to a time interval between the formation of Bjurböle meteorite and a sample object by a relation,

$$\Delta t = 1/\lambda_{129} \ln\{(^{129}I/^{127}I)_{sample}/(^{129}I/^{127}I)_{Bjurböle}\}$$

where λ_{129} denotes the decay constant of ^{129}I. Most Δt values so far determined for meteorites fall in a rather narrow range of ±10 Ma, which corresponds to the initial isotopic ratio of about $(0.5 \sim 1.5) \times 10^{-4}$ (e.g., Swindle & Podosek, 1988).

Iodine initial ratio can be calculated for the Earth from the estimations of I and Xe contents. Whetherill (1975) calculated a value of $(^{129}I/^{127}I)_0 = 3 \times 10^{-6}$, using crustal concentration of I of about 0.4 ppm and atmospheric Xe inventory with 6% radiogenic ^{129}Xe*. The underlying assumption here is that all the Earth's radiogenic ^{129}Xe* is represented by the atmospheric amount and that the iodine now present in the crust (including ocean) is a measure of the ^{127}I associated with the extinct ^{129}I, which generated the ^{129}Xe*.

If we know the total amounts of I (monoisotopic ^{127}I in the present Earth) and of radiogenic ^{129}Xe* in the Earth, we may estimate $(^{129}I/^{127}I)_0$ (= ^{129}Xe*/^{127}I) at the time of the "formation of the Earth." The approach employed by Whetherill for the estimation of an initial iodine ratio $(^{129}I/^{127}I)_0$ in the Earth is essentially the same as a "coherent degassing" assumption proposed by Schwartzmann (1973a, 1973b). He assumed that radiogenic ^{40}Ar in the present atmosphere was produced originally in the mantle from the ^{40}K, which was also "degassed" from the mantle to the crust together with the radiogenic ^{40}Ar. This assumption is reasonable and useful. In support of this assumption, we note that the amount of ^{40}Ar in the atmosphere (1.7×10^{18} moles) is close to the amount of radiogenic ^{40}Ar (ca. 1×10^{18} moles), which can be produced by K in the crust (about 1%; cf. Taylor & McLennan, 1985) over 4.5 Ga. The agreement between the two estimates is in good accordance with a generally accepted conclusion on crustal evolution; that is, ^{40}Ar in the present atmosphere as well as K in the crust were concomitantly (or "coherently") derived from the mantle.

With the use of recent estimates of I content of 1.55 ppm (Déruelle, Dreibus, & Jambon, 1992) in the crust and of ^{129}Xe* excess of 6.8% (Pepin, 1991), we revise the foregoing estimate by Whetherill (1975) to obtain 8.7×10^{-7} for $(^{129}I/^{127}I)_0$. Although there are some uncertainties in the estimates of I content and excess ^{129}Xe*, it seems to be imperative to accept that the ratio $(^{129}I/^{127}I)_0$ in the Earth was smaller by more than an order of magnitude than meteorite values. As a possible explanation, Whetherill (1975) suggested that the Earth formed about 100 Ma later than meteorites. Next we will examine another Xe isotope produced from an extinct nuclide ^{244}Pu, which also carries information on the early evolution of the Earth.

^{244}Pu/U: In contrast to I, the element Pu has no stable isotope, nor any longer-lived isotope than ^{244}Pu. Hence, the abundance of ^{244}Pu must be indexed to an isotope of a different element. The usual choice is ^{238}U. Because of the highly refractory nature and geochemical coherence of Pu and U, it would be safe to assume that ^{244}Pu/^{238}U is fairly constant in most planetary objects including meteorites and planets, and thus that differences in ^{244}Pu/^{238}U are due to radioactive decay (i.e., time). Recent studies of this ratio in meteorites seem to support this view (e.g., Hagee et al., 1990). Most investigators now adopt ^{244}Pu/U = 0.0068 ± 0.001 in the early solar system, the chondrite value of Hudson et al. (1989). But note that this is a difficult measurement, and that the recommended value is based on a single meteorite. Next we will estimate ^{244}Pu/^{238}U ratio in the Earth.

Estimation of extinct ^{244}Pu can be made from extant fission Xe. In the following discussion, we use $(^{136}Xe)_{Pu}$ as a proxy for the entire fission Xe isotopes of ^{244}Pu. The problem is then reduced to estimating $(^{136}Xe)_{Pu}/^{238}$U in the Earth. As with estimating ^{129}Xe*/I, we also employ the "coherent degassing" assumption, in that a crustal value of U-content and atmospheric $(^{136}Xe)_{Pu}$ are used in calculating $(^{136}Xe)_{Pu}/^{238}$U in the Earth. Pepin and Phinney (1978) (also quoted in Pepin, 1991) first showed from multidimensional correlation analyses that terrestrial atmospheric ^{136}Xe contains (4.65 ± 0.3)% of $(^{136}Xe)_{Pu}$. However, Igarashi (1995), with larger database and more refined statistical technique (i.e., a principal component analysis), concluded $(^{136}Xe)_{Pu}/(^{136}Xe)_{total}$ = (2.8 ± 1.3)% in atmospheric Xe. Taking the latter value for $(^{136}Xe)_{Pu}$ and U = 0.91 ($\cong$ ^{238}U) ppm in the crust (Taylor & McLennan, 1985), we obtain ^{244}Pu/^{238}U = 2.6 × 10^{-3} as the best estimate for the Earth. This value is considerably smaller than the average solar ratio of 6.8 × 10^{-3}. A straightforward interpretation of this smaller ratio is that the Earth formed considerably later than meteorites. The latter conclusion is consistent with the conclusion from ^{129}I/^{127}I.

Next, we examine whether or not the formation time interval suggested from ^{129}I/^{127}I and $(^{136}Xe)_{Pu}/^{238}$U is quantitatively consistent. We will also examine how these conclusions from extinct nuclides are affected by other constraints, especially that of missing Xe.

7.7 ^{244}Pu/^{238}U-^{129}I/^{127}I Systematics

As we discussed in Section 7.6, extant nucleogenic Xe isotopes, ^{129}Xe* and $(^{136}Xe)_{Pu}$, have yielded robust information on the early chronology of Earth evolution. In this section, we show that the early chronology can be more conveniently discussed with the use of a combined ^{244}Pu/^{238}U-^{129}I/^{127}I diagram [Figure 7.12(a)]. For a system that has not undergone Pu/U fractionation (this may be reasonable; see Section 7.6), changes in ^{244}Pu/^{238}U and ^{129}I/^{127}I are only due to radioactive and spontaneous fission decays. We can then construct a ^{244}Pu/^{238}U-^{129}I/^{127}I evolution curve [Figure 7.12(a)], which is chosen to go through an average meteorite. As the average meteorite isotopic composition, we choose generally adopted values of ^{244}Pu/^{238}U $\equiv$ 0.0068 and ^{129}I/^{127}I $\equiv$ 10^{-4} (see Section 7.6). Thus chosen values are close to those for Bjurböle meteorite, which is commonly used as a reference meteorite for meteorite formation

7.7 ^{244}Pu/^{238}U–^{129}I/^{127}I Systematics

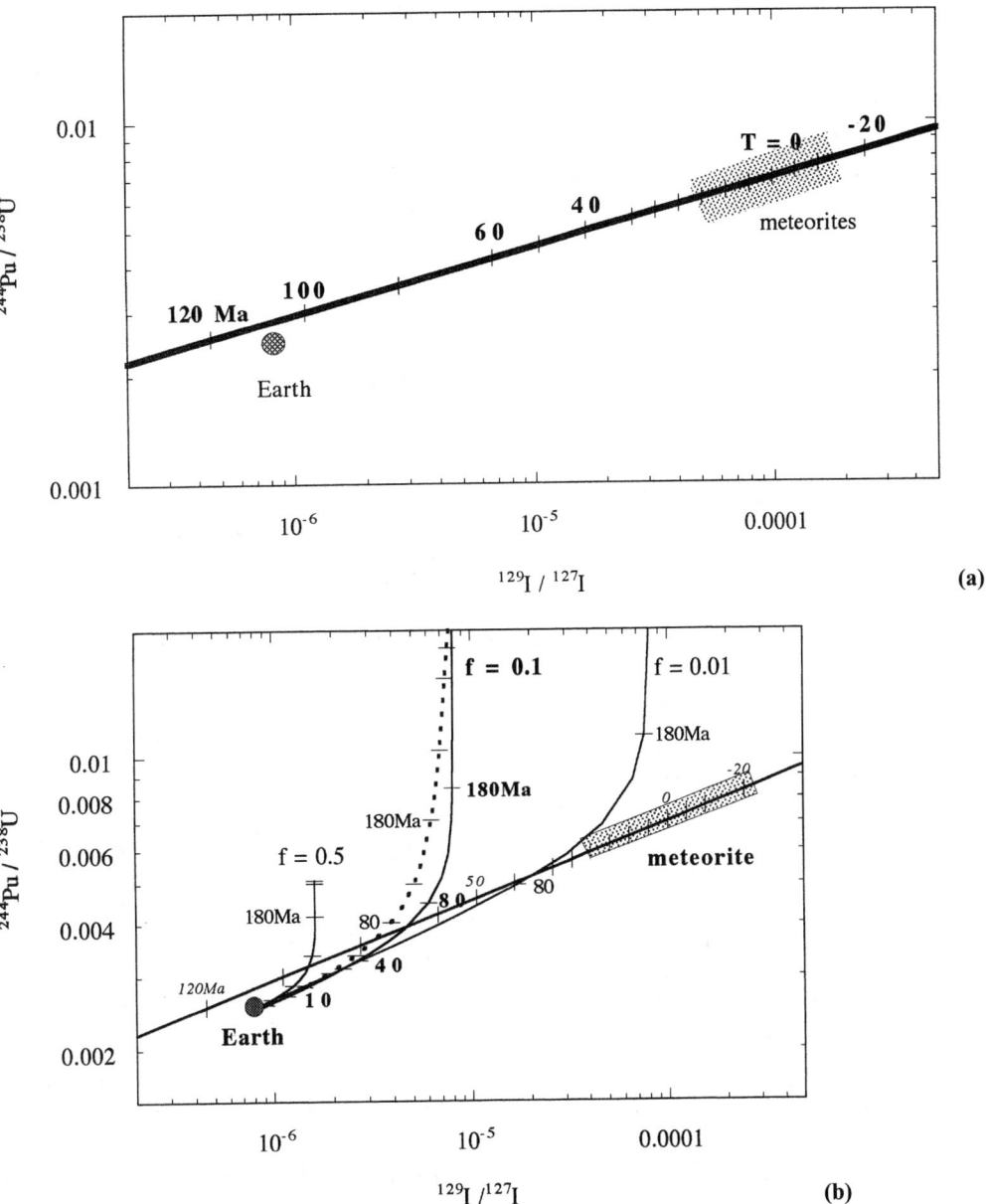

Figure 7.12 (a) Abundance evolution (decay) of ^{129}I/^{127}I and ^{244}Pu/^{238}U as a function of time (T). Fields for early solar system values (meteorites) and initial values for the Earth are indicated. After Ozima and Podosek (1999). (b) Abundance evolution of ^{129}I/^{127}I and ^{244}Pu/^{238}U; the scales, straight line, and meteorites and Earth field are the same as in (a). The curved (solid) lines indicate the effect of assuming that, at some time δt after formation of the Earth, much of the atmosphere's Xe was somehow lost, taking with it radiogenic ^{129}Xe and ^{136}Xe produced up to that time, leaving behind only a fraction f. Each curve corresponds to a different value of f; the ticks on each curve indicate the labeled values of δt. For example, if 90% of the Xe were lost ($f = 0.1$) 60 Ma after formation of the Earth, the initial ^{129}I/^{127}I value must have been nearly tenfold higher, and the ^{244}Pu/^{238}U ratio nearly twofold higher, than the apparent Earth values. The dashed curve illustrates, for $f = 0.1$, the comparable locus for more gradual loss (see text). After Ozima and Podosek (1999).

chronology. Taking the average meteorite point for the origin of time, we put time marks (in Ma units) on the evolution curve. The Earth representative point [$(^{129}I/^{127}I)_0 = 8.1 \times 10^{-7}$, $^{244}Pu/^{238}U = 2.5 \times 10^{-3}$; see Section 7.6] is also shown in Figure 7.12.

Assume that the Earth had a similar primordial Pu/U ratio as that in meteorites, apart from changes due to radioactive/spontaneous fission decays. If we accept this aasumption, we then expect that the Earth point should lie on or close to the evolution curve in Figure 7.12(a). Note that the estimated Earth point is situated very close to the growth curve at $t \approx 110$ Ma. Hence, taking account of uncertainties in the estimations of contents of U, I, and Xe isotopes, we may argue that both meteorites and the Earth are on a single $^{244}Pu/^{238}U$-$^{129}I/^{127}I$ evolution curve; hence, the difference on the curve between the Earth and meteorite representative points suggests that Earth formation postdated Bjurbole meteorite by about 110 Ma, or a formation interval of the Earth relative to Bjurbole is about 110 Ma. Next we examine how this conclusion would need to be changed, if we take account of missing Xe.

As discussed in Sections 7.3 and 7.5, the comparison of relative elemental abundance between Earth and solar noble gas strongly suggests that a considerable fraction of Xe (about 90% of the initial inventory) was lost from the Earth or that Xe is hiding somewhere in the Earth's interior. If we impose this missing Xe as an additional constraint on terrestrial Xe, the conclusions derived from the $^{244}Pu/^{238}U$-$^{129}I/^{127}I$ systematics need to be changed. Qualitatively this is stated as follows. Imposition of the missing Xe as a constraint requires that the primordial terrestrial Xe must have had more radiogenic $^{129}Xe^*$ and fission $(^{136}Xe)_{Pu}$ than those observed in the present atmospheric Xe. This results in moving back the Earth point toward the upper right in Figure 7.12(a) or in decreasing the formation time interval. To demonstrate this quantitatively, we construct a family of curves corresponding to a different degree of missing Xe, which is defined as $f \equiv$ (Amount of currently observed Xe)/(Amount of primordial Xe) [Figure 7.12(b)]. Here, we assumed that the missing Xe event took place instantaneously at a time t (labeled on each curve by ticks), and a fraction f of Xe was lost.

Let us consider a case where that missing Xe is 90% (i.e., $f = 0.1$). To give rise to the observed Earth Xe in Figure 7.12(b), after subtracting the missing Xe from the primordial amount, the primordial Xe must lie somewhere on a curve designated $f = 0.1$. The exact location on the curve depends on the time when the missing Xe took place. If the missing Xe event took place immediately after the formation of the Earth, the corresponding primordial Xe must be the same as the present Earth representative point in Figure 7.12(b), since radiogenic $^{129}Xe^*$ and fission ^{136}Xe had not yet been accumulated in the Earth. Hence, missing Xe, if taken into account, will not change the Earth representative point in Figure 7.12(b). As another extreme case, if we suppose that missing Xe took place a few hundred Ma after the formation of the Earth, the corresponding Earth representative point must lie near the upper right end of the curve (thin lines), which would be considerably above the growth curve (thick line). Therefore, if we further assume that the primordial Xe should lie on the growth curve in accordance with the assumed unfractionated Pu/U ratio, we must conclude from Figure 7.12(b) that the time for the occurrence of missing Xe must be younger than about 80 Ma (time is marked on the curve corresponding to $f = 0.1$). Accordingly, the ^{244}Pu/^{238}U-^{129}I/^{127}I formation time interval of the Earth would be considerably reduced, down to about 70 Ma.

In this discussion, the missing Xe event was assumed to take place instantaneously. However, if the missing Xe event were assumed to be more gradual, for example, Xe loss proceeded as an exponential function of time (i.e., $e^{-t/\tau}$ with some arbitrary time constant τ), the results of this continuous Xe-loss model are similar to the singular event model (Ozima & Podosek, 1999). Allowance for missing Xe permits higher initial abundance or a shorter formation interval, as long as the time scale for loss of Xe is long compared to the lifetime of ^{129}I but not the lifetime of ^{244}Pu.

In concluding this section, it may be interesting to comment on some implications of the I-Pu-Xe chronology on planetary evolution. According to a standard model of planet formation, the growth time of a planet formation increases with the distance from the sun (e.g. Hayashi et al., 1993) because material density in the solar nebula decreases with the distance. Hence, the model predicts that the Earth formed prior to the formation of meteorite parent bodies because the Earth is closer to the sun than to the asteroids, which are generally assumed to be meteorite parent bodies. Figure 7.12(b) suggests that the earlier formation of the Earth than meteorite parent bodies is only possible for $f < 10^{-2}$. However, as we discussed in Section 7.3, such an exhaustive missing Xe is not supported by observational data. To resolve this contradiction, we not only need to constrain more precisely the fraction of radiogenic ^{136}Xe and ^{129}Xe in terrestrial Xe but also to reconsider the standard model in this regard.

7.8 Origin of Terrestrial Noble Gas

To discuss the origin of noble gas in the Earth, we must address at least the following basic questions, that is, (i) inventory or the amount of noble gas in the Earth, (ii) relative elemental abundance, and (iii) isotopic composition. In the first issue, the most difficult problem would be to explain how the Earth could have acquired a substantial amount of He, which seems to be too elusive to be captured in any meaningful amount by solid material. If a satisfactory explanation were given for He trapping, it would be comparatively easier to understand the capture of the heavier noble gases. The second issue requires an explanation for the general fractionation trend of enrichment of the heavier noble gases relative to the solar noble gas, from which the terrestrial noble gas is generally assumed to be derived. The third issue is to explain the characteristic isotopic compositions of noble gases, especially the unique isotopic compositions of Xe and Ne.

Let us discuss the first issue, that is, how terrestrial noble gases, in extremely small but not negligible amounts, were captured by the Earth. Adsorption of noble gases by Earth-accreted dust grains had once been favored as a noble gas capture mechanism by many (e.g., Fanale & Cannon, 1972) because the mechanism can explain the relative elemental abundance pattern (enrichment of heavier noble gases relative to the solar composition) as well as a reasonable amount of absorbed noble gases (except for He; see the following discussion). However, for the dust grains to adsorb a meaningful amount of He, the ambient temperature must be extremely low. In this regard, it may be useful to recall that even activated charcoal, one of the most effective noble gas adsorbers, barely adsorbs He above 20 K. Thus, the required temperature appears to be unrealistically low in the inner planetary region in the early solar nebula, and an alternative mechanism needs to be invoked to explain the capture of He by the Earth. Also, adsorption alone cannot explain nonsolar isotopic compositions of terrestrial noble gases, if the terrestrial noble gases were derived from the solar noble gas. Noble gas adsorption hardly causes isotopic fractionation (see Section 2.9).

As a possible mechanism by which the Earth could have captured a substantial amount of noble gases, including He, implantation of solar wind noble gases has often been assumed. The implantation origin of the terrestrial noble gas assumes that solar emitted noble gases (solar wind) were implanted into small dust grains, which subsequently have accreted to form the Earth. A fairly large amount of the surface-correlated solar He observed in lunar breccia and in some meteorites gives reasonable grounds for this supposition. However, for implantation to be effective, the size of the dust grains or a volume-to-surface ratio should be very small. Also, dust density should be fairly high so that the Earth could have accumulated enough noble gases. However, the existence of such thick dust grains necessarily prevents the penetration of solar corpuscular radiation into the planetary region.

A possible way to circumvent this difficulty would be to appeal to off-disk solar wind implantation. Sasaki (1991) proposed that a dust cloud in the early solar nebula before the formation of the terrestrial planets had an axisymmetric disk shape and was characterized by a power law density distribution in a radial direction from the sun and a Gaussian distribution in a vertical direction from the equatorial plane. In this particular dust distribution, solar corpuscular flux decreases very rapidly on the equatorial plane with distance from the sun, but attenuation of the solar corpuscular flux becomes less important away from the equatorial plane. Consequently, a considerable amount of solar corpuscular flux can be implanted into dust grains at the off-disk region. Taking some permissible values for dust grain distribution and assuming 100 times as intense solar wind flux as the present SW, which had persisted for 10^7 years during the T Tauri stage of the Sun's evolution, they concluded that implantation of the solar-emitted noble gas on dust grains may have supplied enough noble gases into the innermost planets. However, for the mechanism to be effective at the Earth's region, we need a much higher flux intensity or longer duration of the irradiation time, either of which is very unlikely. The attempt by Sasaki seems to illuminate the fundamental difficulty in the implantation model for the origin of terrestrial noble gases rather than to offer a solution.

Another difficulty in the implantation model is that it fails to explain the observed heavy elemental fractionation of the Earth's noble gas relative to the SW component.

7.8 Origin of Terrestrial Noble Gas

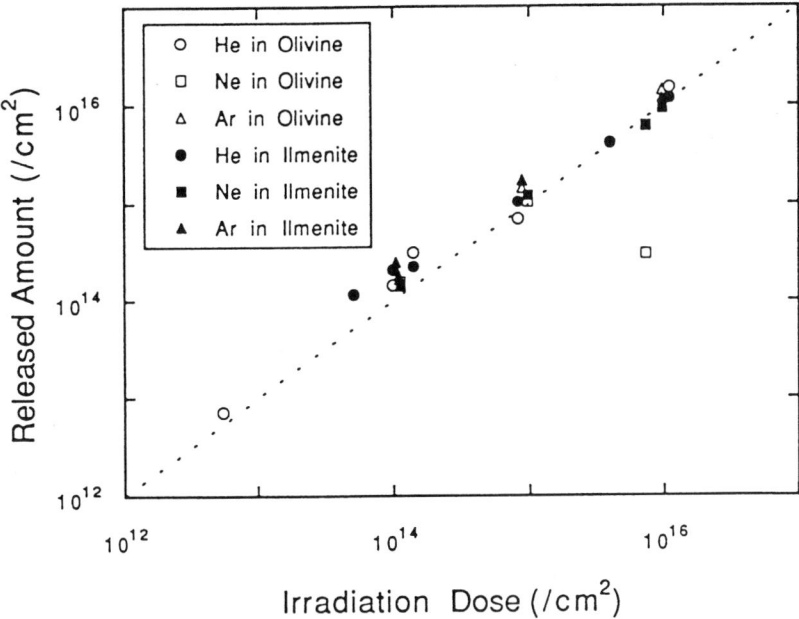

Figure 7.13 Noble gas amounts (labeled as "Released Amount" in the ordinate), which were implanted on target minerals (olivine and ilmenite), are plotted against irradiation dose. Acceleration energies of the implanted noble gases range from 20 to 200 KeV (total energy). After Futagami et al. (1993).

As we show in Figure 7.1, the Earth's noble gas is enriched in the heavier ones relative to the SW noble gases implanted in lunar breccia; Ar/Ne and Kr/Ar ratios are almost two orders of magnitude larger than the ratios in the SW. Laboratory ion implantation experiments by Futagami et al. (1993) seem to rule out such a heavy elemental fractionation. They studied implantation of noble gases (He, Ne, Ar) of moderate energies (20 ~ 200 KeV; total energy) on ilemenite and olivine crystal plate targets. They found that the amount of implanted noble gases are proportional to the irradiation dose over a wide range (Figure 7.13), which covers the observed amount of noble gases in lunar soils.

The result clearly indicates that implantation of noble gas atoms of a moderate energy does not result in a significant elemental fractionation. It is also difficult to attribute the observed elemental fractionation to difference in noble gas retentivity of implanted noble gases. Futagami et al. (1993) studied thermal release of noble gases from the implanted minerals. Figure 7.14 shows an Arrhenius plot constructed from the data for the thermal release experiments. Using a simple-minded extrapolation of the data to 0°C, which may be a realistic approximation for the temperature prevailing around the Earth orbital region in the early solar nebula, they estimated the diffusion coefficients for He, Ne, and Ar to be 10^{-18}, 10^{-21}, and $10^{-25}\,\text{cm}^2\,\text{s}^{-1}$, respectively.

Since the mean penetration depth of a 20 KeV He ion into ilmenite is about 0.2 μm (see Table 2.5), we may conclude from a comparison of a diffusion length

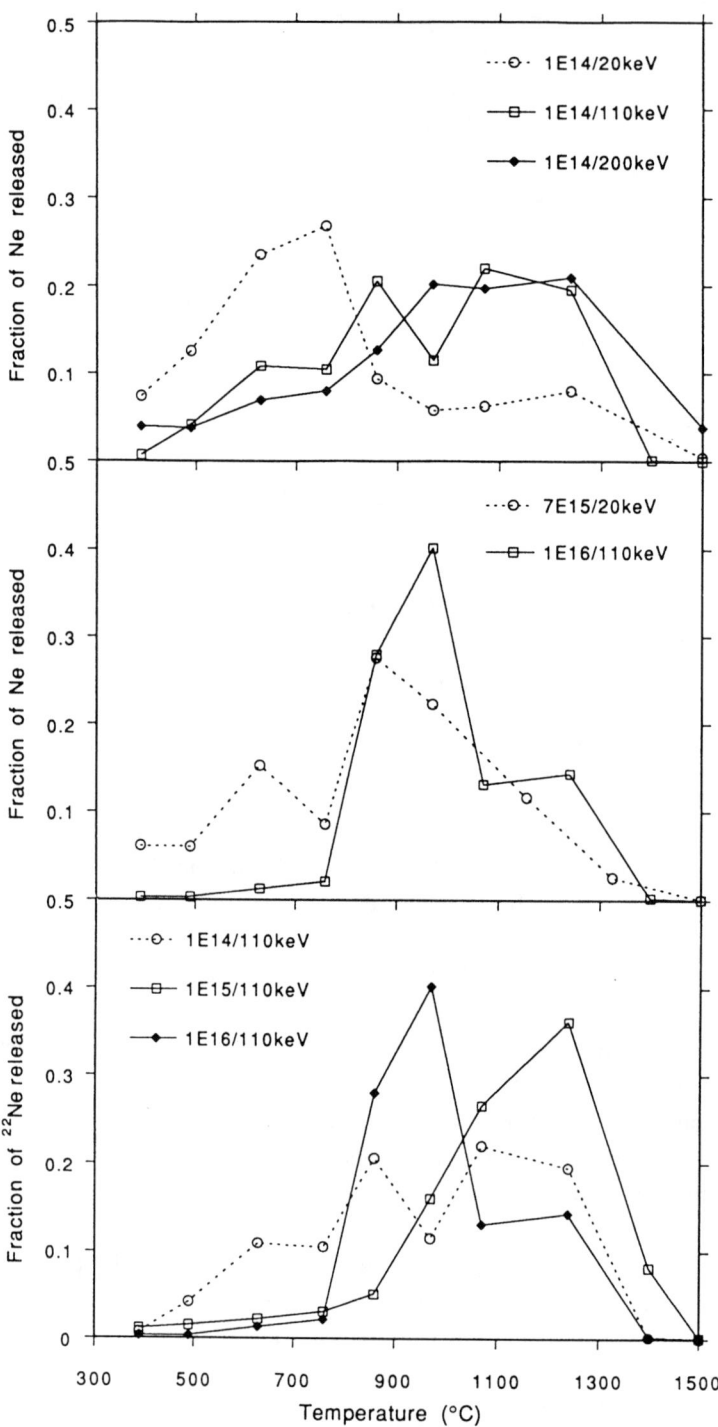

Figure 7.14 Arrhenius plot for noble gas diffusion from noble gas-implanted target minerals. Samples are the same as those in Figure 7.13. After Futagami et al. (1993).

7.8 Origin of Terrestrial Noble Gas

($= \sqrt{2} Dt$, where t is time, and D is the diffusion coefficient) with penetration depth that, except for He and Ne, other noble gas atoms can hardly diffuse out from the target material if the irradiated materials were kept at 0°C for more than 10^7 years. The latter conditions are likely to approximate those in the preaccretionary solar nebula. Therefore, considering all these experimental results, we may conclude that implantation is unlikely to have played a major role in acquiring noble gases by the Earth.

As an alternative mechanism, some authors suggest that a substantial fraction of terrestrial noble gases were supplied by cometary materials (Owen et al., 1992; Porcelli & Wasserburg, 1995b). For example, Owen et al. (1992) suggested that some cometary materials may have supplied a substantial amount of heavy noble gases in the Earth. This suggestion is based on their laboratory experiments on noble gas adsorption of ice at very low temperatures. They suggested that the atmospheric noble gas can be explained as a mixture between this ice (comets) adsorbed noble gas at about 50 K and the mantle noble gas. However, because cometary materials are assumed to have supplied the major fraction of the atmospheric noble gases as a late-accreting veneer, we need to conclude that the noble gases in the cometary materials had the same isotopic compositions of Ar, Kr, and Xe as those in the mantle noble gases; the isotopic composition of the atmospheric noble gases are indistinguishable from the mantle component (see Section 7.2). There are no data for isotopic compositions of noble gases in cometary materials. Hence, it is difficult to judge validity of the cometary hypothesis until the isotopic data are made available.

Any scenario for the origin of terrestrial noble gas crucially depends on a model for Earth evolution. Basically, there are two schools of thought in this regard. One assumes that the formation of the Earth took place before the dissipation of the primordial solar nebula, and another assumes that it occurred after the dissipation (e.g., Levy & Lunine, 1993, and references therein). Based on the former Earth formation scenario, Mizuno, Nakazawa, and Hayashi (1980) showed that a considerable amount of solar noble gases would be dissolved into the Earth during the later stage of its accretion in the presence of a thick primary terrestrial atmosphere. In this scenario, the formation of this thick primary atmosphere is a natural consequence of gravitational attraction of the primordial nebula around the Earth; hence, the noble gas compositions were the same as in the solar nebula. Mizuno et al. showed that the primary atmosphere thus formed had a temperature of 4000 K and a pressure of 900 atm and argued that the molten surface layer of the Earth would be in solubility equilibrium with the thick atmosphere. The amount of noble gases dissolved in the surface layer is then calculated by balancing the diffusive rate of noble gases into the molten layer and the growth rate of the latter layer, which yielded 2 ~ 200 times (depending on the choice of opacity of the primary atmosphere) as much Ne as in the present atmosphere but insignificant amounts of Kr and Xe. As a natural consequence of this scenario, Mizuno et al. predicted that neon in the Earth must have a solar neon isotopic composition because absorption does not fractionate an isotopic composition. Harper and Jacobsen (1996) revised the model by Mizuno et al. They concluded that the noble gas absorption from a thick primordial atmosphere was the principal process for the origin of lighter noble gases (He, Ne) in the Earth. In supportoing this conclusion, they pointed out that the ^{3}He/^{22}Ne ratio in the mantle is also close to the solar ratio.

The specific model for Earth formation is the crucial assumption in this absorption scenario. If Earth formation proceeded in "nebula-free" space as was originally postulated by Safronov (1969) and henceforth advocated by various researchers (e.g., Levy & Lunine, 1993), the absorption scenario must be abandoned. Also, the Mizuno model predicts that Ne as well as other noble gases in the mantle should have solar noble gas isotopic compositions. However, as we discussed in Chapter 5, the currently available observational data on mantle noble gases show nonsolar isotopic compositions, arguing against the Mizuno model.

On the basis of a standard model for terrestrial planet formation (e.g., Hayashi, Nakazawa & Nakagawa, 1993), Ozima and Nakazawa (1980) proposed a different model for the origin of terrestrial noble gases. As discovered independently by Safronov (1969), Hayashi (1972), and Goldreich and Ward (1973), condensed grains in the solar nebula would sediment toward the median plane and, when a critical density (about 10^{-7} g/cm^3) is exceeded, spontaneously aggregate into bodies (planetesimals) of substantial size, about 10^{18} g (radius about 5 km). The process is known as the fragmentation of the solar nebula. The planetesimals would subsequently grow more slowly by mutual collision and aggregation, ultimately leading to both meteorite parent bodies and the larger terrestrial planets. Ozima and Nakazawa pointed out that since lithostatic pressures in planetesimals were rather low (central pressures of about 1 bar for 10 km radius, 100 bar for 100 km, and so on), the planetesimals would be rather porous, and that the pores were filled with nebular gas, assuming that nebular gas has not yet been totally dissipated. Ultimately, the growth of the planetesimals would seal the pores and presumably some of the gas would be captured. The abundance of this natural supply of nebular gases is approximately $\rho_0 \times$ (pore space of a planetesimal), where ρ_0 is nebular density. Therefore, the noble gas abundance relative to solid materials in a planetesimal is $s\rho_0/\rho_p$ where s denotes the porosity and ρ_p, ρ_0 are the densities of a planetesimal and nebula, respectively. The abundance of this natural supply of nebular volatiles would depend on nebular temperature and density as well as on porosity; for plausible values (e.g., $s = 0.33$, $T = 225$ K, $\rho = 3 \times 10^{-8}$ g/cm^3, and mass ratio of solid to gas in nebula = 0.00343) the pore space gas would correspond to a depletion factor of about 10^{-11} to 10^{-12} relative to cosmic abundance. This factor would be the same for all chemical species and clearly could supply only a trivial amount of observed amounts of most volatiles; it is, however, of the same order of magnitude as the Ne depletion in the Earth. The model also predicts that the magnitude of He depletion will be similar to that for Ne in the primitive Earth. In this regard, it is worth noting that this model necessarily leads to trapping of a substantial amount of helium in the Earth's interior.

7.8 Origin of Terrestrial Noble Gas

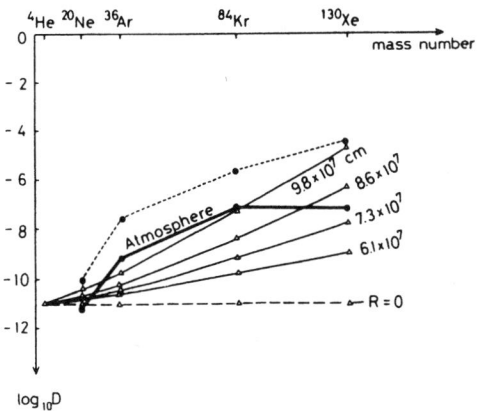

Figure 7.15 Noble gas elemental abundance ($\log_{10}D$) relative to the solar abundance in planetesimals of various sizes are plotted against a mass number. The amount of noble gases is normalized to Si. In the case of atmospheric values (thick line), M_i stands for the noble gases in the atmosphere and Si, in the whole Earth. Note that the mixing of the gravitationally fractionated planetesimal-type (thin lines) and the planetary-type noble gases would approximate the atmospheric pattern. $[D = (M_i/\text{Si})_{\text{atm}}/(M_i/\text{Si})_{\text{solar}}]$. After Ozima and Nakazawa (1980).

When the planetesimal grows further and the radius exceeds a few hundred kilometers, we need to consider concentration of gases in pores due to self-gravitational force. Ozima and Nakazawa showed that gases in pores could be concentrated by the planetesimal's own gravitational field. If diffusion equilibrium could be maintained so that each gas species made its own equilibrium adjustment to the gravitational field, substantial elemental fractionation could result. As illustrated in Figure 7.15, a planetesimal radius of several hundred kilometers would enrich Xe by a few orders of magnitude but have little effect on He and Ne. An important and unique feature of this model is its ability to predict isotopic effects. Along with elemental fractionation in planetesimal gas, there will be concomitant isotopic fractionation favoring heavier isotopes; moreover, the fractionation is strongest for the heaviest elements. Ozima and Nakazawa calculate that a planetesimal radius of about 600 km would produce isotopic fractionation characteristic of the difference between solar and terrestrial Xe.

As already seen, a key question for this model is whether pores remain accessible and gravitational equilibration remains possible until a radius of several hundred kilometers is achieved. Considering this point, Zahnle et al. (1990a) modified the Ozima-Nakazawa model. In their model, only the porous outer layer in a growing planetesimal is assumed to be capable of freely exchanging gases with the

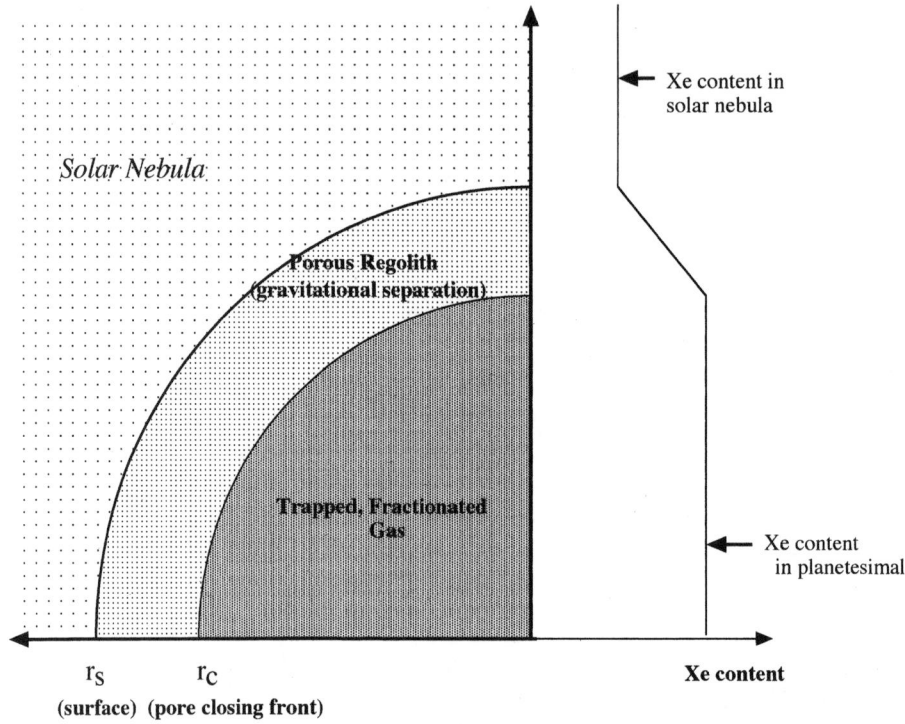

Figure 7.16 Schematic view of a planetesimal immersed in the solar nebula. The deep porous regolith is suffused with adsorbed nebular gases. These fractionate under gravity. Adsorbed gases are trapped when lithostatic pressures (p_c) becomes large enough to close off pores. All gas below the pore-closing front has the same isotopic composition. After Zhanle et al., 1990a.

surrounding solar nebula (Figure 7.16). Since the lunar highland crust is cracked to a depth of 20–25 km, they assumed that a pore-closing pressure would correspond to this depth or to 1–1.2 bars. Although it is difficult to judge how closely the lunar crust simulates the physical conditions in a planetesimal, it would at least be a better analogue than Earth's crust. The isotopic fractionation of Xe at this pore-closing pressure depends on the density and temperature of a planetesimal. Zahnle et al. showed that for $\rho = 2\,\text{g/cm}^3$ and $T = 200\,\text{K}$ about 4%/amu of Xe isotopic fractionation would result, providing a good explanation the observed Earth and Mars Xe isotopic fractionation relative to the solar Xe.

The Ozima-Nakazawa model requires the "right" size of a planetesimal a priori to explain the observed xenon isotopic fractionation; the xenon fractionation is a natural consequence in this pore-closing model. However, in this model the amount of Xe captured by a planetesimal is too small to account for terrestrial Xe, if the similar nebular density ($3 \times 10^{-8}\,\text{g cm}^{-3}$) as in the case of the Ozima-Nakazawa model were assumed. Zahnle et al., therefore, suggested that adsorbed Xe on dust grains prior to the formation of a planetesimal was the primary source: Most of the adsorbed

Xe was released inside a planetesimal and gravitationally fractionated and finally sealed.

Although no attempt was made to explain how noble gases were captured by the Earth, Pepin (1991) proposed a scenario to explain the characteristic terrestrial noble gas isotopic composition, especially of Xe. Taking for granted that the primordial terrestrial noble gases are solar, he proposed that a giant moon-forming impact dissipated noble gases from the Earth's primitive atmosphere almost completely. Only Xe that survived the impact (about 15% of the original amount) underwent severe isotopic fractionation to give rise to the present atmospheric Xe isotopic composition. Lighter noble gases (Kr, Ar, Ne) were subsequently degassed from the Earth's interior, and Ne, the lightest noble gas among them, underwent further isotopic fractionation due to the hydrodynamic escape of H_2 from the atmosphere to give rise to the present atmospheric $^{20}Ne/^{22}Ne$ ratio. The neon isotopic fractionation process here is the same as the one originally proposed by Hunten et al. (1987). The scenario is rather specific in the choices of free parameters such as intensity and duration of the solar EUV radiation, temperature of the primary atmosphere, and time and degree of mantle degassing. It remains to be seen whether the assumed choices of the free parameters are compatible with observations. However, a crucial test of the scenario is in a comparison of noble gas isotopic compositions between the mantle and the atmosphere. Mantle noble gases should differ from the atmospheric noble gases because the latter were degassed from the mantle and subsequently underwent fractionation in the atmosphere.

In summary, there is still no satisfactory theory on the origin of terrestrial noble gases. Attempts to resolve this problem has been hampered by our lack of knowledge of noble gas in the Earth's interior, especially of the noble gas isotopic composition representative of the whole solid Earth. However, a growing number of noble gas data on mantle-derived materials are now unveiling an unequivocal picture of mantle noble gas (Chapter 6). For example, currently available data strongly suggest that mantle noble gases have nearly identical isotopic composition as the atmospheric components, but both are distinctly different from the solar gases (see Chapter 6). This observation, if established, would impose a stringent constraint and greatly narrow the range for possible models.

7.9 Inventory

It is generally assumed that except for He, which is not gravitationally bound to the Earth, the majority of terrestrial noble gases presently reside in two major reservoirs, namely, the atmosphere and the mantle. Noble gases in other reservoirs such as the hydrosphere or the crust are likely to be insignificant in comparison with these other two reservoirs. The latter conclusion is based on a number of observations on noble gas contents in a variety of samples from these reservoirs. However, a considerable amount of Xe may be stored in some unknown region in the Earth's interior including a deeper mantle or the core (cf., Section 7.5).

As a result of the well mixing of the atmosphere in the troposphere where most of the atmospheric mass is contained, isotopic composition as well as elemental composition of atmospheric noble gas are well defined from the measurement of near-surface air noble gases (Table 1.3 and Section 1.6). Although isotopic and elemental compositions of noble gases in the mantle are still a hot issue, currently available data impose some useful constraints on the noble gas inventory at least in the mantle regions, which can be assessed by samples such as MORB or OIB. Mantle noble gas characteristics have mainly been inferred from the analyses of noble gases trapped in various mantle-derived materials. Table 6.1 shows noble gas amounts in these mantle-derived samples. Assuming that the mantle-derived samples represent the noble gas inventory in the mantle, comparison of these data with air noble gas inventory (the last row in the table) indicates that the major fraction of noble gases, especially the lighter noble gases, resides in the atmosphere. This conclusion is further strengthened if we consider that the mantle noble gas inventory deduced from mantle-derived samples is very likely to be overestimated due to comparatively high concentration of noble gas in typical mantle-derived samples such as volcanic glasses and to preferred partition of noble gases in magma (cf. Section 6.2).

From a simple ^{40}Ar inventory in the Earth, we also concluded that a majority of the terrestrial Ar has been degassed from the mantle to the atmosphere (Section 6.6 and 6.8). Hence, the assumption that, except for He and Xe, atmospheric noble gas is a good proxy of the terrestrial noble gas inventory serves as a useful working hypothesis.

He is distinct from the heavier noble gases. At high altitude (around a few hundred kilometers) where temperature is likely to be above 1000 K, the thermal velocity of a He atom would exceed the escape velocity ($v = 11.18$ km). This nonconservative nature of He in the atmosphere can also be argued from the following simple budgetary consideration of radiogenic ^{4}He in the Earth. Taking average crustal values of U = 0.9 ppm (Taylor & McLenann, 1985) and Th/U = 3.3, the ^{4}He production rate in the whole crust will be 5.32×10^{12} cm^3 STP/yr. If thus produced radiogenic ^{4}He were released to the atmosphere in a geologically short time – the assumption seems to be reasonable judging from very poor retentivity of radiogenic He in minerals (cf. Section 2.7) – the amount of radiogenic ^{4}He released to the atmosphere throughout the Earth's history would be 1.44×10^{23} cm^3 STP. This amount is about 10^4 times as much as the present amount of atmospheric ^{4}He (2.07×10^{19} cm^3 STP). Hence, this simple budgetary consideration strongly suggests that ^{4}He was not conserved in the atmosphere but must have been escaping to the interplanetary space. We also infer that a mean residence time of ^{4}He in the atmosphere would be about one million years.

Although the escape of He from the atmosphere must be accepted, a formal application of the classical Jeans thermal escape theory to ^{4}He shows that the temperature

7.9 Inventory

Table 7.2. *Major fluxes in atmospheric He*

Flux Term	^{3}He (atoms cm^{-2} s^{-1})	^{4}He ($\times 10^6$ atoms cm^{-2} s^{-1})
Escape		
Thermal (Jeans)	5.9 (1)	0.06 (2)
Nonthermal	4.2 (2)	3.0 (2)
Sources		
Ocean floor (mantle degassing) (3)	4.0	0.3
Crust (4)	0.086	0.86
Atmosphere (cosmic-ray spallation) (1)	0.6	
Precipitation from magnetosphere	4.0 (1, 5)	0.01 (1)
Cosmic dust (6)	0.8	

Table 7.2 References: 1, Johnson & Axford (1969); 2, Kockart (1973); 3, Craig et al. (1975); 4, estimated from ^{4}He production in the crust (see text); 5, Lind et al. (1979); 6, the estimated ET-^{3}He in abyssal sediment (0.0025 atoms/cm^2 s^{-1}, see Section 5.3) is assumed to correspond to 0.4% of total imported ET-^{3}He into the atmosphere (Anderson, 1993).

at the base of the exosphere required for effective thermal escape of ^{4}He becomes uncomfortably high (cf. Kockart, 1973). Hence, it seems to be imperative to call for a nonthermal escape process to account for the ^{4}He budget in the atmosphere. It is now generally considered that the dominant loss mechanism for ^{4}He is nonthermal and appears to be the escape of photoionized He$^+$ along magnetic field lines (cf. Kockart, 1973).

So far we discussed the case for ^{4}He escape. ^{3}He escape must at least be as effective as ^{4}He; the ^{4}He lifetime against Jeans escape can be shown to be 70 times greater than the ^{3}He lifetime. In the case of ^{4}He, an inflow flux into the atmosphere is essentially the radiogenic ^{4}He from the solid Earth. However, a flux from the interplanetary space may become important in the case of ^{3}He inventory. From He implantation experiment on metal foil collectors in the *Skylab*, a spacecraft operated in a circular orbit at the altitude of 443 km from 1973 to 1974, Lind, Geiss, and Stettler (1979) concluded that the solar wind ^{3}He precipitating from the magnetosphere to the upper atmosphere is comparable to ^{3}He degassing flux from the solid Earth.

In Table 7.2, we summarize major He fluxes in the atmosphere. The He budget seems to be in balance, at least within a factor of two, although there is certainly room for modification of some of the terms or introduction of new ones. However, there is no reason to require the He budget to be in balance. On the 10^6 year scale of the mean residence time, variations in fluxes will be smoothed out, and the present epoch may not be typical. The geological and cosmic ray sources are probably fairly steady on this time scale, but the thermal loss is very sensitive to solar activity, the nonthermal loss is sensitive to the geomagnetic field, and precipitation is sensitive to both. At an extreme, the fluxes may be highly irregular; Sheldon and Kern (1972), for example, have suggested that major loss of ^{4}He and gain of ^{3}He occurs during geomagnetic field reversals. In any case, it is evident that the present composition and abundance of He are not closely related to any major geological reservoir, and so are only local and probably temporary features. In this regard, it may be worth noting that He is totally decoupled from other noble gases in the mantle (cf. Section 6.6).

References

Abe, Y., & Matsui, T. (1985) The formation of an impact-generated H_2O atmosphere and its implications for the early thermal history of the Earth. *J. Geophys. Res.*, **90**, C545–59.

Abe, Y., & Matsui, T. (1986) Early evolution of the Earth: Accretion, atmosphere formation, and thermal history. *J. Geophys. Res.*, **91**, E291–302.

Aeschbach-Hertig, W., Kipfer, R., Hofer, M., Imboden, D. M., Wieler, R., & Signer, P. (1996) Quantification of gas fluxes from the subcontinental mantle: The example of Laacher See, a maar lake in Germany. *Geochim. Cosmochim. Acta*, **60**, 31–41.

Albarède, F. (1978) The recovery of spatial isotope distributions from stepwise degassing data. *Earth Planet. Sci. Lett.*, **39**, 387–97.

Aldrich, L. T., & Nier, A. O. (1948) The occurrence of He_3 in natural sources of helium. *Phys. Rev.*, **74**, 1590–4.

Alexander, Jr., C. E., Lewis, R. S., Reynolds, J. H., & Michel, M. (1971) Plutonium-244: Confirmation as an extinct radioactivity. *Science*, **172**, 837–40.

Allard, P. (1992) The origin of hydrogen, carbon, sulfur, nitrogen and rare gases in volcanic exhalations: Evidence from isotope geochemistry. In *Forecasting Volcanic Events*. H. Tazieff & J. C. Sabroux, Eds., pp. 337–86. Amsterdam: Elsevier.

Allègre, C. J., Hofmann, A., & O'Nions, K. (1996) The argon constraints on mantle structure. *Geophys. Res. Lett.*, **23**, 3555–7.

Allègre, C. J., Sarda, P., & Staudacher, Th. (1993) Speculations about the cosmic origin of He and Ne in the interior of the Earth. *Earth Planet. Sci. Lett.*, **117**, 229–33.

Allègre, C. J., Staudacher, T., & Sarda, P. (1987) Rare gas systematics: Formation of the atmosphere, evolution, and structure of the Earth's mantle. *Earth Planet. Sci. Lett.*, **81**, 127–50.

Allègre, C. J., Staudacher, Th., Sarda, P., & Kurz, M. (1983) Constraints on evolution of Earth's mantle from rare gas systematics. *Nature*, **303**, 762–6.

Amari, S., & Ozima, M. (1985) Search for the origin of exotic helium in deep-sea sediments. *Nature*, **317**, 520–2.

Amari, S., & Ozima, M. (1988) Extra-terrestrial noble gases in deep sea sediments. *Geochim. Cosmochim. Acta*, **52**, 1087–95.

Anders, E. (1988) Circumstellar material in meteorites: Noble gases, carbon and nitrogen. In *Meteorites and the Early Solar System*, J. F. Kerridge & M. S. Matthews, Eds., pp. 927–55. Tuscon: University of Arizona Press.

Anders, E., & Ebihara, M. (1982) Solar system abundances of the elements. *Geochim. Cosmochim. Acta*, **46**, 2363–80.

Anders, E., & Grevesse, N. (1989) Abundances of the elements: Meteoritic and solar. *Geochim. Cosmochim. Acta*, **53**, 197–214.

Anders, E., & Zinner, E. (1993) Interstellar grains in primitive meteorites: Diamond, silicon carbide, and graphite. *Meteoritics*, **28**, 490–514.

Anderson, D. L., Ed. (1991) *Hotspot Handbook*. Caltech Plume Symposium California Inst. Technology.

Anderson, D. L. (1993) Helium-3 from the mantle: Primordial signal or cosmic dust? *Science*, **261**, 170–6.

Andrew, J. N. (1991) Noble gases and radioelements in ground water. In *Applied Groundwater Hydrology*, R. A. Downing & W. B. Wilkinson, Eds., pp. 243–65. Oxford: Oxford University Press.

Andrew, J. N., Fontes, J.-Ch., Michelot, J.-L., & Elmore, D. (1986) In-situ neutron flux, ^{36}Cl production and groundwater evolution in crystalline rocks at Stripa, Sweden. *Earth Planet. Sci. Lett.*, **77**, 49–58.

Andrew, J. N., Hussain, N., & Youngman, M. J. (1989) The in situ production of radio isotopes in rock minerals with particular reference to the Stripa granite. *Geochim. Cosmochim. Acta*, **53**, 1803–15.

Aston, F. W. (1919) Positive ray spectrograph. *Philosophical Magazine*, VI, **38**, 707–14.

Azbel, I. Ya., & Tolstikhin, I. N. (1990) Geodynamics, magmatism, and degassing of the Earth. *Geochim. Cosmochim. Acta*, **54**, 139–54.

Azbel, I. Ya., & Tolstikhin, I. N. (1993) Accretion and early degassing of the Earth: Constraints from modeling of Pu-U-Xe isotopic systematic. *Meteoritics*, **56**, 609–21.

Azuma, S., & Ozima, M. (1993) Investigation of ^{4}He and ^{40}Ar in beryl by laser extraction technique. *Earth Planet. Sci. Lett.*, **90**, 69–76.

Azuma, S., Hiyagon, H., Futagami, T., Syono, Y., & Fukuoka, K. (1994) Impact-induced degassing of noble gases from olivine. In *A.I.P. Conference Proc.*, vol. 341, K. A. Farley, Ed., pp. 270–5. Amer. Institute Physics.

Baker, E. T., & Lupton, J. E. (1992) Changes in submarine hydrothermal ^{3}He/heat ratios as an indicator of magmatic/tectonic activity. *Nature*, **346**, 556–8.

Ballentine, C. J. (1997) Resolving the mantle He/Ne and crustal ^{21}Ne/^{22}Ne in well gases. *Earth Planet. Sci. Lett.*, **152**, 233–49.

Ballentine, C. J., & O'Nions, R. K. (1991) The nature of mantle neon contributions to Vienna Basin hydrocarbon reservoirs. *Earth Planet. Sci. Lett.*, **113**, 553–67.

Barrer, R. M., & Edge, A. J. V. (1967) Gas hydrate containing argon, krypton, and xenon: Kinetics and energetic of formation and equilibrium. *Proc. Roy. Soc., London*, **A300**, 1–24.

Basford, J. R., Dragon, J. C., Pepin, R. O., Coscio, Jr., M. R., & Murthy, V. R. (1973) Krypton and xenon in lunar fined. *Proc. Lunar Planetary Conf.*, **2**, 1915–55.

Becker, L., Bada, J. L., Winans, R. E., Hunt, J. E., Bunch, T. E., & Grench, B. M. (1994) Fullerenes in the 1.85 billion year old Sudbury Impact Crater. *Science*, **265**, 642–5.

Becker, L., Bunch, T. E., & Allamandolla, L. J. (1999) Higher fullerenes in the Allende meteorite, *Nature*, **400**, 227–8.

Becker, L, Poreda, R. J., & Bada, J. F. (1996) Extraterrestrial helium in trapped in fullerenes in the Sudbury impact structure. *Science*, **272**, 249–52.

Becker, R. H., Schlutter, D. J., Rider, P. E., & Pepin, R. O. (1996) Reevaluation of solar wind 36Ar/38Ar ratio. *Lunar Planet. Sci.*, **XXVII**, 83–4.

Becker, R. H., Schlutter, D. J., Rider, P. E., & Pepin, R. O. (1998) An acid-etch study of the Kapoeta achondrite: Implications for the argon-36/argon-38 ratio in the solar wind. *Meteoritics Planet. Sci.*, **33**, 109–13.

Begemann, F. (1994) Indigenous, & extraneous noble gases in terrestrial diamonds. In *Noble Gas Geochemistry and Cosmochemistry*, J. Matsuda, Ed., pp. 217–27. Tokyo: Terra Scientific Publ. Co.

Benkert, J.-P., Baur, H., Pedroni, A., Wieler, R., & Signer, P. (1988) Solar He, Ne and Ar in regolith minerals: All are mixtures of two components. *Lunar Planet. Sci.*, **XIX**, 56–60.

Benkert, J.-P., Baur, H., Signer, P., & Wieler, R. (1993) He, Ne and Ar from the solar wind and solar energetic particles in lunar ilmenites and pyroxenes. *J. Geophy. Res.*, **98**, 13, 147–62.

Ben-Naim, A., & Egel-Thal, M. (1965) Thermodynamics of aqueous solutions of noble gases. III. Effect of electrolytes. *J. Phys. Chem.*, **69**, 3250–3.

Benson, B. B. (1973) Noble gas concentration ratios as paleotemperature indicators. *Geochim. Cosmochim. Acta*, **37**, 1391–5.

Benson, B. B., & Kraus, D. J. (1976) Empirical laws for dilute aqueous solutions of non-polar gases. *J. Chem. Phys.*, **64**, 689–709.

Bernatowicz, T. J. (1981) Noble gases in ultramafic xenoliths from San Carlos, Arizona. *Contrib. Mineral. Petrol.*, **76**, 84–91.

Bernatowicz, T. J., & Fahey, A. J. (1986) Xe isotopic fractionation in a cathode glow discharge. *Geochim. Cosmochim. Acta*, **50**, 445–52.

Bernatowicz, T. J., & Hagee, B. E. (1987) Isotopic fractionation of Kr and Xe implanted in solids at very low energies. *Geochim. Cosmochim. Acta*, **51**, 1599–611.

Bernatowicz, T. J., & Podosek, F. A. (1978) Nuclear components in the atmosphere. In *Terrestrial Rare Gases*, E. C. Alexander, Jr., & M. Ozima, Eds., pp 99–135. Tokyo: Japan Sci. Soc. Press.

Bernatowicz, J. T., Goettel, K., Hohenberg, C. M., & Podosek, F. A. (1979) Anomalous noble gasses in josephinite and related rocks? *Earth Planet. Sci. Lett.*, **43**, 368–84.

Bernatowicz, T., Brannon, J., Brazzle, R., Cowsik, R., Hohenberg, C. M., & Podosek, F. A. (1993) Precise determination of relative and absolute ββ-decay rates of 128Te and 130Te. *Phys. Rev. C*, **47**, 806–25.

Bernatowicz, T., Fraundorf, G., Tang, M., Anders, E., Wopenka, B., Zinner, E., & Fraundorf, P. (1987) Evidence for interstellar SiC in the Murray carbonaceous meteorite. *Nature*, **330**, 728–30.

Bernatowictz, T. J., Kennedy, B. M., & Podosek, F. A. (1985) Xenon in glacial ice and the atmospheric inventory of noble gases. *Geochim. Cosmochim. Acta*, **49**, 2561–4.

Bernatowicz, T. J., Podosek, F. P., Honda, M., Kramer, F. E., & Podosek, F. A. (1983) Xe on shales: The plastic bag experiment (abstract), *Lunar Planet. Sci.*, **XIV**, 31–2.

Bernatowicz, T. J., Podosek, F. A., Honda, M., & Kramer, F. E. (1984) The atmospheric inventory of xenon and noble gases in shales: The plastic bag experiment. *J. Geophys. Res.*, **89B6**, 4597–611.

Beyerle, U., Purtshert, R., Aeschbach-Hertig, W., Imboden, D. M., Loosli, H. H., Wieler, R., & Kipfer, R. (1998) Climate and groundwater recharge during the last glaciation in an ice-covered region. *Science*, **282**, 731–4.

Bieri, R. H. (1971) Dissolved noble gases in marine waters. *Earth Planet. Sci. Lett.*, **10**, 329–33.

Bieri, R. H., & Koide, M. (1972) Dissolved noble gases in the east equatorial and southeast Pacific. *J. Geophys. Res.*, **77**, 1667–76.

Bieri, R. H., Koide, M., & Goldberg, E. D. (1964) Noble gases in sea water. *Science*, **146**, 1035–7.

Bieri, R. H., Koide, M., & Goldberg, E. D. (1966) The noble gas content of Pacific seawaters. *J. Geophys. Res.*, **71**, 5243–65.

Bieri, R. H., Koide, M., & Goldberg, E. D. (1967) Geophysical implications of the excess helium found in Pacific waters. *J. Geophys. Res.*, **72**, 2497–511.

Bieri, R. H., Koide, M., & Goldberg, E. D. (1968) Noble gas contents of marine waters. *Earth Planet. Sci. Lett.*, **4**, 329–40.

Black, D. C. (1972) On the origin of trapped helium, neon, and argon siotopic variations in meteorites – I. Gas-rich meteorites, lunar soil and breccia. *Geochim. Cosmochim. Acta*, **36**, 347–75.

Blander, M., Grimes, W. R., Smith, N. V., & Watson, G. N. (1959) Solubility of the noble gases in moltenfluorides. II. In the LiF-NaF-KF eutectec mixture. *J. Phys. Chem.*, **63**, 1164–7.

Bloch, M. R., Fechtig, H., Gentner, W., Neukum, G., & Schneider, E. (1971) Meteorite impact craters, crater simulations, and the meteoroid flux in the early solar system. *Proc. Second Lunar Science Conf.*, **3**, 2639–52.

Bochsler, P., Geiss, J., & Maeder, A. (1990) The abundance of ^{3}He in the solar wind – A constraint for models of solar evolution. *Solar Phys.* **128**, 203–15.

Bogard, D., Hörz, F., & Johnson, P. (1987) Shock effects and argon loss in samples of the Leedy L6 chondrite experimentally shocked to 29–70 Gpa pressures. *Geochim. Cosmochim. Acta*, **51**, 2035–44.

Bogard, D. D., Rowe, M. W., Manuel, O. K., & Kuroda, P. K. (1965) Noble gas anomalies in the mineral thucolite. *J. Geophys. Res.*, **70**, 703–8.

Boslough, M. B., Ahrens, T. J., Vizgirda, R. H., Becker, R. H., & Epstein, S. (1982) Shock-induced devolatalization of calcite. *Earth Planet. Sci. Lett.*, **61**, 166–70.

Boyd, S. R., Pineau, F., & Javoy, M. (1994) Modeling of the growth of natural diamonds. *Chem. Geol.*, **116**, 29–42.

Brawer, S. A., & White, W. B. (1975) Theory of vibrational spectra of some network and molecular glasses. *Phys. Rev.*, **B11**, 3173–94.

Brenemann, H. H., & Stone, E. C. (1985) Solar coronal and photospheric abundances from solar energetic particle measurements. *Astrophys. J.*, **299**, L57–61.

Broadhurst, C. L., Drake, M. J., Hagee, B. E., & Bernatowicz, T. J. (1990) Solubility and partition of Ar in anorthosite, diopside, forsterite, spinel, & synthetic basaltic liquids. *Geochim. Cosmochim. Acta*, **54**, 299–309.

Broadhurst, C. L., Drake, M. J., Hagee, B. E., & Bernatowicz, T. J. (1992) Solubility and partitioning of Ne, Ar, Kr, and Xe in minerals and synthetic basalt melts. *Geochim. Cosmochim. Acta*, **56**, 709–23.

Broecker, W. S. (1974) *Chemical Oceanography*. New York: Harcourt Brace Javanovich.

Brown, H. (1952) Rare gases and formation of the earth's atmosphere. In *The Atmospheres of the Earth and Planets*, 2nd ed. G. P. Kuiper, Ed., pp. 258–66. Chicago: University of Chicago Press.

Brownlee, D. (1979) Interplanetary dust. *Rev. Geophys. Space Phys.*, **17**, 1735–43.

Brownlee, D. (1985) Cosmic dust: Collection and research. *Ann. Rev. Earth Planet. Sci.*, **XIII**, 147–73.

Brunauer, S., Emmett, P. H., & Teller, E. (1938) Adsorption of gases in multimolecular layers. *J. Amer. Chem. Soc.*, **60**, 309–19.

Burgess, R., Turner, G., & Harris, J. W. (1991) ^{40}Ar-^{39}Ar laser probe studies of clinopyroxene inclusions in eclogitic diamonds. *Geochim. Cosmochim. Acta*, **56**, 389–402.

Burnard, P., Graham, D., & Turner, G. (1997) Vesicle-specific noble gas analyses of "popping rock"; Implications for primordial noble gases in Earth. *Science*, **276**, 568–71.

Busemann, H., Baur, H., & Wieler, R. (1997) Noble gas component "Q" in HF/HCl-resistant residues of unequilibrated chondrites. *Lunar Planet. Sci.*, **XXVIII**, 183–4.

Butler, W. A., Jeffery, P. M., Reynolds, J. H., & Wasserburg, G. J. (1963) Isotopic variations in terrestrial gases. *J. Geophys. Res.*, **68**, 3283–91.

Caffee, M. W., Hohenberg, C. M., Horz, F., Hudson, B., Kennedy, B. M., Podosek, F. A., & Swindle, T. D. (1982) Shock disturbance of the I-Xe system. *J. Geophys. Res.*, **87**, A318–30.

Caffee, M. W., Hudson, G. B., Velsco, C., Alexander, E. C., Jr., Huss, G. R., & Chivas, A. R. (1989) Non-atmospheric noble gases from CO_2 well gases. *Lunar Planet. Sci.*, **XIX**, 154–5.

Caffee, M. W., Hudson, G. B., Velsco, C., Huss, G. R., Alexander, E. C., Jr., & Chivas, A. R. (1999) Primoridial noble gases from Earth's mantle: Identification of a primitive volatile component. *Science*, **285**, 2115–8.

Caldwell, W. A., Ngyuyen, J. H., Pfrommer, B. G., Mauri, F., Louie, S. G., & Jeanloz, R. (1997) Structure, bonding, and geochemistry of xenon at high pressures. *Science*, **227**, 930–3.

Canalas, R. A., Alexander, Jr., E. C., & Manuel, O. K. (1968) Terrestrial abundance of noble gases. *J. Geophys. Res.*, **73**, 3331–4.

Carroll, M. R., & Stolper, E. M. (1991) Argon solubility and diffusion in silica glass: Implications for the solution behavior of molecular gases. *Geochim. Cosmochim. Acta*, **55**, 211–25.

Carroll, M. R., & Stolper, E. M. (1993) Noble gas solubility in silicate melts and glasses: New experimental results for Ar and the relationship between solubility and ionic porosity. *Geochim. Cosmochim. Acta*, **57**, 5039–51.

Carroll, M. R., Draper, D. S., Brooker, R. A., & Kelly, S. (1994) Noble gas solubility in melts and crystals. In *Noble Gas Geochemistry*, J. Matsuda, Ed., pp. 325–41. Tokyo: Terra Scientific Publishing Co.

Carroll, M. R., Sutton, S. R., Rivers, M. L., & Woolum, D. (1993) Krypton solubility and diffusion in silicic melts. *Chem. Geol.*, **109**, 9–28.

Carslaw, H. S., & Jaeger, J. C. (1959) *Conduction of Heat in Solids*. Oxford: Clarendon Press.

Cerling, T. E. (1990) Dating geomorphic surfaces using cosmogenic ^{3}He. *Quat. Res.*, **33**, 148–56.

Cerling, T. E., & Craig, H. (1994) Cosmogenic ^{3}He production rates from 39°N to 46°N latitude, western USA and France. *Geochim. Cosmochim. Acta*, **58**, 249–55.

Cerling, T. E., Poreda, R. E., & Rathburn, S. L. (1994) Cosmogenic ^{3}He and ^{21}Ne age of the Big Lost River flood, Snake River Plain, Idaho. *Geology*, **22**, 227–30.

Cerutti, H. (1974) Die Bestimmung des Argons im Sonnenwind aus Messungen an den Apollo-SWC-Folien. Dissertation, University of Bern.

Chamorro-Perez, E., Gillet, P., & Jambon, A. (1996) Argon solubility in silicate melts at very high pressures. Experimental set-up and preliminary results for silica and anorthite melts. *Earth Planet. Sci. Lett.*, **145**, 97–107.

Chamorro-Perez, E., Gillet, P., Jambon, A., Badro, & McMillan, P. (1998) Low argon solubility in silicate melts at high pressure. *Nature*, **393**, 352–5.

Chaussidon, M., & Jambon, A. (1994) Boron content and isotopic composition of oceanic basalts: Geochemical and cosmochemical implications. *Earth Planet. Sci. Lett.*, **121**, 277–91.

Clarke, W. B., Beg, M. A., & Craig, H. (1969) Excess 3He in the sea: Evidence for terrestrial primordial helium. *Earth Planet. Sci. Lett.*, **6**, 213–20.

Clarke, W. B., Beg, M. A., & Craig, H. (1970) Excess ^{3}He at the North Pacific Geosecs station. *J. Geophys. Res.*, **75**, 7676–8.

Classen, H. H. (1966) *The Noble Gases*. Boston: D. C. Heath.

Clayton, R. N., Grossman, L., & Mayeda, T. K. (1973) A component of primitive nuclear composition in carbonaceous meteorites. *Science*, **182**, 485–8.

Cook, G. A. (1961) Introduction and general survey. In *Argon, Helium, and Rare Gases: I. History, Occurrence and Properties*, G. A. Cook, Ed., pp. 1–15. New York: Interscience.

Craig, H., & Clarke, W. B. (1970) Oceanic ^{3}He: Contribution from cosmogenic tritium. *Earth Planet. Sci. Lett.*, **9**, 289–96.

Craig, H., & Lupton, J. (1976) Primordial neon, helium and hydrogen in oceanic basalts. *Earth Planet. Sci. Lett.*, **31**, 369–85.

Craig, H., & Weiss, R. F. (1968) Argon concentration in the ocean: a discussion. *Earth Planet. Sci. Lett.*, **5**, 175–82.

Craig, H., & Weiss, R. F. (1971) Dissolved gas saturation anomalies and excess helium in the ocean. *Earth Planet. Sci. Lett.*, **10**, 289–96.

Craig, H., Clarke, F. R., & Beg, M. A. (1975) Excess ^{3}He in deep water on the East Pacific Rise. *Earth Planet. Sci. Lett.*, **26**, 125–32.

Craig, H., Lupton, J. E., & Horibe, Y. (1978) A mantle helium component in circum-Pacific volcanic gases: Hakone, the Marianas, and Mt. Lassen. In *Terrestrial Rare Gases*, E. C. Alexander, Jr., & M. Ozima, Eds., pp. 3–16. Tokyo: Center for Academic Press.

Craig, H., Weiss, R. F., & Clarke, W. B. (1967) Dissolved gases in the equatorial and south Pacific Ocean. *J. Geophys. Res.*, **72**, 6165–81.

Crank, J. (1975) *The Mathematics of Diffusion*, 2nd ed. Oxford: Clarendon Press.

Damon, P. E., & Kulp, J. L. (1957) Determination of radiogenic helium in zircon by stable isotope dilution technique. *Trans. Amer. Geophys. Union*, **38**(6), 945–53.

Damon, P. E., & Kulp, J. L. (1958) Excess helium and argon in beryl and other minerals. *Amer. Mineralogists*, **43**, 433–59.

Davies, G. F. (1984) Geophysical and isotopic constraints on mantle convection: An interim synthesis. *J. Geophys. Res.*, **89**, 6017–40.

Davies, G. F. (1990) Mantle plumes, mantle stirring and hotspot chemistry. *Earth Planet. Sci. Lett.*, **99**, 94–109.

Davies, G. F. (1999) Geophysically constrained mantle mass flow and the ^{40}Ar budget: A degassed lower mantle? *Earth Planet. Sci. Lett.*, **166**, 149–62.

Davis, P. K. (1977) Effects of shock pressure on ^{40}Ar-^{39}Ar radiometric age determinations. *Geochim. Cosmochim. Acta*, **41**, 195–205.

Dean, J. A., Ed. (1985) Lange's handbook of chemistry, p. 1856. New York: McGraw-Hill.

Dempster, A. J. (1918) A new method of positive ray analysis. *Phys. Rev.*, **11**, 316–25.

Déruelle, B., Dreibus, G., & Jambon, A. (1992) Iodine abundances in oceanic basalts: Implications for Earth dynamics. *Earth Planet. Sci. Lett.*, **108**, 217–27.

Dodson, A., Kennedy, B. M., & DePaolo, D. J. (1997) Helium and neon isotopes in the Imnaha basalt, Columbia River Basalt Group: Evidence for a Yellowstone plume source. *Earth Planet. Sci. Lett.*, **150**, 443–51.

Drescher, J., Kirsten, T., & Schäffer, K. (1998) The rare gas inventory of the continental crust, recovered by the KTB Continental Deep Drilling Project. *Geochim. Cosmochim. Acta*, **154**, 247–63.

Drozd, R. J., & Podosek, F. P. (1976) Primordial ^{129}Xe in meteorites. *Earth Planet. Sci. Lett.*, **31**, 15–30.

Drozd, R. J., Hohenberg, C. M., & Morgan, C. J. (1974) Heavy rare gases from Rabbit Lake (Canada) and the Oklo Mine (Gabon): Natural spontaneous chain reactions in old uranium deposits. *Earth Planet. Sci. Lett.*, **23**, 28–33.

Dubey, V. S., & Holmes, A. (1929) Estimates of the ages of the Whin Sill and the Cleveland Dyke by the helium method. *Nature*, **123**, 794–5.

Dunai, T. J., & Baur, H. (1995) Helium, neon, and argon systematics of the European subcontinental mantle: Implications for its geochemical evolution. *Geochim. Cosmochim. Acta*, **59**, 2767–83.

Dymond, J., & Hogan, L. (1973) Noble gas abundance patterns in deep-sea basalts – primordial gases from the mantle. *Earth Planet. Sci. Lett.*, **20**, 131–9.

Eberhardt, P., Eugster, O., & Marti, K. (1965) A redetermination of the isotopic composition of atmospheric neon. *Z. Naturforsch.*, **20a**, 623–4.

Eberhardt, P., Geiss, J., Graf, H., Grögler, N., Krähenbühl, U., Schwaller, H., Schwarzmüller, J., & Stettler, A. (1970) Trapped solar wind noble gases, exposure age and K/Ar age in Apoollo 11 lunar fine material. *Proc. Apollo 11 Lunar Science Conf.*, 1037–70.

Eberhardt, P., Geiss, J., Graf, H., Grögler, N., Mendia, M. D., Mörgeli, M. U., Schwaller, H., & Stettler, A. (1972) Trapped solar wind noble gases in Apollo 12 lunar fines 12001 and Apollo 11 breccia 10046. *Proc. Third Lunar Science Conf.*, **2**, 1821–56.

Eikenberg, J., Signer, P., & Wieler, R. (1993) U-Xe, U-Kr, and U-Pb systematics for dating uranium minerals and investigations of the production of nucleogenic neon and argon. *Geochim. Cosmochim. Acta*, **57**, 1053–69.

Elliot, T., Ballentine, C. J., O'Nions, R. K., & Ricchiuto, T. (1993) Carbon, helium, neon and argon isotopes in a Po Basin (northern Italy) natural gas field. *Chem. Geol.*, **106**, 429–40.

Esser, B. K., & Turekian, K. K. (1988) Accretion rate of extraterrestrial particles determined from osmium isotope systematics of Pacific pelagic clay and manganese nodules. *Geochim. Cosmochim. Acta*, **52**, 1383–8.

Fabryka-Martin, J., Bentley, H., Elmore, D., & Airey, P. L. (1985) Natural iodine-129 as an environmental tracer. *Geochim. Cosmochim. Acta*, **49**, 337–47.

Fanale, F. P. (1971) A case for catastrophic early degassing of the Earth. *Chem. Geol.*, **8**, 79–105.

Fanale, F. P., & Cannon, W. A. (1971) Physical absorption of rare gas on terrigenous sediments. *Earth Planet. Sci. Lett.*, **11**, 362–86.

Fanale, F. P., & Cannon, W. A. (1972) Origin of planetary rare gas: The possible role of adsorption. *Geochim. Cosmochim. Acta*, **36**, 319–28.

Fanale, F. P., Cannon, W. A., & Owen, T. (1978) Mars: Regolith adsorption and the relative concentrations of atmospheric noble gases. *Geophys. Res. Lett.*, **5**, 77–80.

Farley, K. A. (1995) Cenozoic variation in the flux of interplanetary dust recorded by ^{3}He in a deep-sea sediment. *Nature*, **376**, 153–6.

Farley, K. A., & Neroda, E. (1998) Noble gases in the Earth's mantle. *Ann. Rev. Earth Planet. Sci.*, **26**, 189–218.

Farley, K. A., & Poreda, R. J. (1991) Rare gas composition of Samoan ultramafic xenoliths. *EOS Trans. Amer. Geophys. Union*, **72**, 536–7.

Farley, K. A., & Poreda, R. J. (1993) Mantle neon and atmospheric contamination. *Earth Planet. Sci. Lett.*, **114**, 325–39.

Farley, K. A., Maier-Reimer, E., Schlosser, P., & Broecker, W. S. (1995) Constraints on mantle ^{3}He fluxes and deep-sea circulation from an oceanic general circulation model. *J. Geophys. Res.*, **100**(B3), 3829–39.

Farrer, H., & Tomlinson, R. H. (1962) Cumulative yields of the heavy fragments in ^{235}U thermal neutron fission. *Nuclear Physics*, **34**, 367–81.

Farrer, H., Fickel, H. R., & Tomlinson, R. H. (1962) Cumulative yields of light fragments in ^{235}U thermal neutron fission. *Canad. J. Phys.*, **40**, 1017–26.

Fields, P. R., Friedman, A. M., Milsted, J., Lerner, J., Stevens, C. M., Metta, D., & Sabine, W. K. (1966) Decay properties of plutonium-244, and comments on its existence in nature. *Nature*, **212**, 131–4.

Fisher, D. E. (1985a) Noble gas data from oceanic island basalts do not require an undepleted mantle source. *Nature*, **316**, 716–18.

Fisher, D. E. (1985b) Radiogenic rare gases and the evolutionary history of the depleted mantle. *J. Geophys. Res.*, **90**(B2), 1801–7.

Fisher, D. E. (1986) Rare gases in MORB. *Geochim. Cosmochim. Acta*, **50**, 2531–41.

Fisher, D. E. (1970) Heavy rare gases in a Pacific seamount. *Earth Planet. Sci. Lett.*, **9**, 331–5.

Foland, K. A. (1974) ^{40}Ar diffusion in homogeneous orthoclase and an interpretation of Ar diffusion in K-feldspars. *Geochim. Cosmochim. Acta*, **38**, 151–66.

Fraundorf, P., Brownlee, D. E., & Walker, R. M. (1982) Laboratory studies of interplanetary dust. In *Comets*, L. L. Wilkning, Ed., pp. 383–409. Tucson: University of Arizona Press.

Fredriksson, K., & De Carli, P. (1964) Shock emplaced argon in a stony meteorite: 1. Shock experiment and petrology of sample. *J. Geophys. Res.*, **69**, 1403–6.

Frick, U., & Chang, S. (1977) Ancient carbon and noble gas fractionation. *Proc. Eighth Lunar Science Conf.*, **1**, 263–72.

Frick, U., Mack, R., & Chang, S. (1979) Noble gas trapping and fractionation during synthesis of carbonaceous matter. *Proc. Tenth Lunar Planetary Science Conf.*, **2**, 1961–73.

Fukumoto, H., Nagao, K., & Matsuda, J. (1986) Noble gas studies on the host phase of high ^{3}He/^{4}He ratios in deep sea sediments. *Geochim. Cosmochim. Acta*, **50**, 2245–53.

Funk, H., Podosek, F. A., & Rowe, M. W. (1967) Fissiogenic Xe in the Renazzo and Murray meteorites. *Geochim. Cosmochim. Acta*, **31**, 1721–32.

Futagami, T., Ozima, M., Nagai, S., & Aoki, Y. (1993) Experiments on thermal release of implanted noble gases from minerals and their implantations for noble gases in lunar soil grains. *Geochim. Cosmochim. Acta*, **57**, 3177–94.

Galer, S. J. G., & O'Nions, R. K. (1985) Residence time of thorium, uranium and lead in the mantle and implications for mantle convection. *Nature*, **316**, 778–82.

Gazis, C., & Ahrens, T. J. (1991) Solution and shock-induced exsolution of argon in vitreous carbon. *Earth Planet. Sci. Lett.*, **104**, 337–49.

Geiss, J., & Reeves, H. (1972) Cosmic and solar system abundances of deuterium and helium-3. *Astron. Astrophys.*, **18**, 126–32.

Geiss, J., Buehler, F., Cerutti, H., Eberhardt, P., & Filleaux, C. H. (1972) Solar wind composition experiments. *Apollo 15 Preliminary Scientific Report*, NASA SP-289, pp. 15-1–15-7.

Göbel, R., Ott, U., & Begemann, F. (1978) On the trapped noble gases in ureilites. *J. Geophys. Res.*, **83**, 855–67.

Gold, T., & Soter, S. (1982) Abiogenic methane and the origin of petroleum. *Energy Explore. Exploit.*, **1**, 89–104.

Goldreich, P., & Ward, W. R. (1973) The formation of planetesimals. *Astrophys. J.*, **183**, 1051–61.

Graf, Th., Kohl, C. P., Marti, K., & Nishiizumi, K. (1991) *Geophys. Res. Lett.*, **18**, 203–6.

Graham, D., & Sarda, P. (1991) Reply to comments by T. M. Gerlach on "Mid-ocean ridge popping rocks: Implications for degassing at ridge crests." *Earth Planet. Sci. Lett.*, **105**, 568–73.

Hagee, B., Bernatowicz, T. J., Podosek, F. A., Johnson, M. L., Burnet, D. S., & Tatsumoto, M. (1990) Actinide abundances in ordinary chondrites. *Geochim. Cosmochim. Acta*, **54**, 2847–58.

Hamano, Y., & Ozima, M. (1978) Earth atmosphere evolution model based on Ar isotopic data. In *Terrestrial Rare Gases*. Jr. E. C. Alexander, & M. Ozima, Eds., pp. 155–72. Tokyo: Center for Academic Pub. Japan.

Hammond, S., Baker, E., Bernard, E., Massoth, G., Fox, C., Feely, R., Embley, R., Rona, P., & Cannon, G. (1991) NOAA's vents program targets oceanic hydrothermal effects. *EOS*, **72**, 561–6.

Harper, C. L., & Jacobsen, S. B. (1996) Noble gases and Earth's accretion. *Science*, **272**, 1814–18.

Hart, R., Hogan, L., & Dymond, J. (1985) The closed-system approximation for evolution of argon and helium in the mantle, crust and atmosphere. *Chem. Geol.*, **52**, 45–73.

Hart, S. R. (1984) He diffusion in olivine. *Earth Planet. Sci. Lett.*, **70**, 297–302.

Hart, S. R., Dymond, J., & Hogan, L. (1979) Preferential formation of the atmosphere-sialic crust system from the earth. *Nature*, **278**, 156–9.

Hayashi, C. (1972) Origin of the solar system (in Japanese). *Report of the 5th Symposium on the Moon and planets*, pp. 13–18. Tokyo: The Space and Aeronautical Research Institute, University of Tokyo.

Hayashi, C., Nakazawa, K., & Nakagawa, Y. (1993) Planetary accretion in the solar gravitational field. In *Protostars & Planets III*, E. H. Levy & J. T. Lunine, Eds., pp. 1089–107. Tucson: University of Arizona Press.

Hayatsu, A., & Waboso, C. E. (1985) The solubility of rare gases in silicate melts and implications for K-Ar dating. *Chem. Geol.*, **52**, 97–102.

Henderson, P. (1982) *Inorganic Geochemistry*. Oxford: Pergamon Press.

Henneke, E. W., Manuel, O. K., & Sabu, D. D. (1975) Double beta decay of 128Te. *Phys. Rev. C*, **11**, 1378–84.

Hilton, D. R., & Craig, H. (1989) A helium isotope transect along the Indonesian archipelago. *Nature*, **342**, 906–8.

Hilton, D. R., Hammerschmidt, K., Loock, G., & Friedrichsen, H. (1993) Helium and argon isotopic systematics of the central Lau Basin and Valu Fa Ridge: Evidence of crustal/mantle interactions in a back-arc basin. *Geochim. Cosmochim. Acta*, **57**, 2819–41.

Hilton, D. R., McMurty, G. M., & Goff, F. (1998) Large variations in vent fluid $CO_2/^3He$ ratios signal rapid changes in magma chemistry at Loihi seamount, Hawaii. *Nature*, **396**, 359–62.

Hiyagon, H. (1981) Preliminary studies on partition of rare gases between crystals and melts. MA Thesis, University of Tokyo, Tokyo.

Hiyagon, H. (1994a) Constraints on rare gas partition coefficients from analysis of olivine-glass from a picritic mid-ocean ridge basalt – Comments. *Chem. Geol.*, **112**, 119–27.

Hiyagon, H. (1994b) Retention of solar He and Ne in IDPs in deep sea sediment. *Science*, **263**, 1257–9.

Hiyagon, H., & Kennedy, B. M. (1992) Noble gases in CH$_4$-rich gas fields, Alberta, Canada. *Geochim. Cosmochim. Acta*, **56**, 1569–89.

Hiyagon, H., & Ozima, M. (1986) Partition of noble gases between olivine and basalt melt. *Geochim. Cosmochim. Acta*, **50**, 2045–57.

Hiyagon, H., Ozima, M., Marty, B., Zashu, S., & Sasaki, H. (1992) Noble gases in submarine glasses from mid-oceanic ridges and Loihi seamount: Constraints on the early history of the Earth. *Geochim. Cosmochim. Acta*, **56**, 1301–16.

Hofmann, A. W. (1997) Mantle geochemistry: The message from oceanic volcanism. *Nature* **385**, 219–29.

Hofmann, A. W., & Hart, S. R. (1978) An assessment of local and regional isotopic equilibrium in the mantle. *Earth Planet. Sci. Lett.*, **38**, 44–62.

Honda, M., McDougall, I., Patterson, D. B., Doulgeris, A., & Clague, D. A. (1991a) Possible solar noble gas component in Hawaiian basalts. *Nature*, **349**, 149–51.

Honda, M., McDougall, I., Patterson, D. B., Doulgeris, A., & Clague, D. A. (1991b) Terrestrial primordial neon – Reply. *Nature*, **352**, 388.

Honda, M., McDougall, I., Patterson, D. B., Doulgeris, A., & Clague, D. A. (1993a) Noble gases in submarine pillow basalt glasses from Loihi and Kilauea, Hawaii: A solar component in the Earth. *Geochim. Cosmochim. Acta*, **57**, 859–74.

Honda, M., Ozima, M., Nakada, Y., & Onaka, T. (1979) Trapping of rare gases during the condensation of solids. *Earth Planet. Sci. Lett.*, **43**, 197–200.

Honda, M., Patterson, D. B., McDougall, I., & Falloon, T. J. (1993b) Noble gases in submarine pillow basalt glasses from the Lau Basin. *Earth Planet. Sci. Lett.*, **120**, 135–48.

Honda, M., Reynolds, J. H., Roedder, E., & Epstein, E. (1987) Noble gases in diamonds: Occurrence of solar helium and neon. *J. Geophys. Res.*, **92**, 12507–21.

Horn, M. K., & Adams, J. A. S. (1966) Computer-derived geochemical balances and element abundances. *Geochim. Cosmochim. Acta*, **30**, 279–97.

House, M. A., Farley, K. A., & Kohn, B. P. (1999) An empirical test of helium diffusion in apatite: Borehole data from the Otway basin, Australia. *Earth Planet. Sci. Lett.*, **170**, 463–74.

Hudson, B., Flynn, G. J., Fraundorf, P., Hohenberg, C. M., & Shirck, J. (1981) Noble gases in stratospheric dust particles: Confirmation of extraterrestrial origin. *Science*, **211**, 383–6.

Hudson, G. B., Kennedy, B. M., Podosek, F. A., & Hohenberg, C. M. (1989) The early solar system abundance of ^{244}Pu as inferred from the St. Severin chondrite. *Proc. 19th Lunar Planetary Science Conf.*, pp. 547–57.

Hughes, D. W. (1978) Meteors. In *Cosmic Dusts*, J. A. M. McDonnell, Ed., pp. 125–83. New York: John Wiley.

Huneke, J. C., & Smith, S. P. (1976) The realities of recoil: ^{39}Ar recoil out of small grains and anomalous age patterns in ^{39}Ar-^{40}Ar dating. *Proc. Seventh Lunar Science Conf.*, 1987–2008.

Hünemohr, H. (1989) Edelgase in U- und Th-reichen mineralen und die Bestimmung der ^{21}Ne-dicktarget-ausbeute der ^{18}O(α, n)^{21}Ne-Kernreaktion in Bereich 4.0–8.8 MeV. Ph.D. Thesis. Johannes-Gutenberg University, Mainz.

Hunten, D. M., Pepin, R. O., & Walker, J. C. G. (1987) Mass fractionation in hydrodynamic escape. *Icarus*, **69**, 532–49.

Huss, G. (1997) The survival of presolar grains in solar system bodies. In *Astrophysical Implications of the Laboratory Study of Presolar Materials*, T. J. Bernatowicz & E. K. Zinner, Eds., pp. 721–48. AIP Conf. Proc. 402.

Huss, G., & Lewis, R. S. (1994) Noble gases in presolar diamonds I: Three distinct components and their implications for diamond origins. *Meteoritics*, **29**, 791–810.

Huss, G. R., Lewis, R. S., & Hemkin, S. (1996) The "normal planetary" noble gas component in primitive chondrites: Compositions, carrier, and metamorphic history. *Geochim. Cosmochim. Acta*, **60**, 3311–40.

Hyde, E. K. (1974) *The nuclear properties of the heavy Elements III, Fission Phenomena*. Englewood Cliffs, NJ: Prentice-Hall.

Hyman, H. H., Ed. (1963) *Noble Gas Compounds*. Chicago: The University of Chicago Press.

Ida, S., Canup, R. M., & Stewart, G. R. (1997) Lunar accretion from an impact-generated disk. *Nature*, **389**, 353–7.

Igarashi, G. (1995) Primitive Xe in the Earth. *AIP Conf. Proc.*, **341**, 70–80.

Igarashi, G., Ozima, M., Ishibashi, J., Gamo, T., Sakai, H., Nojiri, Y., & Kawai, T. (1992) Mantle helium flux from the bottom of Lake Mashu, Japan. *Earth Planet. Sci. Lett.*, **108**, 11–18.

Inghram, M. G., & Reynolds, J. H. (1950) Double beta-decay of Te130. *Phys. Rev.*, **78**, 822–3.

Irwin, J. I., & Reynolds, J. H. (1995) Multiple stages of fluid trapping in the Stripa granite indicated by laser microprobe analysis of Cl, Br, I, K, U, and nucleogenic plus radiogenic Ar, Kr, and Xe in fluid inclusions. *Geochim. Cosmochim. Acta*, **59**, 355–69.

Jambon, A., Weber, H., & Braun, O. (1986) Solubility of He, Ne, Ar, Kr, and Xe in a basalt melt in the range 1250–1600°C. Geochemical implications. *Geochim. Cosmochim. Acta*, **50**, 401–8.

Jean-Baptiste, P., Bougault, H., Vangriesheim, A., Charlou, J. L., Radford-Knoery, J., Fouquet, Y., Needham, D., & German, C. (1998) Mantle ^{3}He in hydrothermal vents and plume of the Lucky Strike site (MAR 37°17′N) and associated geothermal heat flux. *Earth Planet. Sci. Lett.*, **157**, 69–77.

Jeffrey, P. M., & Reynolds, J. H. (1961) Origin of excess Xe129 in stone meteorites. *J. Geophys. Res.*, **66**, 3582–3.

Jenkins, W. J., & Clarke, W. B. (1976) The distribution of ^{3}He in the western Atlantic Ocean. *Deep Sea Res.*, **23**, 481–94.

Jenkins, W. J., Beg, M. A., Clarke, W. B., Wangersky, P. J., & Craig, H. (1972) Excess 3He in the Atlantic Ocean. *Earth Planet. Sci. Lett.*, **16**, 122–6.

Jenkins, W. J., Edmond, J. M., & Corliss, J. B. (1978) Excess ^{3}He and ^{4}He in Galapagos submarine hydrothermal waters. *Nature*, **272**, 156–8.

Jephcoat, A. P. (1998) Rare gas solids in the Earth's deep interior. *Nature*, **393**, 355–8.

Jephcoat, A. P., Mao, H.-K., Finger, L. W., Cox, D. E., Hemley, R. J., & Zha, C.-S. (1987) Pressure-induced structural phase transition in solid Xenon. *Phys. Rev. Lett.*, **59**(23), 2670–3.

Jephcoat, A. P., & Besedin, S. P. (1996) Temperature measurement and melting determination in the laser-heated diamond anvil cell. *Phil. Trans. Roy. Soc. A*, **354**, 133–60.

Jessberger, E. K., & Ostertag, R. (1982) Shock effects on the K-Ar system of plagioclase feldspar and the age of inclusions from North-eastern Minnesota. *Geochim. Cosmochim. Acta*, **46**, 1465–71.

Johnson, H. E., & Axford, W. I. (1969) Production and loss of ^{3}He in the earth's atmosphere. *J. Geophys. Res.*, **74**, 2433–8.

Jokipii, J. R. (1964) The distribution of gases in the protoplanetary nebula. *Icarus*, **3**, 248–52.

Kamijo, K., Hashizume, K., & Matsuda, J. (1998) Noble gas constraints on the evolution of the atmosphere-mantle system. *Geochim. Cosmochim. Acta*, **62**, 2311–21.

Kellog, L. H. (1992) Mixing in the mantle. *Rev. Earth Planet. Sci.*, **20**, 365–88.

Kellog, L. H., & Wasserburg, G. J. (1990) The role of plumes in mantle helium fluxes. *Earth Planet. Sci. Lett.*, **99**, 276–89.

Kellog, L. H., Hager, B. H., & van der Hilst, R. D. (1999) Compositional stratification in the Deep Mantle. *Science*, **283**, 1881–4.

Kennedy, B. M., Hiyagon, H., & Reynolds, J. H. (1990) Crustal neon: A striking uniformity. *Earth Planet. Sci. Lett.*, **98**, 277–86.

Kennedy, B. M., Lynch, M. A., Reynolds, J. H., & Smith, S. P. (1985) Intensive sampling of noble gases in fluids at Yellowstone: I. Early overview of the data; regional patterns. *Geochim. Cosmochim. Acta*, **49**, 1251–61.

Kester, D. R. (1975) Dissolved gases other than CO_2. In *Chemical Oceanography*, vol. 1, 2nd ed., J. P. Reily & G. Skiro, Eds., pp. 498–556. New York: Academic Press.

Kipfer, R., Aeschbach-Hertig, W., Baur, H., Imboden, D. M., & Signer, P. (1994) Injection of mantle type helium into Lake Van (Turkey): The clue for quantifying deep water renewal. *Earth Planet. Sci. Lett.*, **125**, 357–70.

Kirsten, T. (1968) Incorporation of rare gases in solidifying enstatite melts. *J. Geophys. Res.*, **73**, 2807–10.

Kirsten, T., Heusser, E., Kaether, D., Oehm, J., Pernicka, E., & Richter, H. (1986) New geochemical double beta decay measurements on various selenium ores and remarks concerning tellurium isotopes. In *Nuclear Beta Decays and Neutrino*, T. Kotani, E. Ejiri, & E. Takasugi, Eds., pp. 81–92. Singapore: World Scientific.

Klots, C. E., & Benson, B. B. (1963) Isotopic effect in the solution of oxygen and nitrogen in distilled water. *J. Chem. Phys.*, **38**, 890–2.

Kockart, G. (1973) Helium in the terrestrial atmosphere. *Space Sci. Rev.*, **14**, 723–57.

König, H. (1963) Über die Löslichkeit der Edelgase in Meerwasser. *Z. Naturforsch.*, **18**, 363–7.

Kothari, B. K., Marti, K., Niemeyer, S., Regnier, S., & Stephens, J. R. (1979) Noble gas trapping during condensation: A laboratory study (abstract). In *Lunar and Planetary Science*, vol. X, pp. 682–4. Houston: Lunar Planetary Institute.

Krylov, A. Ya, Mamyrin, B. A., Silin, Yu. I., & Khabarin, L. V. (1973) Helium isotopes in ocean sediments. *Geochem. Internat.*, **10**, 202–5.

Kunz, J. (1999) Is there solar argon in the Earth's mantle? *Nature*, **399**, 649–50.

Kunz, J., Staudacher, T., & Allègre, C. J. (1998) Plutonium-fission Xe found in Earth's mantle. *Science*, **280**, 877–80.

Kuroda, P. K. (1956) On the nuclear physical stability of the uranium minerals. *J. Chem. Phys.*, **25**, 781–800.

Kuroda, P. K., & Sherill, R. D. (1977) Xenon and krypton isotope anomalies in the Besner Mine, Ontario, thucolite. *Geochem. J.*, **11**, 9–19.

Kurz, M. (1986) In situ production of terrestrial cosmogenic helium and some application to geochronology. *Geochim. Cosmochim. Acta*, **50**, 2855–62.

Kurz, M. D. (1993) Mantle heterogeneity beneath oceanic islands: Some inferences from isotopes. *Proc. Roy. Soc. London.*, **A342**, 91–103.

Kurz, M. D., & Jenkins, W. J. (1981) The distribution of helium in oceanic basalt glasses. *Earth. Planet. Sci. Lett.*, **53**, 41–54.

Kurz, M. D., Colodner, D., Trull, T. W., Moor, R. B., & O'Brien, K. (1990) Cosmic ray exposure dating with in situ produced cosmogenic ^3He: Results from young Hawaiian lava flows. *Earth Planet. Sci. Lett.*, **97**, 177–89.

Kusakabe, M., & Matsubaya, O. (1986) Volatiles in magmas, volcanic gases, and thermal waters. *Bull. Volcanol. Soc. Japan*, **30**, S267–83.

Kyser, T. K., & Rison, W. (1982) Systematics of rare gas isotopes in basaltic lavas and ultramafic xenoliths. *J. Geophys. Res.*, **87**, 5611–30.

Lal, D. (1988) In situ-produced cosmogenic isotopes in terrestrial rocks. *Ann. Rev. Earth Planet. Sci.*, **16**, 355–88.

Lal, D. (1989) An important source of ^{4}He (and ^{3}He) in diamonds. *Earth Planet. Sci. Lett.*, **96**, 1–7.

Lal, D. (1991) Cosmic ray labeling of erosion surfaces: *in situ* nuclide production rates and erosion models. *Earth Planet. Sci. Lett.*, **104**, 424–39.

Lal, D., Nishiizumi, K., Klein, J., Middleton, R., & Craig, H. (1987) Cosmogenic ^{10}Be in Zaire alluvial diamonds: Implications for ^{3}He contents of diamonds. *Nature*, **328**, 139–41.

Lal, D., Wacker, J. F., Poreda, R., & Craig, H. (1987) Diffusion of helium in diamonds and implications for primordial helium in the Earth (abstract). *Meteoretics*, **22**, 437–8.

Lambert, G., Le Cloarec, M. F., & Pennisi, M. (1988) Volcanic out-put of SO_2 and trace metals: a new approach. *Geochim. Cosmochim. Acta*, **52**, 39–42.

Lange, M. A., & Ahrens, T. J. (1982) Impact-induced dehydration of serpentine and the evolution of planetary atmosphere. *Proc. Lunar Planet. Sci. Conf. 13, J. Geophys. Res.*, **87**, A451–6.

Lange, M. A., & Ahrens, T. J. (1986) Shock-induced CO_2 loss from $CaCO_3$: Implications for early planetary atmospheres. *Earth Planet. Sci. Lett.*, **77**, 409–18.

Lange, M. A., Lambert, P., & Ahrens, T. J. (1985) Shock effects on hydrous minerals and implications for carbonaceous meteorites. *Geochim. Cosmochim. Acta*, **49**, 1715–26.

Langmuir, I. (1918) The adsorption of gases on plane surface of glass, mica and platinum. *J. Amer. Chem. Soc.*, **40**, 1361–403.

Levy, E. H., & Lunine, J. I., Eds. (1993) *Protostars and planets*, III, p. 1596. Tucson: University of Arizona Press.

Lewis, J. S., & Prinn, R. G. (1984) Planets and their atmospheres – Origin and evolution. New York: Academic Press.

Lewis, R. S., Srinivasan, B., & Anders, E. (1975) Host phase of a strange xenon component in Allende. *Science*, **190**, 1251–62.

Lewis, R. S. (1975) Rare gases in separated whitlockite from the St. Severin chondrite: Xenon and krypton from fission extinct ^{244}Pu. *Geochim. Cosmochim. Acta*, **39**, 417–32.

Leya, I., & Wieler, R. (1999) Nucleogenic production of Ne isotopes in Earth's crust and upper mantle induced by alpha particles from the decay of U and Th. *J. Geophys. Res.*, **104**(7), 15439–50.

Lin, W. J., Manuel, O. K., Cumming, J. L., Krstic, R. I., & Thorpe, R. I. (1988) Geochemically measured half-lives of ^{82}Se and ^{130}Te. *Nucl. Phys.*, **A481**, 484–93.

Lind, D. L., Geiss, J., & Stettler, W. (1979) Solar and terrestrial noble gases in magnetospheric precipitation. *J. Geophys. Res.*, **84**, 6435–42.

Lippolt, H. J., & Weigel, E. (1988) ^{4}He diffusion in ^{40}Ar-retentive minerals. *Geochim. Cosmochim. Acta*, **52**, 1449–58.

London, F. (1930) Zur Theorie und Systematik der Molekularkräfte. *Z. Phys.*, **63**, 245–79.

Lunine, J. I., & Stevenson, D. J. (1985) Thermodynamics of clathrate hydrate at low and high pressures with application to the outer solar system. *Astrophys. J. Suppl.*, **58**, 493–531.

Lupton, J. E. (1979) Helium-3 in the Guaymas Basin: Evidence for injection of mantle volatiles in the Gulf of California. *J. Geophys. Res.*, **84**, 7446–52.

Lupton, J. E., & Craig, H. (1981) A major helium-3 source at 15°S on the East Pacific Rise. *Science*, **214**, 13–18.

Lupton, J. E., & Craig, H. (1975) Excess ^{3}He in oceanic basalts: Evidence for terrestrial primordial helium. *Earth Planet. Sci. Lett.*, **26**, 133–9.

Lupton, J. E., Baker, E. T., & Massoth, G. J. (1989) Variable ^{3}He/heat ratios in submarine hydrothermal systems: Evidence from two plumes over the Juan de Fuca ridge. *Nature*, **337**, 161–4.

Lupton, J. E., Weiss, R. F., & Craig, H. (1977a). Mantle helium in red sea brines. *Nature*, **266**, 244–6.

Lupton, J. E., Weiss, R. F., & Craig, H. (1977b). Mantle helium in hydrothermal plumes in the Galapagos Rift. *Nature*, **267**, 603–4.

Luther, L. C., & Moore, W. J. (1964) Diffusion of helium in silicon, germanium and diamond. *J. Chem. Phys.*, **41**, 1018–26.

Lux, G. (1987) The behavior of noble gases in silicate liquids: Solution, diffusion, bubbles, and surface effects, with applications to natural samples. *Geochim. Cosmochim. Acta*, **51**, 1549–60.

Macedo, C. R., Costa, C. V., Ferreira, J. T., & Reynolds, J. H. (1977) Rare-gas dating, III. Evaluation of a double spiking procedure for potassium-argon dating. *Earth Planet. Sci. Lett.*, **34**, 411–18.

Mamyrin, B. A., Anufriev, G. S., Kamenskii, I. L., & Tolstikhin, I. N. (1970) Determination of isotopic composition of atmospheric helium. *Geochem. Internat.*, **7**, 498–505.

Mamyrin, B. A., Tolstikhin, I. N., Anufriev, G. S., & Kamensky, I. L. (1969) Anomalous isotopic composition of helium in volcanic gases. *Dokkl. Akad. Nauk SSSR*, **184**, 1197–9. (In Russian.)

Manuel, O. K. (1991) Geochemical measurements of double-beta decay. In *Nuclear Beta Decays and Neutrino*, T. Kotani, E. Ejiri, & E. Takasugi, Eds., pp. 71–80. Singapore: World Scientific.

Manuel, O. K., Hennecke, E. W., & Sabu, D. D. (1972) Xenon in carbonaceous chondrites. *Nature Phys. Sci.*, **240**, 99–101.

Markov, M. S., Shchipansky, A. A., Shukolyukov, Yu. A., & Verkhovsky, A. B. (1990) Distribution and origin of noble gases in magnetites from Precambrian banded iron ores. In *Ancient banded iron formations (regional presentation)*, pp. 281–4. Athens: Theophrastus Publications, S.A.

Marti, K. (1967) Isotopic composition of trapped krypton and xenon in chondrites. *Earth Planet. Sci. Lett.*, **3**, 243–8.

Marti, K. (1969) Solar-type xenon: A new isotopic composition of xenon in the Pesyanoe meteorite. *Earth Planet. Sci. Lett.*, **3**, 243–8.

Marti, K., & Craig, H. (1987) Cosmic-ray produced neon and helium in the summit lavas of Maui. *Nature*, **325**, 335–7.

Marty, B. (1989) Neon and xenon isotopes in MORB: Implications for the earth-atmosphere evolution. *Earth Planet. Sci. Lett.*, **94**, 45–56.

Marty, B. (1995) Nitrogen content of the mantle inferred from N_2-Ar correlation in oceanic basalts. *Nature*, **377**, 326–9.

Marty, B., & Jambon, A. (1987) C/^{3}He in volatile fluxes from the solid Earth: Implications for carbon geodynamics. *Earth Planet. Sci. Lett.*, **83**, 16–26.

Marty, B., & Le Cloarec, M.-F. (1992) Helium-3 and CO_2 fluxes from subaerial volcanoes estimated from polonium-210 emissions. *J. Volcanol. Geotherm. Res.*, **53**, 67–72.

Marty, B., & Lussiez, P. (1993) Constraints on rare gas partition coefficients from analysis of olivine-glass from a picritic mid-ocean ridge basalt. *Chem. Geol.*, **106**, 1–7.

Marty, B., & Ozima, M. (1986) Noble gas distribution in oceanic basalt glasses. *Geochim. Cosmochim. Acta*, **50**, 1093–7.

Marty, B., Torgersen, T., Meynier, V., O'Nions, R. K., & de Marsily, C. (1993) Helium isotope fluxes and groundwater ages in the Dogger Aquifer, Paris Basin. *Water Resources Res.*, **29**(4), 1025–35.

Marty, B., Zashu, S., & Ozima, M. (1983) Two noble gas components in a mid-Atlantic Ridge basalt. *Nature*, **302**, 238–40.

Masarik, J., & Reedy, R. C. (1995) Terrestrial cosmogenic-nuclide production systematics calculated from numerical simulations. *Geochim. Cosmochim. Acta*, **136**, 381–95.

Matsuda, J., & Marty, B. (1995) The $^{40}Ar/^{36}Ar$ ratio of the undepleted mantle: A reevaluation. *Geophys. Res. Lett.*, **22**, 1937–40.

Matsuda, J., & Matsubara, K. (1989) Noble gases in silica and their implications for the terrestrial "missing" Xe. *Geophys. Res. Lett.*, **16**, 81–4.

Matsuda, J., Fukunaga, K., & Ito, K. (1991) Noble gas studies in vapor-growth diamonds: Comparison with shock-produced diamonds and the origin of diamonds in ureilites. *Geochim. Cosmochim. Acta*, **55**, 2011–23.

Matsuda, J., Matsubara, K., Yajima, H., & Yamamoto, K. (1989) Anomalous Ne enrichment in obsidians and Darwin glass: Diffusion of noble gases in silicate-rich glasses. *Geochim. Cosmochim. Acta*, **53**, 3025–33.

Matsuda, J., Murota, M., & Nagao, K. (1990) He and Ne isotopic studies on the extraterrestrial material in deep-sea sediments. *J. Geophys. Res.*, **95**, 7111–17.

Matsuda, J., Sudo, M., Ozima, M., Ito, K., Ohtaka, O., & Ito, E. (1993) Noble gas partitioning between metal and silicate under high pressures. *Science*, **259**, 788–90.

Matsuda, J., Fukunaga, K., & Ito, K. (1991) Noble gas studies in vapor-growth diamonds: Comparison with shock-produced diamonds and the origin of diamonds in ureilites. *Geochim. Cosmochim. Acta*, **55**, 2011–23.

Matsui, T., & Abe, Y. (1986) Impact-induced oceans on Earth and Venus. *Nature*, **322**, 526–8.

Matsumoto, T., Honda, M., McDougall, I., Yatsevich, I., & O'Reilly, S. Y. (1997) Plume-like neon in a metasomatic apatite from the Australian lithospheric mantle. *Nature*, **388**, 162–4.

Matsumoto, T., Maruo, K., Tsuchiyama, A., & Matsuda, J. (1996) Occlusion of noble gases (He, Ne, Ar, Kr, Xe) into synthetic magnetite at 500–1300°C. *Earth Planet. Sci. Lett.*, **141**, 315–24.

Matsuo, S., & Miyake, Y. (1966) Gas composition in ice samples from Antarctica. *J. Geophys. Res.*, **71**, 5235–41.

Maurette, M., Olinger, C., Christophe Michel-Levy, M. C., Kurat, G., Pourchet, M., Brandstätter, F., & Bourot-Denise, M. (1991) A collection of diverse micrometeorites recovered from 100 tons of Antarctic blue ice. *Nature*, **351**, 44–7.

Mazor, E. (1972) Paleotemperatures and other hydrological parameters deduced from noble gases dissolved in ground waters; Jordan Rift Valley, Israel. *Geochim. Cosmochim. Acta*, **36**, 1321–36.

Mazor, E. (1975) Atmospheric and radiogenic noble gases in thermal waters: Their potential applications to prospecting and steam production studies, *Proc. 2nd UN Symp. Development and Use of Geothermal Resources.* San Francisco, pp. 793–802.

Mazor, E., & Fournier, R. O. (1973) More on noble gases in Yellowstone National Park hot waters. *Geochim. Cosmochim. Acta*, **37**, 515–25.

Mazor, E., & Verhagen, B. Th. (1983) Dissolved ions, stable and radioactive isotopes and noble gases in thermal waters of South Africa. *J. Hydrol.*, **63**, 315–29.

Mazor, E., & Wasserburg, G. J. (1965) Helium, neon, argon, krypton and xenon in gas emanations from Yellowstone and Lassen volcanic National Parks. *Geochim. Cosmochim. Acta*, **29**, 443–54.

Mazor, E., Heymann, D., & Anders, E. (1970) Noble gases in carbonaceous chondrites. *Geochim. Cosmochim. Acta*, **34**, 781–824.

Mazor, E., Wasserburg, G. J., & Craig, H. (1964) Rare gases in Pacific Ocean water. *Deep Sea Res.*, **11**, 922–32.

McConville, P., & Reynolds, J. H. (1989) Cosmogenic helium and volatile-rich fluid in Sierra Leone alluvial diamonds. *Geochim. Cosmochim. Acta*, **53**, 2365–75.

McConville, P., Reynolds, J. H., Epstein, S., & Roedder, E. (1991) Implanted ^{3}He, and Xe in further studies of diamonds from Western Australia. *Geochim. Cosmochim. Acta*, **55**, 1977–89.

McKenzie, D. (1979) Finite deformation during fluid flow. *Geophys. J. Roy. Astr. Soc.*, **58**, 689–715.

McKenzie, D., & Richter, F. M. (1981) Parametrized thermal convection in a layered region and the thermal history of the Earth. *J. Geophys. Res.*, **86**, 11667–80.

Meier-Reimer, E., Mikolajewicz, U., & Hasselmann, K. (1993) Mean circulation of the Hamburg LSG OGCM and its sensitivity to the thermocline surface forcing. *J. Phys. Oceanogr.*, **23**, 731–57.

Merrihue, C. (1964) Rare gas evidence for cosmic dust in modern Pacific red clay. *Ann. N. Y. Acad. Sci.*, **119**, 351–67.

Merrihue, C., & Turner, G. (1966) Potassium-argon dating by activation with fast neutrons. *J. Geophys. Res.*, **71**, 2852–7.

Merrill, R. T., & McElhinny, M. W. (1996) *The Magnetic Field of the Earth: Paleomagnetism, the Core, and the Deep Mantle*. New York: Academic Press.

Meshik, A. P., Kehm, K., & Hohenberg, C. M. (2000) Anomalous Xenon in zone 13 Okelobondo. *Geochim. Cosmochim. Acta*, **64**, 1651–61.

Meshik, A. P., Shukolyukov, Yu. A., & Jessberger, E. K. (1995) Chemically fractionated fission Xe on the Earth and in meteorites. In *Nuclei in the Cosmos*, pp. 603–6. New York: American Institute of Physics.

Messenger, S., & Walker, R. M. (1997) Evidence for molecular cloud material in meteorites and interplanetary dust. In *Astrophysical Implications of the Laboratory Study of Presolar Materials*, T. J. Bernatowicz & E. K. Zinner, Eds., pp. 545–66. AIP Conf. Proc. 402.

Mirtov, B. A. (1961) *Gaseous Composition of the Atmosphere and its Analysis* (in Russian). Moscow.

Mizuno, H., Nakazawa, K., & Hayashi, C. (1980) Dissolution of the primordial rare gases into the molten earth's material. *Earth Planet. Sci. Lett.*, **50**, 202–10.

Moreira, M., Kunz, J., & Allègre, C. J. (1997) Rare-gas systematics in the upper mantle: A complete study of a popping rock. *Seventh Annual V. M. Goldschmidt Conference*, p. 145.

Moreira, M., Kunz, J., & Allègre, C. (1998) Rare gas systematics in Popping Rock: Isotopic and Elemental Compositions in the Upper Mantle. *Science*, **279**, 1178–81.

Moreira, M., Staudacher, T., Sarda, P., Schilling, J.-G., & Allègre, C. J. (1995) A primitive plume neon component in MORB: the Shona ridge-anomaly, South Atlantic (51–52°S). *Earth Planet. Sci. Lett.*, **133**, 367–77.

Morgan, W. J. (1972) Plate motions and deep mantle convection. *Geol. Soc. Amer. Memoir*, **132**, 7–22.

Morris, J. D., Leemann, W. P., & Tera, F. (1990) The subducted component in island arc lavas: Constraints from the Be isotopes and B-Be systematics. *Nature*, **344**, 31–6.

Morrison, P., & Pine, J. (1955) Radiogenic origin of helium isotopes in rock. *Ann. N.Y. Acad. Sci.*, **62**, 71–9.

Morrison, T. J., & Johnstone, N. B. (1954) Solubilities of the inert gases in water. *J. Chem. Soc.*, **3**, 3441–6.

Murer, Ch. A., Baur, H., Signer, P., & Wieler, R. (1997) Helium, neon, and argon abundances in the solar wind: In vacuo etching of meteoritic iron-nickel. *Geochim. Cosmochim. Acta*, **61**, 1303–14.

Navon, O., Hutcheon, D., Rossman, G. R., & Wasserburg, G. J. (1988) Mantle-derived fluids in diamond micro-inclusions. *Nature*, **335**, 784–9.

Nichols, R. H., Jr., Hohenberg, C. M., & Olinger, C. T. (1994) Implanted solar helium, neon, and argon in individual lunar ilmenite grains: Surface effects and temporal variation in the solar wind composition. *Geochim. Cosmochim. Acta*, **58**, 1031–42.

Niedermann, S., Bach, W., & Erzinger, J. (1997) Noble gas evidence for a lower mantle in MORBs from the southern East Pacific Rise: Decoupling of helium and neon isotope systematics. *Geochim. Cosmochim. Acta*, **61**, 2697–715.

Niedermann, S., Graf, Th., & Marti, K. (1993) Mass spectroscopic identification of cosmic-ray-produced neon in terrestrial rocks with multiple neon components. *Earth Planet. Sci. Lett.*, **118**, 65–73.

Niedermann, S., Graf, Th., Kim, J. S., Kohl, C. P., Marti, K., & Nishiizumi, K. (1994) Cosmic-ray-produced ^{21}Ne in terrestrial quartz: The neon inventory of Sierra Nevada quartz separates. *Geochim. Cosmochim. Acta,* **125**, 341–55.

Nief, G. (1960) Isotopic abundance ratios reported for reference samples stocked at the National Bureau of Standards, F. Mohler, Ed. *NBS Technical Notes 51*.

Niemann, H. B., Atreya, S. K., Carignan, G. R., Donahue, T. M., Haberman, J. A., Harpold, D. N., Hartle, R. E., Hunten, D. M., Kasprzak, W. T., Mahaffy, P. R., Owen, T. C., Spencer, N. W., & Way, S. H. (1996) The Galileo probe mass spectrometer: Composition of Jupiter's atmosphere. *Science*, **272**, 846–9.

Nier, A. O. (1950a) A redetermination of the relative abundances of the isotopes of carbon, nitrogen, oxygen, argon and potassium. *Phys. Rev.*, **77**, 789–93.

Nier, A. O. (1950b) A redetermination of the relative abundances of the isotopes of neon, krypton, rubidium, xenon and mercury. *Phys. Rev.*, **79**, 450–4.

Nier, A. O., & Schlutter, D. J. (1990) Helium and neon isotopes in stratospheric particles. *Meteoritics*, **25**, 263–7.

Nier, A. O., Schlutter, D. J., & Brownlee, D. E. (1990) Helium and neon isotopes in deep Pacific Ocean sediments. *Geochim. Cosmochim. Acta*, **54**, 173–82.

Nittler, L. R. (1997) Presolar oxide grains in meteorites. In *Astrophysical Implications of the Laboratory Study of Presolar Materials*, T. J. Bernatowicz & E. K. Zinner, Eds., pp. 59–84. AIP Conf. Proc. 402.

Olinger, C. T., Maurette, M., Walker, R. M., & Hohenberg, C. M. (1990) Neon measurements of individual Greenland sediment particles: proof of an extraterrestrial origin and comparison with EDX and morphological analyses, *Earth Planet. Sci. Lett.*, **100**, 77–93.

O'Nions, R. K. (1987) Relationships between chemical and conventive layering in the Earth. *J. Geol. Soc. London*, **144**, 259–74.

O'Nions, R. K., & Oxburgh, E. R. (1983) Heat and helium in the Earth. *Nature*, **306**, 429–31.

O'Nions, R. K., & Oxburgh, E. R. (1988) Helium, volatile fluxes and the development of continental crust. *Earth Planet. Sci. Lett.*, **90**, 331–47.

O'Nions, R. K., & Tolstikhin, I. N. (1994) Behaviour and residence times of lithophile and rare gas tracers in the upper mantle. *Earth Planet. Sci. Lett.*, **124**, 131–8.

O'Nions, R. K., & Tolstikhin, I. N. (1996) Limits on the mass flux between lower and upper mantle and stability of layering. *Earth Planet. Sci. Lett.*, **139**, 213–22.

O'Nions, R. K., Carter, S. R., Evensen, N. M., & Hamilton, P. J. (1983) Upper-mantle geochemistry. In *The Sea*, vol. 7, C. Emiliani, Ed., pp. 49–71. New York: Wiley-Interscience.

Owen, T., Bar-Nun, A., & Kleinfeld, I. (1991) Noble gases in terrestrial planets: Evidence for cometary impacts? In *Comets in the Post-Halley Era*, R. L. Newburn, Jr., et al., Eds., vol. 1, pp. 429–37. Kluwer Academic Publishers.

Owen, T., Bar-Nun, A., & Kleinfeld, I. (1992) Possible cometary origin of heavy noble gases in the atmospheres of Venus, Earth, and Mars. *Nature*, **358**, 43–6.

Oxburgh, E. R. (1991) Helium, plumes, and mantle reservoirs. In *Hotspot Handbook*, D. L. Anderson, Ed. Pasadena: Caltech.

Oxburgh, E. R., & O'Nions, R. K. (1987) Helium loss, tectonics, and the terrestrial heat budget. *Science*, **237**, 1583–8.

Ozima, M. (1989) Gases in diamonds. *Ann. Rev. Earth Planet. Sci.*, **17**, 361–84.

Ozima, M. (1990) Comment on "An important source of ^{4}He (and ^{3}He) in diamonds" by D. Lal. *Earth Planet. Sci. Lett.*, **101**, 107–9.

Ozima, M., & Alexander, Jr., E. C. (1976) Rare gas fractionation patterns in terrestrial samples and the earth-atmosphere evolution model. *Rev. Geophys. Space Phys.*, **14**, 385–90.

Ozima, M., & Igarashi, G. (2000) The Primordial noble gases in the Earth: A key constraint on Earth evolution models. *Earth Planet. Sci. Lett.*, **176**, 219–32.

Ozima, M., & Kudo, K. (1972) Excess argon in submarine basalts and an earth-atmosphere evolution model. *Nature Phys. Sci.*, **239**, 23–4.

Ozima, M., & Nakazawa, K. (1980) Origin of rare gases in the Earth. *Nature*, **284**, 313–6.

Ozima, M., & Podosek, F. A. (1983) *Noble Gas Geochemistry*. p. 367. Cambridge, UK: Cambridge University Press.

Ozima, M., & Podosek, F. A. (1999) Formation age of the earth from ^{129}I/^{127}I and ^{244}Pu/^{238}U systematics and the missing Xe. *J. Geophys. Res.*, **104**(B11), 25493–9.

Ozima, M., & Takigami, Y. (1980) Activation energy for thermal release of Ar from some DSDP submarine rocks. *Geochim. Cosmochim. Acta*, **44**, 141–4.

Ozima, M., & Tatsumoto, M. (1997) Radiation-induced diamond crystallization: Origin of carbonados and its implications on meteorite nano-diamonds. *Geochim. Cosmochim. Acta*, **61**, 369–76.

Ozima, M., & Wada, N. (1993) Noble gases in atmospheres. *Nature*, **361**, 693.

Ozima, M., & Zahnle, K. (1993) Mantle degassing and atmospheric evolution: Noble gas view. *Geochem. J.*, **27**, 185–200.

Ozima, M., & Zashu, S. (1983) Noble gases in submarine pillow volcanic glasses. *Earth Planet. Sci. Lett.*, **62**, 24–40.

Ozima, M., & Zashu, S. (1988) Solar-type Ne in Zaire cubic diamonds. *Geochim. Cosmochim. Acta*, **52**, 19–25.

Ozima, M., & Zashu, S. (1991) Noble gas state in the ancient mantle as deduced from noble gases in coated diamonds. *Earth Planet. Sci. Lett.*, **105**, 13–27.

Ozima, M., Podosek, F. A., & Igarashi, G. (1985) Terrestrial xenon isotope constraints on the early history of the Earth. *Nature*, **315**, 471–4.

Ozima, M., Takayanagi, M., Zashu, S., & Amari, S. (1984) High ^{3}He/^{4}He ratio in ocean sediments. *Nature*, **311**, 448–50.

Ozima, M., Wieler, R., Marty, B., & Podosek, F. A. (1998) Comparative studies of solar, Q-gas and terrestrial noble gases, and implications on the evolution of the solar nebula. *Geochim. Cosmochim. Acta*, **62**, 301–14.

Ozima, M., Zashu, S., & Nitoh, O. (1983) ^{3}He/^{4}He ratio, noble gas abundance and K-Ar dating of diamonds – An attempt to search for the record of early terrestrial history. *Geochim. Cosmochim. Acta*, **47**, 2217–24.

Ozima, M., Zashu, S., Takigami, Y., & Turner, G. (1989) Origin of the anomalous ^{40}Ar-^{39}Ar age of Zaire cubic diamonds: Excess ^{40}Ar in pristine mantle fluids. *Nature*, **337**, 226–9.

Patterson, D. B., & Farley, K. A. (1998) Extraterrestrial ^{3}He in seafloor sediments: Evidence for correlated 100 kyr periodicity in the accretion rate of interplanetary dusts, orbital parameters, and Quaternary climate. *Geochim. Cosmochim. Acta*, **62**, 3669–82.

Patterson, D. B., Honda, M., & McDougall, I. (1990) Atmospheric contamination: A possible source for heavy noble gases from Loihi seamount, Hawaii. *Geophys. Res. Lett.*, **17**, 705–8.

Pedroni, A., & Begemann, F. (1994) On unfractionated solar gases in the H3-6 meteorite Acfer 111. *Meteoritics*, **29**, 632–42.

Pepin, R. O. (1991) On the origin and early evolution of terrestrial planet atmospheres and meteoritic volatiles. *Icarus*, **92**, 2–79.

Pepin, R. O. (1997) Evolution of earth's noble gases: Consequence of assuming hydrodynamic loss driven by giant impact. *Icarus*, **126**, 148–56.

Pepin, R. O. (1998) Isotopic evidence for a solar argon component in the Earth's mantle. *Nature*, **394**, 684–7.

Pepin, R. O., & Signer, P. (1965) Primordial rare gases in meteorites. *Science*, **149**, 253–65.

Pepin, R. O., & Phinney, D. (1978) Components of xenon in the solar system. Unpublished preprint, University of Minnesota.

Pepin, R. O., Becker, R. H., & Rider, P. E. (1995) Xenon and krypton isotopes in extraterrestrial regolith soils and in the solar wind. *Geochim. Cosmochim. Acta*, **59**, 4997–5022.

Pepin, R. O., Reynolds, J. H., & Turner, G. (1964) Shock emplaced argon in a stony meteorite, 2: A comparison with natural argon in its diffusion. *J. Geophys. Res.*, **29**, 1406–11.

Phinney, D. (1972) ^{36}Ar, Kr and Xe in terrestrial materials. *Earth Planet. Sci. Lett.*, **16**, 413–20.

Phinney, D., Tennyson, J., & Frick, U. (1978) Xenon in CO_2 well gas revisited. *J. Geophys. Res.*, **83**, 2313–19.

Pierotti, R. A. (1963) The solubility of gases in liquids. *J. Phys. Chem.*, **67**, 1840–5.

Pinti, D. L., Marty, B., & Andrew, J. N. (1997) Atmosphere-derived noble gas evidence for the preservation of ancient waters in sedimentary basins. *Geology*, **25**, 111–14.

Podosek, F. A. (1970) The abundance of ^{244}Pu in the early solar system. *Earth Planet. Sci. Lett.*, **8**, 183–7.

Podosek, F. A. (1972) Gas retention chronology of Petersburg and other meteorites. *Geochim. Cosmochim. Acta*, **36**, 755–72.

Podosek, F. A. (1978) Isotopic structure in solar system materials. *Ann. Rev. Astron. Astrophys.*, **16**, 293–334.

Podosek, F. A. (1999) A couple of uncertain age. *Science*, **283**, 1863–4.

Podosek, F. A., & Nichols, Jr., R. H. (1997) Short-lived radionuclides in the solar nebula. In *Astrophysical Implications of the Laboratory Study of Presolar Materials*, J. T. Bernatowicz & E. K. Zinner, Eds., pp. 617–47. *AIP Proc. Conf.*, 402.

Podosek, F. A., & Ozima, M. (1999) The Xenon Age of the Earth. In *Origin of Earth and Moon*, R. M. Canup & K. J. Righter, Eds., pp. 63–72. Tucson: University of Arizona Press.

Podosek, F. A., Bernatowicz, T. J., & Kramer, F. E. (1981a) Adsorption and excess fission xenon. *Proc. Lunar Planet. Sci.*, **12B**, 891–901.

Podosek, F. A., Bernatowicz, T., & Kramer, F. E. (1981b) Adsorption of xenon and krypton on shales. *Geochim. Cosmochim. Acta*, **45**, 2401–15.

Podosek, F. A., Brannon, J. C., Bernatowicz, J. T., Brazzle, R., Grauch, R., Cowsik, R., & Hohenberg, C. M. (1994) Geochronology of tellurium ores and the double-beta decay lifetime of ^{130}Te. In *Noble Gas Geochemistry and Cosmochemistry*, J. Matsuda, Ed., pp. 89–113, Tokyo: Terra Scientific Publishing Co.

Podosek, F. A., Honda, M., & Ozima, M. (1980) Sedimentary noble gases. *Geochim. Cosmochim. Acta*, **44**, 1875–84.

Podosek, F. A., Huneke, J. C., Burnett, D. S., & Wasserburg, G. J. (1971) Isotopic composition of xenon and krypton in the lunar soil and in the solar wind. *Earth Planet. Sci. Lett.*, **10**, 199–216.

Polak, B. G., Kononov, V. I., Tolstikhin, I. N., Mamyrin, B. A., & Khabarin, L. V. (1975) The helium isotopes in thermal fluids. In *Terrestrial and Chemical Problems of Thermal Waters*, Publication No. 119, A. I. Johnson, Ed., pp. 15–29. Grenoble: International Association of Hydrological Science.

Poole, J. C., McNeil, G. W., Langman, S. R., & Dennis, F. (1997) Analysis of noble gases in water using a quadrupole mass spectrometer in static mode. *Appl. Geochem.*, **12**, 707–14.

Porcelli, D., & Wasserburg, G. J. (1995a) Mass transfer of xenon through a steady-state upper mantle. *Geochim. Cosmochim. Acta*, **59**, 1991–2007.

Porcelli, D., & Wasserburg, G. J. (1995b) Mass transfer of helium, neon, argon, and xenon through a steady-state upper mantle. *Geochim. Cosmochim Acta*, **59**, 4921–37.

Poreda, R. J., & Cerling, T. E. (1992) Cosmogenic neon in recent lavas from the western United States. *Geophys. Res. Lett.*, **19**, 1863–6.

Poreda, R., & Craig, H. (1989) Helium isotope ratios in circum-Pacific volcanic arcs. *Nature*, **338**, 473–8.

Poreda, R. J., & Farley, K. A. (1992) Rare gases in Samoan xenoliths. *Earth Planet. Sci. Lett.*, **113**, 129–44.

Poreda, R., & Radicati di Brozolo, F. (1984) Neon isotope variations in Mid-Atlantic Ridge basalts. *Earth Planet. Sci. Lett.*, **69**, 277–89.

Potter, R. W., II, & Clynne, M. A. (1978) The solubility of noble gases He, Ne, Ar, Kr, and Xe in water up to the critical point. *J. Sol. Chem.*, **7**, 837–44.

Pyykkö, P. (1997) Strong closed-shell interactions in inorganic chemistry. *Chem. Rev.*, **97**, 597–636.

Ragettli, R. A., Hebeda, E. H., Signer, P., & Wieler, R. (1994) Uranium-xenon chronology: Precise determination of $\lambda_{sf}^{136}Y_{sf}$ for spontaneous fission of ^{238}U. *Earth Planet. Sci. Lett.*, **128**, 653–70.

Rajan, R. S., Brownlee, D. E., Tomandle, D., Hodge, P. W., Farrar, H., & Britten, R. A. (1977) Detection of stratospheric particles gives evidences of extraterrestrial origin. *Nature*, **267**, 133–4.

Rayleigh, J. W. S. (1933) Beryllium and helium I. The helium contained in beryls of different geologic age. *Proc. Roy. Soc. London Ser. A*, **142**, 370–81.

Reiss, H. R., Frisch, H. L., Helfand, E., & Lebowitz, J. L. (1960) Aspects of the statistical thermodynamics of real fluids. *J. Chem. Phys.*, **32**, 119–24.

Revelle, R., & Suess, H. E. (1962) Interchange of properties between sea and atmosphere. In *The Sea*, vol. 1, M. N. Hill, Ed., New York: Wiley (Interscience).

Reynolds, J. H. (1956) High-sensitivity mass spectrometer for noble gas analysis. *Rev. Scientific Instruments*, **27**, 928–34.

Reynolds, J. H. (1960) Determination of the age of the elements. *Phys. Rev. Lett.*, **4**, 8.

Reynolds, J. H., & Turner, G. (1964) Rare gases in chondrite Renazzo. *J. Geophys. Res.*, **69**, 3263–81.

Richardson, S. H., Gurney, J. J., Earlang, A. J., & Harris, J. W. (1984) Origin of diamonds in old enriched mantle. *Nature*, **310**, 198–202.

Richardson, S. H., Harris, J. W., & Gurney, J. J. (1993) Three generations of diamonds from old continental mantle. *Nature*, **366**, 256–8.

Rison, W. (1980a) Isotopic fractionation of argon during stepwise release from shungite. *Earth Planet. Sci. Lett.*, **47**, 383–90.

Rison, W. (1980b) Isotopic studies of the rare gases in igneous rocks: implications for the mantle and the atmosphere, PhD Thesis, University of California, Berkeley.

Robert, F., Gautier, D., & Dubrulle, B. (2000) The solar system D/H ratio: Observations and theories. *Space Sci. Rev.* **92**, 201–24.

Rocholl, A., Heusser, E., Kirsten, T., Oehm, J., & Richter, H. (1996) A noble gas profile across a Hawaiian mantle xenolith: Mantle-derived cognate and accidental noble gas components and evidence for anomalous krypton isotopes. *Geochim. Cosmochim. Acta*, **60**, 4773–83.

Roselieb, K., Rammensee, W., Bütter, H., & Rosenhauer, M. (1992) Solubility and diffusion of noble gases in vitreous albite. *Chem. Geol.*, **96**, 241–66.

Roselieb, K., Rammensee, W., Bütter, H., & Rosenhauer, M. (1995) Diffusion of noble gases in melts of the system SiO_2-$NaAlSi_2O_6$. *Chem. Geol.*, **120**, 1–13.

Ross, S., & Olivier, J. P. (1964) *On Physical Adsorption*. New York: Wiley (Interscience).

Rubey, W. W. (1951) Geologic history of sea water. *Geolog. Soc. Amer. Bull.*, **62**, 1111–48.

Safronov, V. S. (1969) Evolution of the protoplanetary cloud and formation of the earth and planets (translated into English). *NASA, TFF-677*.

Saito, K., Alexander, E. C., Jr., Dragon, J. C., & Zashu, S. (1984) Rare gases in cyclosilicates and cogenetic minerals. *J. Geophys. Res.*, **89**, 7891–901.

Sano, Y. (1986) Helium flux from the solid Earth. *Geochem. J.*, **20**, 227–32.

Sano, Y., & Wakita, H. (1987) Island arc tectonics of New Zealand manifested in helium isotope ratios. *Geochim. Cosmochim. Acta*, **51**, 1855–69.

Sano, Y., Wakita, H., & Huang, C.-W. (1986) Helium flux in a continental land area estimated from $^3He/^4He$ ratio in northern Taiwan. *Nature*, **323**, 55–7.

Sano, Y., Wakita, H., Makide, Y., & Tominaga, T. (1989) A ten-year decrease in the atmospheric helium isotopic ratio possibly caused by human activity. *Geophys. Res. Lett.*, **16**, 1371–4.

Sarda, P., & Graham, D. (1990) Mid-ocean ridge popping rocks: Implications for degassing at ridge crests. *Earth Planet. Sci. Lett.*, **97**, 268–89.

Sarda, P., Moreira, M., & Staudacher, Th. (1999) Argon-lead isotope correlation in mid-Atlantic Ridge basalts. *Science*, **283**, 666–9.

Sarda, P., Staudacher, Th., & Allègre, C. J. (1988) Neon isotopes in submarine basalts. *Earth Planet. Sci. Lett.*, **91**, 73–88.

Sasaki, S. (1991) Off-disk penetration of ancient solar wind. *Icarus*, **91**, 29–38.

Sasaki, S., & Nakazawa, K. (1988) Origin of isotopic fractionation of terrestrial Xe: Hydrodynamic fractionation during escape of the primordial H_2-He atmosphere. *Earth Planet. Sci. Lett.*, **89**, 323–34.

Saunders, M., Cross, R. J., Hugo, A., Jiménez-Vázquez, R., Shimshi, R., & Khong, A. (1996) Noble gas atoms inside fullerenes. *Science*, **271**, 1693–7.

Saunders, M., Hugo, A., Jiménez-Vázquez, R., Cross, R. J., & Poreda, R. J. (1995) Stable compounds of helium and neon: He@C_{60} and Ne@C_{60}. *Science*, **259**, 1428–31.

Schäfer, J. M., Ivy-Ochs, S., Wieler, R., Leya, I., Baur, H., Denton, G. H., & Schlüchter, C. (1999) Cosmogenic noble gas studies in the oldest landscape on earth: Surface exposure ages of the Dry Valleys, Antarctica. *Earth Planet. Sci. Lett.*, **167**, 215–26.

Schlosser, P., Stute, M., Dorr, H., Sonntag, C., & Munnich, K. O. (1988) Tritium/^{3}He dating of shallow groundwater. *Earth Planet. Sci. Lett.*, **89**, 353–62.

Schwartzmann, D. W. (1973a) Argon degassing model of the Earth. *Nature Phys. Sci.*, **245**, 20–1.

Schwartzmann, D. W. (1973b) Argon degassing and the origin of the sialic crust. *Geochim. Cosmochim. Acta*, **37**, 2479–95.

Sears, D. W. G., & Dodd, R. T. (1988) Overview and classification of meteorites. In *Meteorites and the Early Solar System*, J. F. Kerridge & M. S. Matthews, Eds., pp. 3–31. Tucson: University of Arizona Press.

Segre, E. (1952) Spontaneous fission. *Phys. Rev.*, **86**, 21–8.

Sekiya, M., Nakazawa, K., & Hayashi, C. (1980) Dissipation of the Primordial terrestrial atmosphere due to irradiation of the solar EUV. *Progr. Theoret. Phys.*, **64**, 1968–85.

Sheldon, W. R., & Kern, J. W. (1972) Atmospheric helium and geomagnetic field reversals. *J. Geophys. Res.*, **77**, 6194–201.

Shibata, T., Takahashi, E., & Matsuda, J. (1996) Noble gas solubility in binary Cao-SiO_2 system. *Geophys. Res. Lett.*, **23**, 3139–42.

Shibata, T., Takahashi, E., & Matsuda, J. (1998) Solubility of neon, argon, krypton, and xenon in binary and ternary silicate systems: A new view on noble gas solubility. *Geochim. Cosmochim. Acta*, **62**, 1241–53.

Shibata, T., Takahashi, E., & Ozima, M. (1994) Noble gas partition between basaltic melt and olivine crystals at high pressure. In *Noble Gas Geochemistry and Cosmochemistry*, J. Matsuda, Ed., pp. 343–54. Tokyo: Terra Scientific Publishing Co.

Shukolyukov, Yu. A., Ashkinadze, G. Sh., & Verkhovskii, A. B. (1976) Anomalous isotopic composition of xenon and krypton in minerals of the natural nuclear reactor. *Soviet Atomic Energy*, **41**(1), 663–6.

Shukolyukov, Yu. A., Jessberger, E. K., Meshik, A. P., Dang Vu Minh, & Jordan, J. L. (1994) Chemically fractionated fission-xenon in meteorites and on the Earth. *Geochim. Cosmochim. Acta*, **58**, 3075–92.

Shukolyukov, Yu. A., Meshek, A. P., Krylov, D. P., & Pravdivsteva, O. V. (1994) Current status of Xe_s-Xe_f dating. In *Noble Gas Geochemistry*, J. Matsuda, Ed., pp. 125–46. Tokyo: Terra Scientific Publishing Co.

Signer, P., & Suess, H. (1963) Rare gases in the sun, in the atmosphere, and in meteorites. In *Earth Science and Meteorites*, J. Geiss & E. D. Goldberg, Eds., pp. 241–72. Amsterdam: North Holland Publ. Co.

Signer, P., Baur, H., & Wieler, R. (1993) Closed system stepped etching; An alternative to stepped heating. In *Proceedings of the Alfred O. Nier Symposium on Inorganic Mass Spectrometry*, D. J. Rokop, Ed., pp. 181–202. U.S. Department of Energy.

Sigurgeirsson, Th. (1962) Dating recent basalt by the potassium argon method. Report of Physical Laboratory of the University of Iceland. (In Icelandic.)

Sill, G., & Wilkening, L. (1978) Ice clathrates as a possible source of the atmospheres of the terrestrial planets. *Icarus*, **33**, 13–22.

Smith, S. P., & Kennedy, B. M. (1983) The solubility of noble gases in water and NaCl brine. *Geochim. Cosmochim. Acta*, **47**, 503–15.

Smith, S. P., & Reynolds, J. H. (1981) Excess ^{129}Xe in a terrestrial sample as measured in a pristine system. *Earth Planet. Sci. Lett.*, **54**, 236–8.

Srinivasan, B. (1976) Barites: Anomalous xenon from spallation and neutron-induced reactions. *Earth Planet. Sci. Lett.*, **31**, 129–41.

Srinivasan, B., Alexander, Jr., E. C., & Manuel, O. K. (1971) Iodine-129 in terrestrial ores. *Science*, **173**, 327–8.

Stacey, F. D. (1992) *Physics of the Earth*, 3rd ed. Brisbane: Brookfield Press.

Staudacher, T., & Allègre, C. J. (1982) Terrestrial xenology. *Earth Planet. Sci. Lett.*, **60**, 389–406.

Staudacher, T., & Allègre, C. J. (1988) Recycling of oceanic crust and sediments: The noble gas subduction barrier. *Earth. Planet. Sci. Lett.*, **89**, 173–83.

Staudacher, T., & Allègre, C. (1991) Cosmogenic neon in ultramafic nodules from Asia and in quartzite from Antarctica. *Earth Planet. Sci. Lett.*, **107**, 87–102.

Staudacher, T., & Srada, P. (1987) Comment on "Radiogenic rare gases and the evolutionary history of the depleted mantle" by D. E. Fisher. *J. Geophys. Res.*, **92**(B3), 2808–12.

Staudacher, T., Jessberger, E. K., Dominik, B., Kirsten, T., & Shaffer, O. A. (1982) ^{40}Ar-^{39}Ar ages of rocks and glasses from the Nordlinger Ries Crater and the temperature history of impact breccias. *J. Geophys.*, **51**, 11.

Staudacher, T., Jessberger, E. K., Dorflinger, D., & Kiko, J. (1978) A refined ultrahigh-vacuum furnace for rare gas. *J. Phys. E: Sci. Instrum.*, **11**, 781–4.

Staudacher, Th. (1987) Upper mantle origin for Harding County well gases. *Nature*, **325**, 605–7.

Staudacher, Th., Kurz, M. D., & Allègre, C. J. (1986) New Noble-gas data on glass samples from Loihi seamount and Hualalai and dunite samples from Loihi and Réunion Island. *Chem. Geol.*, **56**, 193–205.

Staudacher, Th., Sarda, P., & Allègre, C. J. (1990) Noble gas systematics of Reunion Island, Indian Ocean. *Chem. Geol.*, **89**, 1–17.

Staudacher, Th., Sarda, Ph., Richardson, S. H., Allègre, C. J., Sagna, I., & Dmitriev, L. V. (1989) Noble gases in basalt glasses from a Mid-Atlantic ridge topographic high at 14°N geodynamic consequences. *Earth Planet. Sci. Lett.*, **96**, 119–33.

Steiger, R. H., & Jäger, E. (1977) Subcommission on geochronology: Convention on the use of decay constants in geo- and cosmochronology, *Earth Planet. Sci. Lett.*, **36**, 359–62.

Stevenson, D. J. (1985) Partitioning of noble gases at extreme pressures within planets. *Lunar Planet. Sci.*, **XVI**, 821–2.

Stone, E. C. (1989) Solar abundances as derived from solar energetic particles. *AIP Conf. Proc.*, **183**, 72–90.

Strutt, R. J. (1908) Helium and radioactivity in rare and common minerals. *Proc. Roy. Soc. London Ser. A*, **80**, 572–94.

Stuart, F. M. (1994) Comments on "Speculations about the cosmic origin of He and Ne in the interior of the Earth." *Earth Planet. Sci. Lett.*, **122**, 170–4.

Stuart, F. M., Harrop, P. J., Knott, S., & Turner, G. (1999) Laser extraction of helium isotopes from Antarctic micrometeorites: Source of He and implications for the flux of extraterrestrial ^{3}He to earth. *Geochim. Cosmochim. Acta*, **63**, 2653–65.

Stute, M., & Schlosser, P. (1993) Principles and applications of the noble gas paleothermometer. In *Climate change in continental isotopic records. Geophysical Monograph*, **78**, 89–100.

Stute, M., Forster, H., Frischkorn, H., Serejo, A., Clark, J. F., Schlosser, P., Broecker, W. S., & Bonati, G. (1995) Cooling of tropical Brazil (5°C) during the last glacial maximum. *Science*, **269**, 379–83.

Stute, M., Schlosser, P., Clark, J. F., & Broecker, W. S. (1992) Paleotemperature in the southwestern United States derived from noble gases in ground water. *Science*, **256**, 1000–3.

Stute, M., Sonntag, C., Deák, J., & Schlosser, P. (1992) Helium in deep circulating ground water in the Great Hungarian Plain: Flow dynamics and crustal and mantle helium flux. *Geochim. Cosmochim. Acta*, **56**, 2051–67.

Sudo, M., Ohtaka, O., & Matsuda, J. (1994) Noble gas partitioning between metal and silicate under high pressures: The case of iron and peridotite. In *Noble Gas Geochemistry and Cosmochemistry*, J. Matsuda, Ed., pp. 217–27. Tokyo: Terra Scientific Publ. Co.

Suess, H. (1949) Die Haufigkeit der Edelgase auf der Erde und in Kosmos. *J. Geol.*, **57**, 600–7.

Suess, H. E., Wänke, H., & Wlotzka, F. (1964) On the origin of gas-rich meteorites. *Geochim. Cosmochim. Acta*, **28**, 595–607.

Swindle, T. D., & Podosek, F. A. (1988) Iodine-xenon dating. In *Meteorites and the Early Solar System*, J. F. Kerridge & M. S. Matthews, Eds., pp. 1127–46. Tucson: University of Arizona Press.

Swindle, T. D., Caffee, M. W., Hohenberg, C. M., Lindstrom, M. M., & Taylor, G. J. (1991) Iodine-xenon studies of petrographically and chemically characterized Chainpur chondrules. *Geochim. Cosmochim. Acta*, **55**, 861–80.

Takagi, J., Hampel, W., & Kirsten, T. (1974) Cosmic-ray muon-induced ^{129}I in tellurium ores. *Earth Planet. Sci. Lett.*, **24**, 141–50.

Takaoka, N. (1972) An interpretation of general anomalies of xenon and the isotopic composition of primitive xenon. *Mass Spectrometry*, **20**, 287–302.

Takaoka, N., & Mizutani, Y. (1987) Tritiogenic ^{3}He in ground water in Takaoka. *Earth Planet. Sci. Lett.*, **85**, 74–8.

Takaoka, N., & Nagao, K. (1980) Rare-gas studies of Cretaceous deep-sea basalts. *Initial Report of the Deep Sea Drilling Project*, LI, LII, LIII, pp. 1121–6.

Takaoka, N., Motomura, Y., & Nagao, K. (1996) Half-life of ^{130}Te double-β decay measured with geologically qualified samples. *Phys. Rev. C*, **53**, 1557–61.

Takayanagi, M., & Ozima, M. (1987) Temporal variation of ^{3}He/^{4}He ratio recorded in deep-sea sediment cores. *J. Geophys. Res.*, **92**(B12), 12531–8.

Talwani, M., Windisch, C. C., & Langseth, Jr., M. G. (1971) Reykjanes Ridge crest: A detailed geophysical study. *J. Geophys. Res.*, **76**, 473–517.

Taylor, S. R., & McLennan, S. M. (1985) *The Continental Crust: Its Composition and Evolution*. Oxford: Blackwell Scientific Publications.

Teitsma, A., & Clarke, W. B. (1978) Fission xenon isotope dating. *J. Geophys. Res.*, **83**, 5443–53.

Thomson, J. J. (1912) Further experiments on positive rays. *Philosophical Magazine*, **24**, 209–53.

Tolstikhin, I. N. (1975) Helium isotopes in the earth's interior and in the atmosphere: A degassing model of the earth. *Earth Planet. Sci. Lett.*, **26**, 88–96.

Tolstikhin, I. N. (1978) A review: Some recent advances in isotope geochemistry of light noble gases. In *Terrestrial Noble Gases*, E. C. Alexander, Jr. & M. Ozima, Eds., pp. 33–62. Tokyo: Japan Scientific Societies Press.

Tolstikhin, I. N., & O'Nions, R. K. (1994) The Earth's missing xenon: A combination of early degassing and of rare gas loss from the atmosphere. *Chem. Geol.*, **115**, 1–6.

Tolstikhin, I. N., Lehmann, B. E., Loosli, H. H., Kamensky, L. L., Nivin, V. A., Orlov, S. P., Ploschansky, L. M., Tokarev, I. V., & Gannibal, M. A. (1999) Radiogenic helium isotope fractionation: The role of tritium as ^{3}He precursor in geochemical application. *Geochim. Cosmochim. Acta*, **63**, 1605–11.

Toramaru, A. (1989) Vesciculation process and bubble size distributions in ascending magmas with constant velocities. *J. Geophys. Res.*, **94**, 17523–42.

Torgersen, T. (1980) Controls of pore-fluid concentration of ^{4}He and ^{222}Rn and the calculation of ^{4}He/^{222}Rn ages. *J. Geochem. Expl.*, **13**, 57–75.

Torgersen, T. (1992) Controls of pore-filled concentration of ^{4}He and ^{232}Rn and the calculation of ^{4}He/^{222}Rn ages. *J. Geochem. Expl.*, **13**, 57–75.

Torgersen, T., & Clarke, W. B. (1985) Helium accumulation in groundwater, I: An evaluation of sources and the continental flux of crustal ^{4}He in the Great Altesian Basin, Australia. *Geochim. Cosmochim. Acta*, **49**, 1211–18.

Torgersen, T., & Clarke, W. B. (1992) Geochemical constraints on formation fluid ages, hydrothermal heat flux, and crustal mass transport mechanisms at Cajon Pass. *J. Geophys. Res.*, **97**(B4), 5031–8.

Torgersen, T., & O'Donnel, J. (1991) The degassing flux from the solid Earth: Release by fracturing. *Geophys. Res. Lett.*, **18**, 951–4.

Toyoda, S., & Ozima, M. (1988) Investigation of excess ^{4}He and ^{40}Ar in beryl by laser extraction technique. *Earth Planet. Sci. Lett.*, **90**, 69–76.

Trieloff, M., Kunz, J., Clague, D., Harrison, D., & Allègre, C. J. (2000) The nature of pristine noble gases in mantle plumes. *Science*, **288**, 1036–8.

Trull, T. (1994) Influx and age constraints on the recycled cosmic dust explanation for high ^{3}He/^{4}He ratios at hot spot volcanoes. In *Noble Gas Geochemistry and Cosmochemistry*, J. Matsuda, Ed., pp. 77–88. Tokyo: Terra Publication.

Trull, T. W., & Kurz, M. D. (1993) Experimental measurements of ^{3}He and ^{4}He mobility in olivine and clinopyroxene at magmatic temperatures. *Geochim. Cosmochim. Acta*, **57**, 1313–24.

Trull, T. W., Brown, E. T., Marty, B., Raisbeck, G. M., & Yiou, F. (1995) Cosmogenic ^{10}Be and ^{3}He accumulation in Pleistocene beach terraces in Death Valley, California, U.S.A.: Implications for cosmic-ray exposure dating of young surfaces in hot climates. *Chem. Geol.*, **119**, 191–207.

Trull, T. W., Kurz, M. D., & Jenkins, W. J. (1991) Diffusion of cosmogenic ^{3}He in olivine and quartz: Implications for surface exposure dating. *Earth Planet. Sci. Lett.*, **103**, 241–56.

Trull, T., Nadeau, S., Pineau, F., Polve, M., & Javoy, M. (1993) C-He systematics in hotspot xenoliths: Implications for mantle carbon contents and carbon recycling. *Earth Planet. Sci. Lett.*, **118**, 43–64.

Turekian, K. K. (1964) Degassing of argon and helium from the earth. In *The Origin and Evolution of Atmosphere and Ocean*, P. J. Brancazio & A. G. W. Cameron, Eds., pp. 74–82. New York: Wiley.

Turner, G. (1988) Hydrothermal fluids and argon isotopes in quartz and veins and cherts. *Geochim. Cosmochim. Acta*, **52**, 1443–8.

Turner, G., Miller, J. A., & Grastry, R. L. (1966) Thermal history of the Bruderheim meteorite. *Earth Planet. Sci. Lett.*, **1**, 155–62.

Tyburczy, J. A., Frisch, B., & Ahrens, T. J. (1986) Shock-induced volatile loss from a carbonaceous chondrite: Implications for planetary accretion. *Earth Planet. Sci. Lett.*, **80**, 201–7.

Valbracht, P. J., Honda, M., Matsumoto, T., Mattielli, N., McDougall, I., Ragettli, R., & Weis, D. (1996) Helium, neon and argon isotope systematics in Kerguelen ultramafic xenoliths: Implications for mantle source signatures. *Earth Planet. Sci. Lett.*, **138**, 29–38.

Valbracht, P. J., Honda, M., Staudigel, H., McDougall, I., & Trost, A. P. (1994) Noble gas partitioning in natural samples: Results from coexisting glass and olivine phenocrysts in four Hawaiian submarine basalts. In *Noble Gas Geochemistry*, J. Matsuda, Ed., pp. 371–81. Tokyo: Terra Scientific Publishing Co.

Valbracht, P. J., Staudacher, T., Malahoff, A., & Allègre, C. J. (1997) Noble gas systematics of deep rift zone glasses from Loihi Seamount, Hawaii. *Earth Planet. Sci. Lett.*, **150**, 399–411.

Van der Hilst, R. D., & Karason, H. (1999) Compositional heterogeneity in the bottom 1000 kilometers of Earth's mantle: Toward a hybrid convection model. *Science* **283**, 1885–8.

Van der Hilst, R. D., Widiyantoro, S., & Engdahl, E. R. (1997) Evidence for deep mantle circulation from global tomography. *Nature*, **386**, 578–84.

Van Soest, M. C., Hilton, R. D., & Kreulen, R. (1998) Tracing crustal and slab contributions to arc magmatism in the Lesser Antilles island arc using helium and carbon relationships in geothermal fluids. *Geochim. Cosmochim. Acta*, **62**, 3323–35.

Verniani, F. (1966) Total mass of the earth's atmosphere. *J. Geophys. Res.*, **71**, 835–91.

Virgo, D., Mysen, B. O., & Kushiro, I. (1980) Anionic constitution of 1-atmosphere silicate melts: Implications for the structure of igneous melts. *Science*, **208**, 1371–3.

Von den Driesch, H., & Jung, P. (1980) An investigation on the solubility of helium in nickel. *High Temperatures-High Pressures*, **12**, 635–41.

Von Weizäcker, C. F. (1937) Über die Möglichkeit eines dualen-Zerfalls von Kalium. *Physik. Z.*, **38**, 623–4.

Wacker, J. F., & Anders, E. (1984) Trapping of xenon in ice: Implications for the origin of the Earth's noble gases. *Geochim. Cosmochim. Acta*, **48**, 2373–80.

Wada, N., & Matsuda, J. (1998) Noble gas study of cubic diamonds from Zaire: Information of their source mantle. *Geochim. Cosmochim. Acta*, **62**, 2335–45.

Walker, J. C. G. (1977) *Evolution of Atmosphere*. New York: Macmillan.

Walker, J. F., Parrington, J. R., & Feiner, F. (1989) *Nuclides and Isotopes*, 4th ed. San Jose: General Electric Company.

Wasserburg, G. J., & Hayden, R. J. (1955) Ar40-K40 dating. *Geochim. Cosmochim. Acta*, **7**, 51–60.

Weis, D. (1996) Helium, neon and argon isotope systematics in Kerguelen ultramafic xenoliths: Implications for mantle source signatures. *Earth Planet. Sci. Lett.*, **138**, 29–38.

Weiss, R. F. (1970a) The solubility of nitrogen, oxygen and argon in water and seawater. *Deep Sea Res.*, **17**, 721–35.

Weiss, R. F. (1970b) Helium isotopic effect in solution in water and seawater. *Science*, **168**, 247–8.

Weiss, R. F. (1971a) Solubility of helium and neon in water and seawater. *J. Chem. Eng. Data*, **16**, 235–41.

Weiss, R. F. (1971b) The effect of salinity on the solubility of argon in seawater. *Deep Sea Res.*, **18**, 225–30.

Weiss, R. F., Lonsdale, P., Lupton, J. E., Bainbridge, A. E., & Craig, H. (1977) Hydrothermal plumes in the Galapagos Rift. *Nature*, **267**, 600–3.

Wells, S. G., McFadden, L. D., Poths, J., & Olinger, C. T. (1995) Cosmogenic ^{3}He surface exposure dating of stone pavement: Implications for landscape evolution in deserts. *Geology* **23**, 613–16.

Wescott, M. R. (1966) Loss of argon from biotite in a thermal metamorphism. *Nature*, **210**, 83–4.

Whetherill, G. W. (1953) Spontaneous fission yields from uranium and thorium. *Phys. Rev.*, **92**, 907–12.

Whetherill, G. W. (1954) Variations in the isotopic abundances of neon and argon extracted from radioactive minerals. *Phys. Rev.*, **96**, 679–83.

Whetherill, G. W. (1975) Radiometric chronology of the early solar system. *Ann. Rev. Nucl. Sci.*, **25**, 283–328.

White, B. S., Brearley, M., & Montana, A. (1989) Solubility of Argon in silicate liquids at high pressures. *Amer. Mineralogist*, **74**, 513–29.

Wieler, R. (1994) "Q-Gasses" as local primordial noble gas component in primitive meteorites. In *Noble Gas Geochemistry and Cosmochemistry.* J. Matsuda, Ed., pp. 31–41. Tokyo: Terra Scientific Publ. Co.

Wieler, R., & Baur, H. (1994) Krypton and xenon from the solar wind and solar energetic particles in two lunar ilmenites of different antiquity. *Meteoritics*, **29**, 570–80.

Wieler, R., & Baur, H. (1995) Fractionation of Xe, Kr, and Ar in the solar corpuscular radiation deduced by closed system etching of lunar soils. *Astrophys. J.*, **453**, 987–97.

Wieler, R., Anders, E., Baur, H., Lewis, R. S., & Signer, P. (1991) Noble gases in "phase Q": Closed-system etching of an Allende residue. *Geochim. Cosmochim. Acta*, **55**, 1709–22.

Wieler, R., Anders, E., Baur, H., Lewis, R. S., & Signer, P. (1992) Characterization of Q-gases and other noble gas components in the Murchison meteorite. *Geochim. Cosmochim. Acta*, **56**, 2907–21.

Wieler, R., Baur, H., & Signer, P. (1986) Noble gases from solar energetic particles revealed by closed system stepwise etching of lunar soil minerals. *Geochim. Cosmochim. Acta*, **50**, 1997–2017.

Wiens, R. C., & Pepin, R. O. (1988) Laboratory shock emplacement of noble gases, nitrogen, and carbon dioxide into basalt, and implications for trapped gases in shergottite EETA 79001. *Geochim. Cosmochim. Acta*, **52**, 295–307.

Wiens, R. C., Huss, G. H., & Burnett, D. S. (1999) The solar oxygen isotopic composition: Predictions and implications for solar nebula processes. *Meteoritics Planet. Sci.*, **34**, 99–107.

Wiens, R., Lal, D., Rison, W., & Wacker, J. F. (1994) Helium isotopic diffusion in natural diamonds. *Geochim. Cosmochim. Acta*, **58**, 1747–57.

Wilhelm, E., Battino, R., & Wilcock, R. J. (1977) Low-pressure solubility of gases in liquid water. *Chem. Rev.*, **77**, 219–62.

Wilson, G. B., & McNeill, G. W. (1997) Noble gas recharge temperatures and the excess air component. *Appl. Geochem.*, **12**, 742–62.

Wolfer, W. G., van Siclen, C. D., Foiles, S. M., & Adams, J. B. (1989) Helium solubility in solid and liquid nickel. *Acta Metal.*, **37**, 579–85.

Wood, D. S., & Caputi, R. (1966) Solubility of Kr and Xe in fresh and sea water. U.S. Naval Radiological Defense Laboratory, TR-988.

Woolum, D. S. (1988) Solar-system abundances and processes of nucleosynthesis. In *Meteorites and the Early Solar System*, J. F. Kerridge & M. S. Matthews, Eds., pp. 995–1020. Tucson: University of Arizona Press.

Yatsevich, I., & Honda, M. (1997) Production of nucleogenic neon in the Earth from natural radioactive decay. *J. Geophys. Res.*, **B102**(5), 10291–8.

Young, B. G., & Thode, H. G. (1960) Absolute yields of the xenon and krypton isotopes in U238 spontaneous fission. *Canad. J. Phys.*, **38**, 1–9.

Young, D. M., & Crowell, A. D. (1962) *Physical Adsorption of Gases*. London: Butterworth.

Zadnik, M. G., & Jeffery, P. M. (1985) Radiogenic neon in an archaean anorthosite. *Chem. Geol.*, **52**, 119–25.

Zadnik, M. G., Smith, C. B., Ott, U., & Begemann, F. (1987) Crushing of a terrestrial diamond: $^3He/^4He$ higher than solar. *Meteoritics*, **22**, 540–1.

Zahnle, K., & Kasting, J. F. (1986) Mass fractionation during transonic escape and implications for loss of water from Mars and Venus. *Icarus*, **74**, 62–97.

Zahnle, K., Kasting, J. F., & Pollack, J. B. (1988) Evolution of a steam atmosphere during Earth's accretion. *Icarus*, **74**, 62–97.

Zahnle, K., Pollack, J. B., & Kasting, J. (1990a) Xenon fractionation in porous planetesimals. *Geochim. Cosmochim. Acta*, **54**, 2577–86.

Zahnle, K., Kasting, J. F., & Pollack, J. B. (1990b) Mass fractionation of noble gases in diffusion-limited hydrodynamic hydrogen escape. *Icarus*, **84**, 502–27.

Zaikowski, A., & Schaeffer, O. A. (1979) Solubility of noble gases in serpentine: Implications for meteoritic noble gas abundance. *Earth Planet. Sci. Lett.*, **45**, 141–5.

Zartman, R. E., Wasserburg, G. J., & Reynolds, J. H. (1961) Helium, argon and carbon in some natural gases. *J. Geophys. Res.*, **66**, 277–306.

Zashu, S., & Hiyagon, H. (1995) Degassing mechanisms of noble gases from carbonado diamonds. *Geochim. Cosmochim. Acta*, **59**, 1321–8.

Zhang, Y., & Xu, Z. (1995) Atomic radii of noble gas elements in condensed phases. *Amer. Mineralogists*, **80**, 670–5.

Zhang, Y., & Zindler, A. (1989) Noble gas constraints on the evolution of the Earth's atmosphere. *J. Geophys. Res.*, **94**(B10), 13719–37.

Zindler, A., & Hart, S. R. (1986) Chemical geodynamics. *Ann. Rev. Earth Planet. Sci.*, **14**, 493–571.

Zinner, E. (1997) Presolar materials in meteorites: An overview. In *Astrophysical Implications of the Laboratory Study of Presolar Materials*, T. J. Bernatowicz & E. K. Zinner, Eds., pp. 3–26. AIP Conf. Proc., 402.

Index

Note: (f) and (t) indicate figure and table respectively.

α-particle range, 150
α-reactions, 153
absorption (noble gas), *see solution*
acid-leaching, 74
adsorption (noble gas); 33–42
   adsorption potential, 37
   BET model, 36
   chemical (chemisorption), 34
   Langmuir model, 36–7
   physical, 34
   sticking (life) time, 36, 39, 55, 56, 129
   temperature dependence, 37–8
air contamination, 41
   in mantle rocks, 189
   in MORB, 162–3
   in xenolith, 168
   in water, 120
air, 10–12
   elemental composition, 11(t)
   noble gas isotopes, 12(t)
Apollo lunar samples, 227
aquifer, 120, 147
   discharged area, 120, 138
   recharged area, 120, 138
   turnover time of, 139
Ar
   discovery, 2

^{40}Ar
   inventory in the Earth, 219, 250
   origin of atmospheric, 3–4
^{40}Ar-^{39}Ar (dating) method, 69–74
^{40}Ar/^{36}Ar
   in the mantle, 182–4
   primordial, 223–4
atmosphere, 12–15
   evolution, 205
   excess volatiles, 13
   gravitational capture, 13
   juvenile, 9
   primary, 13, 245
   secondary, 3, 13
authegenic minerals, 124

B (boron), 190
^{10}Be, 190
batch melting, 198
big bang, 15, 81
BIF (banded iron formation), 157
Bjurböle (meteorite), 240
Bruderheim (meteorite), 73
Bunsen coefficient, 103

C-flux, 211
C-C bond, 61
C-He systematics, 211–12

C/^{3}He, 211
C/^{4}He, 211
^{14}C age, 140
carbonado (diamond), 167
CFF (chemically fractionated fission)-Xe, 155
CH$_4$, 209
chemisorption, 34, 42; *see also* adsorption
chondrite, 88
clathrate, 60–2
   decomposition temperature, 61
   ice-noble gas, 61
   ice-methane, 61
   in solar nebula, 61
CMB (core mantle boundary), 216
CO$_2$, 13, 196, 209
CO$_2$ vesicles, 161, 162, 192
CO$_2$ well Xe, 203–4
coherent degassing, 237
cometary materials, 245
comet, 219
component (noble gas), 7–8
   in-situ, 8
   mantle, 163
   nuclear, 8
   planetary, 8
   primordial, 8
   radiogenic, 8
   solar, 8
   trapped, 8
continuous (mantle) degassing, 191
core, 55
correlation line, 27–28; *see also* mixing line
cosmic dusts,
   earth accreting, 241
   elemental composition, 82
   fallout rate, 130–2
   He (Ne) diffusion from, 133
cosmic elemental abundance, 81–3, 227
cosmic isotopic composition, 82, 82(t)
cosmic ray, 23, 140
   attenuation range in air, 23
   attenuation range in rocks, 23
   exposure age, 145, 146
cosmic rock, 82, 83
cosmogenic noble gases, 140–7, 141(t), 142(t)
cosmogenic nuclides, 23
"crustal He," 147, 151
crustal He flux, 209
"crustal Ne," 150, 151
CSSE (closed system step etching) method, 74–5

cubic diamond, 165
cyclo-silicate, 157–9

D (deuterium), 86
D/H, 81, 86, 87
degassing (mantle) coefficient, 197
depleted (upper) mantle, 164
desorption, 42
deuterium burning, 86
diamond, 94, 161, 165–7
   α-particle injection in, 166
   fluid inclusions, 165
   noble gas diffusion, 67–9, 166
   presolar, 96
   Xe in, 186
diffusion (noble gas), 63–9
   activation energy, 64
   characteristic length, 64
   in cyclo-silicate, 159
   effect of perthitization, 64
   grain boundary, 64
   multi-activation process in, 67
   in silicate melts, 68(f)
   in various materials, 65(t), 66(f)
discontinuity (670-km), 214, 215
double beta decay, 22
double spike, 80
dunite, 168

early (mantle) degassing, 199
earth evolution, 245
earth formation, 237–9
elemental fractionation, 9; *see also* isotopic fractionation
emanation (noble gas), 111
erosion rate, 146
EUV (extreme ultra violet) radiation, 232, 233
"excess air" in ground water, 120
excess ^{40}Ar, 8
excess ^{3}He in sea water, 113–14, 115, 116
excess He in water, 109, 139
excess volatile, 13
excess ^{129}Xe, 154, 155, 203
excess ^{131}Xe, 154
exosphere, 14, 76
"exosphere" (Rubey's), 211
extinct nuclides, 21
extrasolar component, 92
extraterrestrial He, 130, 132
extraterrestrial noble gas, 135
extraterrestrial particles, 132

fission Xe, 186, 235–8
fluid inclusions
    in diamond, 165
    in granite, 153
    in xenolith, 168
formation interval, 237
fractionation
    in earth accretion, 196
    in impact shock, 195
    in ion implantation, 243
fragmentation of solar nebula, 246
Freundlich isotherm, 37
fullerene, 62–3
    in Allende meteorite, 137
    $^3He/^4He$ in, 61

GAB (great artesian basin), 138
Galapagos rift, 117
GCM (global circulation model), 206
GCR (galactic cosmic ray), 140
geomagnetic field, 140
geomagnetic field reversal, 252
geomorphology, 146
GHP (great Hungarian plain), 139
giant impact, 236

H_2O, 13, 196
    photodissociation of, 13
He
    atmospheric inventory, 251
    atmospheric residence time, 14, 250
    discovery, 2
    juvenile He, 14, 117–19, 208
    non-thermal escape, 14, 251
^3He
    abundance in the mantle, 174(f)
    anthropogenic, 145
    in cosmic dusts, 130(f), 131(f), 132
    cosmogenic, 145
    degassing, 205–12
    excess, 24
    in sediments, 124(t), 125(t)
    primordial, 115, 116
    spallation-induced, 167
^3He flux, 131
^3He precipitation, 251
^3He production, 148
^4He/heat, 208, 212–16
    theoretical value, 214
$^3He/^4He$, 178–80, 179(f)
    anthropologic change, 14
    crustal, 178, 179(f)

in deep-sea sediments, 129
in diamond, 167
in lunar ilmenite, 136
in mantle, 180(f)
nucleogenic, 148
primordial, 87
solar, 87
$^4He/^{40}Ar$, 155–7, 156(f)
    in diamond, 167
    in MORB, 176
    in OIB, 176
    in water, 113
He-flux, 118, 138, 251(t)
He/H, 81
heat flow, 117
heat of solution, 48
Henry's law (adsorption), 35–6
Henry's law (solution), 42–6
hydrodynamic flow of H_2, 231, 234
hydrogen burning, 86
hydroquinone, 61
hydrous minerals, 127

^{129}I, 146, 186
^{129}I–^{129}Xe (dating) method, 69–74
^{129}I–^{129}Xe systematics, 200–2
$^{129}I/^{127}I$, 236–7
    in chondrite, 202
    in primordial earth, 202, 240
I-Pu-Xe systematics, 204–5
ice, 111, 112, 245
IDP (inter-planetary dust particles), 85, 135, 136, 137
ilmenite, 58
impact shock, 192–5
    devolatalization (degassing), 193, 194, 195
    disturbance on K-Ar, 193
    noble gas trapping, 57
implantation, 58, 58(t)
    of ion, 57, 243
    of SEP, 74
    of SW, 74, 135
incompatibility
    of elements, 52
    of noble gases, 54
interstellar medium, 97
ionic porosity, 46, 47(f)
isochron, 28
isotopic fractionation, 9, 74–80
    cosmic, 82
    instrumental, 79
    in implantation, 56

isotopic fractionation (*cont.*)
  in mechanical shock, 59
  of Ne, 230–3
  in solution, 105
  in van der Waals bond, 76
  of Xe, 79

Jeans escape, 14, 76, 250
Jupiter, 13, 84(t)
juvenile ^{3}He, 113–19
  injection, 116
juvenile ^{4}He, 112–13
  flux, 112
juvenile gas (volatiles), 53, 209

K in the mantle, 215
K/T boundary, 136
K/U, 216
kimberlite, 168
komatiite, 196
Kr
  discovery, 2
  at high pressures, 33
KTB (German deep drilling project), 154

Langmuir isotherm, 35–6
layered mantle, 187–90, 216
less-depleted (lower?) mantle, 164
lherzolite, 168
lunar ilmenite, 136
lunar soils (breccia), 42, 58, 220

"magma ocean," 193, 196
magnetite, 129
  in cosmic dust, 135
  in deep-sea sediments, 136
"mantle He," 147, 161
  emanation, 207
  in tectonic extension area, 209
"mantle Ar," 223, 224
mantle convection, 214, 215
  He barrier, 214–15
mantle degassing, 53
  degree of, 196
mantle dichotomy, 189, 212–16
"mantle fluids," 165
"mantle Ne," 29, 161
mantle plume, 161
"mantle Xe," 201
Mars, 13, 220, 221
  atmosphere, 60, 84(t), 193

mass spectrometer
  noble gas, 7, 9, 189
  Reynolds-type, 3
  thermal, 9
mass transfer
  in the mantle, 198
mechanical shock
  in noble gas trapping, 58–9
meteoric water, 168
meteorite, 23, 80, 88, 97
methane, 61
mfl (mass fractionation line), 90
Milankovitch cycle, 120
"missing heat," 216
"missing Xe," 92, 123, 228, 229(f), 233–5, 240
  fraction of, 240
mixing diagram, 26–7
mixing line, 27; *see also* correlation line
molecular cloud, 82
MORB, 161, 164, 172
  glassy margin, 161, 162
  noble gas abundance, 170–1(t), 172(f), 173(f)
  noble gas isotopic composition, 178(t), 187(t)
  source of, 176
  vescicularity, 192
MORB-type He, 164
muon, 141, 142

N_2 in water, 106
natural nuclear chain reaction, 155
NBO/Si (nonbridging oxygen per silicon), 46
Ne
  cosmogenic, 74
  discovery, 2
  isotopic anomaly, 94
  outgassing (degassing) rate of, 133
  in sediments, 124(t), 125(f)
  in solar system, 75(f)
^{20}Ne/^{22}Ne, 92
  in air, 92
  in earth, 221–3
  in mantle, 180–2, 221(f)
  in SW, 230
^{21}Ne (cosmogenic), 145
^{21}Ne* (nucleogenic), 177, 180–2
^{21}Ne*/^{4}He
  in MORB, 176
  in OIB, 176

Ne fractionation, 219, 230–3
  in the atmosphere, 219
Ne-E, 90, 96(f)
Ne-E(H), 92
Ne-E(L), 92
Ne/He
  in mantle, 177
  in solar system, 84(t)
neutron
  natural neutron detector, 149
  production rate, 147
neutron capture, 141, 153
neutron flux, 149, 154
  in air, 142, 143(f)
  in the crust, 143(f)
noble gas
  in air, 12(t), 10–12
  atomic size, 32
  chemical interaction, 30, 32
  geochemical characteristics, 4–5
  inventory, 10, 12, 249–52
  in meteoric water, 110–12
  physical properties, 31(t)
  polarizability, 32
  in sea water, 106–10, 107(t), 110(t)
  in Sun, 85, 86
  under high pressure, 31–3
noble gas partition
  in mantle degassing, 234
  in the mantle, 199
noble gas trapping, 55–60, 59(t), 61
Novo Urei (meteorite), 225, 236
nuclear absorption mean free path, 140
nuclear reaction
  anthropologic, 24
  noble gas production, 24, 24(t)
nuclear test, 144
nucleosynthesis, 24, 94

O_2, 13
OIB, 161, 172
  noble gas, isotopic composition, 187(t)
  noble gas abundance, 170–1(t)
  source of, 176
OKLO phenomena, 24, 155
olivine, 53, 58
ordinate-intercept diagram, 162

paleoatmosphere, 124
paleotemperature, 168
partition (noble gas), 52–5

nonequilibrium, 76
partition (distribution) coefficient, 9, 54(t)
penetration depth of ion, 243
PI component, 93
planet formation, 241
"planetary (meteorite) Xe," 224
"planetary (type) noble gas," 85, 87–93
"planetary Ne," 230
planetesimal, 192, 193, 196, 219, 236, 246, 247
$^{210}Po/^{3}He$, 211
polymerization of silicate melt, 46
popping rock, 57, 161, 204
presolar grains, 93, 94
presolar noble gas, 83
primary atmosphere, 245
"primitive Xe," 225
primordial He, 215
primordial noble gas, 217, 218–27
primordial terrestrial Xe, 226, 236
Pu fission, 22(t)
^{244}Pu, 70
^{244}Pu fission Xe, 22, 200, 201, 203–205, 235
$^{244}Pu/^{238}U$
  in Earth, 237–8
  in solar system, 203

Q (noble gas component), 84(t), 90, 93, 94, 220, 221, 224

R_a, 179
^{222}Rn, 20, 138, 139
^{226}Ra, 138
radiogenic ^{129}Xe, 146
rate equation, 197–8
Rayleigh fractionation (distillation), 77–8, 92, 228
recharge zone, 138
recoil range of fission Xe, 70
recycling of air, 14, 15
red sea brines, 112
Reynolds, J. H., x
Rubey, W. W., 13

salinity, 103
SCR (solar cosmic ray), 140
sediments, 123–9
  noble gas contents in, 124(t), 125–7(t), 128(f)
sedimentation rate, 132

SEP (solar energetic particles), 74, 84(t), 90, 136
serpentine, 124, 126, 193
Setchenow relation, 103
shale (clay), 38–39
Shergottite (meteorite), 60
shock implantation of noble gas, 193, 194
silicate melt
  noble gas solubility, 43–47
  three-dimensional network, 42
silicon carbide, 94
SiO_2 network, 43
SNC (shergottite-nakhlite-chassignite) meteorites, 193
SO_4, 209
solar (type) noble gas, 83–7, 84(t), 220, 226
solar He, 130
solar nebula, 227, 228
solar Ne, 245
solar wind (SW), 74, 83, 84(t)
solar wind implantation, 242
solar Xe, 224
solution (noble gas), 42–52
  in air, 12(t)
  in brine, 103, 104(f)
  dependence on atomic radius, 49, 52
  in ice, 111
  in metallic iron, 51
  pressure dependence, 48–9
  salinity dependence, 104(f), 105
  in solid, 43
  temperature dependence, 48, 120
  in various melts and solids, 44–45(t)
  in water, 51, 98–103, 99(t), 100(f), 102(t)
spallation, 22, 23, 141, 167
spike, 113
stellar wind, 82
Stripa granite, 149, 153
subduction, 132, 135, 190
SUCOR (surface correlated) Xe, 224
Sudbury impact crater, 137
Sun, 85, 86

surface tension, 49
SW (solar wind), 220, 227–30
  noble gas abundance, 84(t)
  noble gas isotopic composition of, 88–9(t)

T Tauri stage, 242
Te, 154
tektite horizon, 136
tritium (^3H), 114, 115, 144

^{238}U, 212
U-fission, 22, 152
U-Xe, 225, 226

van der Waals force, 4, 32, 34, 60
van der Waals radius, 31(t), 32
van't Hoff equation, 103
veneer materials, 219
vescicularity, 192
volcanic glass, 173

Whetherill reaction, 24

$^{129}Xe^*$ (radiogenic), 186, 188, 199, 225, 235–8
  in the atmosphere, 236
Xe
  concentration in rocks, 3
  discovery, 2
  isotopic anomaly, 94
  primordial Xe, 187
  solid phase of, 235
  in SW, 224, 236
  under high pressure, 235
  Xe-F bond, 30
Xe isotopic fractionation in planetesimals, 247–9
Xe solubility in planets, 51
Xe-closure, 23, 36
Xe-H, 92
Xe-HL, 96
xenolith, 168–9